ELECTRONIC DRAFTING

Printed Circuit Design

Electronic Drafting

PRINTED CIRCUIT DESIGN

Robert S. Villanucci
Alexander W. Avtgis
William F. Megow

All at Wentworth Institute of Technology

Macmillan Publishing Company
NEW YORK

Collier Macmillan Publishers
LONDON

Macmillan Publishing Company
866 Third Avenue, New York, New York 10022

Collier Macmillan Canada, Inc.

Library of Congress Cataloging in Publication Data

Villanucci, Robert S.
  Electronic drafting.

  Bibliography
  Includes index.
  1. Electronic drafting.   2. Printed circuits—Design
and construction.   I. Avtgis, Alexander W.
II. Megow, William F.   III. Title.
TK7866.V55   1985   621.381′73′0221   84-4394
ISBN 0-02-423050-2

Printing: 1 2 3 4 5 6 7 8   Year: 5 6 7 8 9 0 1 2

ISBN 0-02-423050-2

TO THE MEMORY AND HONOR OF OUR PARENTS:

Annette, Nazzareno, Pelagia, William, Ludolf, and Irene

PREFACE

The electronic technician in modern high-technology industry requires a much broader proficiency in electronic drafting than is currently offered in most traditional drafting courses. Many of these courses provide general information on electronic symbols, reading and drawing schematic diagrams, and developing both pictorial and isometric views of subassemblies. Even though these topics will always be a vital part of an electronic technician's training, they do not address the recent demands of industry. More training needs to be provided in the area of printed circuit board design and artwork preparation in electronic drafting courses. Most often, this topic is given minimal attention due to the unavailability of a meaningful and comprehensive textbook which not only presents the traditional electronic drafting topics but also provides modern printed circuit board layout techniques. This book is designed to fill this void.

The demands of industry have steadily been shifting away from the traditional "electronic drafting" requirements into "drafting design." More than ever, skilled electronic technicians with a solid background in printed circuit board design are needed. In order for the design technician to better communicate with the circuit design engineer, a more comprehensive background in basic electronic drafting, electronic circuit theory, and some knowledge of printed circuit board fabrication is required. This knowledge also allows the printed circuit design technician to communicate design specifications more meaningfully to the board manufacturer.

This book has been designed to fulfill the needs of a modern electronic drafting course. The material presented is directed toward educational institutions such as vocational-technical schools, technical institutes and junior colleges, as well as industrial and military training programs. It is intended that this book not only serve as a text in electronic drafting courses, but also as a useful reference to practicing technicians, technologists, and engineers.

In order to clearly demonstrate drafting techniques and to provide detailed information, over 250 figures and drawings have been included in addition to 36 Appendices. End-of-chapter exercises allow the students to practice the design techniques presented in this book.

The material presented is divided into two general categories. Chapters 1 through 5 present the topics normally treated in a traditional electronic drafting course, i.e., drafting tools, materials, equipment, line construction, lettering, symbols, circuits and some mechanical drawing techniques including dimensioning. Chapters 6 through 14 provide a detailed discussion of specifications, methods and procedures required to design, lay out and develop complete documentation packages necessary to produce single-sided, double-sided and multilayer printed circuit boards. Chapter 15 presents an overview of modern computer-aided design (CAD) systems for printed circuits.

Chapter 1 provides information on basic drafting equipment, tools, and materials together with explanations of their proper use. Chapter 2 covers conventional mechanical drafting techniques, such as line construction, line codes and conventions, lettering, and dimensioning. Chapter 3 is a complete overview of electronic symbols, such as resistors, capacitors, transistors, and integrated circuits (including a brief description of each), and the drawing techniques of these, including the use of electronic symbol templates. Chapter 4 discusses the reading and drawing of electronic circuit schematics and converting rough drafts to finished schematic diagrams for both digital and analog circuits. Chapter 5 covers other types of drawings, such as multiviews, pictorials, and simple two-dimensional views of components, hardware, and printed circuit board outlines. Chapter 6 begins the discussion of printed circuit board layout, starting with the basic concept of the printed circuit board and its use in electronic packaging, including general information as to the types of board designs: single-sided, double-sided, and multilayer. Chapter 7 provides information on the essential elements of laying out a simple single-sided printed circuit board using a basic analog circuit as a vehicle. Included is information on grid systems, terminal pad areas, determining lead access hole spacing, drawing component and device body outlines, positioning components, and routing conductor paths, as well as coding, keying, and labeling. Chapter 8 extends the concept of the single-sided design to more complex layouts, including reduction scales, corner brackets, mounting holes, and additional information regarding terminal pad selection as well as conductor path width and spacing determination. Chapter 9 provides basic information on how to lay out and draw a double-sided analog circuit. The concept of feed-through holes is covered as well as registration techniques. Chapter 10 treats the design of a double-sided digital circuit, including the layout of gold fingers (edge-board contacts) for connector insertion, as well as the location and coding of fingers. Decoupling techniques are also presented. Chapter 11 develops a multilayer printed circuit board with one power inner layer and one ground inner layer. Included are thermal relief pads as well as layer coding and identification techniques. Chapter 12 covers artwork master taping methods for single-sided, double-sided, and multilayer boards, including the tools, drafting aids, materials, and methods of constructing complete taped artwork masters used for processing printed circuit board phototools. Chapter 13 describes the techniques for the layout of component marking artworks for producing screens used for labeling printed circuit boards. Chapter 14 covers the construction of artworks for solder masking. Chapter 15 is an overview of computer-assisted digitizing systems which are used to produce more accurate artworks by means of photoplotting. Also included is a discussion of computer-aided design and drafting (CAD) and how it affects traditional manual layout and design techniques.

We wish to express our gratitude to our colleagues, Professors Robert F. Coughlin and Frederick F. Driscoll, for their technical assistance in the preparation of this book and to the Air Force Contracts Section at Wentworth for their support. We also acknowledge the expert photographic work of Mr. Thomas E. Naylor, Jr. Our thanks are extended to Mr. Dennis Sokol of Altron, Incorporated, for helping us to obtain technical information on the state of the art manufacturing of printed circuit boards. We are also grateful to the many drafting manufacturers who provided us with photographs of their equipment with a special thanks to Bishop Graphics, Inc. for furnishing us with the many drafting aids which are shown throughout this book. For the patient and skillful typing of the manuscript, we express our appreciation to Mrs. Phyllis Wolff.

Finally, we acknowledge the understanding and encouragement displayed by our families, who again believe that this is the last one.

Robert S. Villanucci
Alexander W. Avtgis
William F. Megow

Boston, Massachusetts

CONTENTS

3

4

5

Detail Drawings for Electronics   111

6

Printed Circuit Boards   131

7

Single-Sided Printed Circuit Board Design: The Preliminary Sketch   147

11

The Design of Multilayer Printed Circuit Boards 279

12

Taped Artwork Masters 295

13

Component Marking Artwork 325

ELECTRONIC DRAFTING

Printed Circuit Design

1

Use of Basic Drafting Equipment

Upon completion of this chapter on the use of basic drafting equipment, the student should be able to:

1. Know the various types and sizes of available drafting paper and the criteria for selecting the proper type for a specific drawing.

2. Know the various types of drafting pencils available and the differences in lead size and hardness.

3. Properly maintain a good point on a drafting pencil.

4. Know the various types of available erasers and their proper use.

5. Use the technical fountain pen.

6. Correctly use the T-square and drafting board.

7. Use a parallel ruling blade and drafting machine.

8. Properly use drafting triangles.

9. Understand the techniques for using basic drafting equipment in preparing a drawing.

10. Correctly use protractors.

11. Know the various types of scales available and how they are read.

12. Properly use the compass to draw precise circles.

13. Know the various types of templates available and their proper use.

14. Use French curves and sweeps to form irregular shapes.

1-0

Introduction

The electronics industry requires the use of many types of drawings to convey specific technical information. Engineers, technicians, assembly workers, and others involved in the field of electronics rely heavily on graphical presentations, such as *circuit schematic diagrams, mechanical assembly drawings, pictorial parts assembly drawings,* and *printed circuit board design drawings* to communicate plans, designs, and ideas. For these individuals, an understanding of how to interpret and use these forms of technical information is essential. One of the most effective means of gaining this knowledge is to learn how to present graphical information in conventionally adopted forms and in sufficient detail so that it will be easily read and understood. The first step toward this goal requires that you become familiar with the use of various types of drafting equipment.

This chapter provides the information on basic drafting equipment and supplies and their proper use for producing all forms of electronic drawings.

1-1

Drafting Paper

The types and sizes of drafting paper discussed in this section are those employed almost exclusively in the development of all kinds of electronic drawings. The choice of the type and size of paper used for a specific drawing is dictated by universally adopted standards and by the paper's suitability in meeting the requirements of the drawing.

The types of drafting paper used in the electronics industry can be generally classified as (1) *detail paper*, (2) *vellum*, (3) *cloth*, and (4) *polyester film*.

Detail drafting paper is a high-quality, heavy-weight paper manufactured from pure rag stock. The weight of this type of paper is described in units of *pound (lb) basis*, available in 70-, 80-, and 110-lb basis. Detail paper is normally buff or light cream in color and is used primarily for pencil drawings that are not to be reproduced. This paper is very durable, has superior erasing qualities, and does not become brittle or discolored with age.

Drafting vellum (often referred to as tracing paper) is a rag stock material that has been treated with oils or waxes to make it more transparent. This paper is used for tracing with either pencil or ink. Untreated tracing paper is also available and is preferred for ink drawings. These untreated vellums are more durable than the treated type. The surfaces of vellum allow for multiple erasures without concern for excessive abrasion of the paper. Because this material is transparent, the drawings can be easily reproduced in standard blueprint machines.

Drafting cloth (or tracing cloth) is a thin transparent muslin fabric treated with a starch or plastic to make its surface suitable for drafting with the use of either pencil or ink. This material is extremely durable and is preferred for drawings that are to be maintained for an extended period of time. Drawings made on drafting cloth lend themselves easily to reproduction. Even though this material is ideal for electronic drafting applications, it is relatively expensive compared to other drafting media.

Drafting film is a superior material for pencil or ink drawings. This film is made by bonding a mat surface to one side of a clear Mylar polyester sheet, resulting in a dull surface on one side (the drawing side) and a high-gloss surface on the other. Typical film thicknesses are 0.004 and 0.007 in. Drafting film has exceptional erasing qualities, leaving no ghost marks. This material is extremely durable with high resistance to bending, cracking, and tearing. Drafting film finds wide use in producing component layout drawings for printed circuit boards because of its excellent dimensional stability, that is, its ability to maintain its dimensions under temperature and/or humidity changes. This stability is the result of the film being heat treated during its manufacture. Polyester drafting film is stable in length or width typically to 0.0011% per 1% change in relative humidity (RH) and 0.0017% per 1°C change in temperature (*T*). The expected change in either film dimension can be approximated by the following typical equation *:

$$\Delta \text{in.} = [1.7 \times 10^{-5}(\Delta°C) + 1.1 \times 10^{-5}(\Delta\%RH)] \times L \qquad (1-1)$$

where Δ in. is the resultant change in dimension (inches), $\Delta°C$ is the change in temperature (degrees Celsius), $\Delta\%RH$ is the change in relative humidity, and L is the known length in inches. If the change in width is also required, L may be replaced with W (width in inches). An example using Equation 1–1 follows. Assume that a 20-in.-long piece of polyster drafting film (0.007 in. thick) is to be moved from one work area, where the temperature is 15°C and the relative humidity is 30%, to another work area, where the temperature is 25°C and the humidity is 50%. What changes in film dimension will result? Since both RH and *T* have increased, the film will also increase in size. The change in temperature is $\Delta°C = 25°C - 15°C = 10°C$. The change in relative humidity is $\Delta\%RH = 50\% - 30\% = 20\%$. Substituting these values in Equation 1–1, we obtain

$$\Delta\text{in.} [1.7 \times 10^{-5}(10) = 1.1 \times 10^{-5}(20)] \times 20$$
$$[1.7 \times 10^{-4} + 2.2 \times 10^{-4}] \times 20$$
$$[0.00039] \times 20 = 0.0078 \text{ in.}$$

The resultant change in film length for this specific film is an increase of approximately 0.008 in. [8 mils or 2 millimeters (mm)], which is acceptable even in particularly critical work.

Because large sheets of polyester film are stored in tightly rolled tubes, they should be unrolled and laid flat for several days to allow them to stabilize to their normal dimensions. Some polyester films are provided with a colored grid (usually blue so as not to be reproduced in blueprinting) to facilitate drawing.

All of the drafting papers above are available in *flat* or *roll* forms. Flat sizes of standard drafting sheets are designated by a conventional system of letters for the English system and a combination of a number and letter for the metric system. The English system designations are A through F, and for the metric system, A10 through 2A. Table 1–1a shows the length and width of sizes A through F flat drafting paper. Appendices I and II show the comparison between

---

* Manufacturers' specifications should be consulted for the exact coefficients. Consult Federal Specifications LF519C and IPC-D-310A.

TABLE 1-1 Standard Available Drawing Sheet Sizes

(a) Flat Sizes Designated by Letters A Through F

Letter Size Designation	Vertical Width (in.)	Horizontal Length (in.)
A	8.5	11.0
B	11.0	17.0
C	17.0	22.0
D	22.0	34.0
E	34.0	44.0
F	28.0	40.0

(b) Available Sizes Not Designated by a Letter

Paper Type	Vertical Width (in.)	Horizontal Length (in.)
Flat	9.0	12.0
Flat	11.0	15.0
Flat	12.0	18.0
Flat	15.0	22.0
Flat	18.0	24.0
Flat	22.0	30.0
Flat	24.0	36.0
Roll	24.0	—
Roll	36.0	—

(c) Roll Sizes Designated by Letters G Through K

Letter Size Designation	Vertical Width (in.)	Horizontal Length (in.)	
		Minimum	Maximum
G	11.0	22.5	90.0
H	28.0	44.0	143.0
J	34.0	55.0	176.0
K	40.0	55.0	143.0

the English and metric sizes. There are nondesignated sizes in between those listed which are also available. See Table 1–1b.

As the name *flat type* implies, these sheets are intended to be stored flat. However, they may also be folded into standard $8\frac{1}{2} \times 11$-in. pamphlets for filing purposes. See Appendix III for methods of folding different-size paper.

The standard-size sheets described are readily adaptable to microfilm frame sizes for reproduction. The size of the paper selected should be compatible with the drawing. The paper should be of sufficient size to provide space for complete detail without crowding, yet not so large as to result in excessive unused areas.

Roll-size papers are stored rolled in tubes, owing to their long horizontal lengths. The dimensions of roll sizes G through K are shown in Table 1–1c. These large sheets are also compatible to microfilm filing systems. Note the alignment arrowheads shown in Fig. 1–1.

Standard drawing sheets of both the flat and roll sizes are typically provided with preprinted borders and information blocks. Although the border widths are standardized for each paper size, all paper sizes do not have the same border dimension. For example, the top and bottom border width for C-size paper is 0.75 in. and is 0.50 in. for D-size paper. The information blocks will vary in size, content, and position from one manufacturer to another. A typical drafting sheet format with standard information blocks is shown in Fig. 1–1. Typically, the title block is located at the lower right corner of the sheet and the *revision block* in the upper right corner. If a *supplementary block* is required, it is usually positioned immediately to the left of

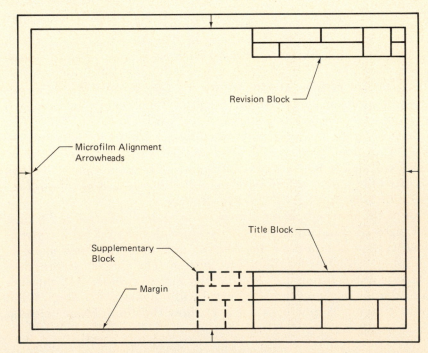

FIGURE 1–1 Basic drafting sheet format showing margins, title block, revision blocks, supplementary block, and microfilm alignment arrowheads.

the title block. These major blocks are subdivided into smaller blocks to accommodate entry of a variety of essential information. The title block includes such information as company name, drawing number and title, date of completion, draftsperson's name, issue date, approval date and initials, drawing scale, sheet numbers, and drawing sheet letter size designation as shown in Table 1–1. The revision block would typically contain information on the revision level (usually the next letter in alphabetical sequence to the previous letter used), description of the change, and approval date and initials of the person authorizing the revision. Supplementary blocks may be required to provide additional information, such as tolerances, material types or finishes, parts lists, or general notes.

The type and amount of information contained in the informational blocks will vary depending on the complexity and format of the drawing and the organization's requirements. There should be sufficient information so as to make the drawing clear and complete, but should not include excessive information which could lead to confusion.

1-2

Drafting Pencils

There is a variety of pencils for use by the draftsperson. The three common types of pencil construction are shown in Fig. 1–2. The standard *wood-cased pencil* is shown in Fig. 1–2a. This pencil is approximately 7 in. long and formed into a ¼-in. hexagonal shape. It is

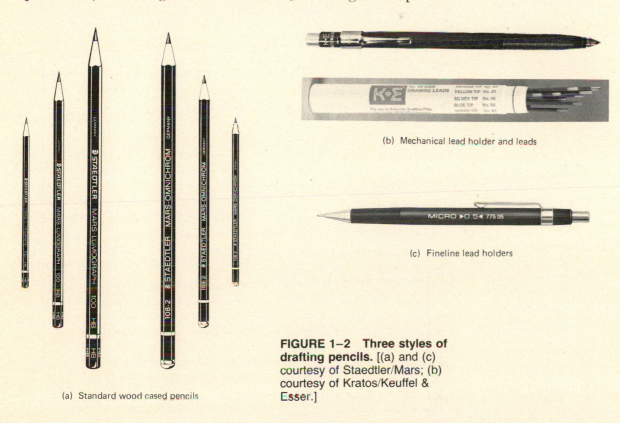

(a) Standard wood cased pencils

(b) Mechanical lead holder and leads

(c) Fineline lead holders

FIGURE 1–2 Three styles of drafting pencils. [(a) and (c) courtesy of Staedtler/Mars; (b) courtesy of Kratos/Keuffel & Esser.]

not fitted with an eraser. The lead grade is stamped along the side near one end. The wood-cased pencils are inexpensive but have the disadvantage of changing their "feel" (weight) as they are sharpened and worn down.

A *mechanical lead holder*, shown in Fig. 1–2b, allows for interchanging separate pieces of lead. These mechanical pencils are constructed of a metal tube lead holder surrounded by a plastic casing and accept standard lead diameters of approximately 0.08 in. (2 mm). The lead is installed into the holder by depressing the button at the opposite end of the pencil holder. This opens the holder jaws and allows the lead to be inserted into the metal tube. Once the lead has been inserted a sufficient amount to leave the desired length of lead exposed, the button is released and the jaws clamp the lead firmly into place. As the point dulls and the lead wears down, holding the pencil vertically and depressing the button will allow more lead to become exposed for sharpening. The standard mechanical pencils can be used with all lead grades. In addition, the "feel" of the pencil remains unchanged as the lead wears down since it is only the lead that is sharpened.

The main objection to the wood-cased pencil and mechanical lead holder is that the lead has to be constantly sharpened for uniform linework throughout the drawing. (Pencil-point sharpening is discussed in the following section.) This objection is overcome with the use of the mechanical pencil shown in Fig. 1–2c. This pencil is similar to the mechanical pencil just described except that it was made to accept microthin leads having diameters of 0.3, 0.5, 0.7, and 0.9 mm. Because of the extremely thin sizes of lead used, they require no sharpening, yet maintain uniform linework. As the lead is worn down, a new portion of lead is exposed by simply depressing the end button. This feature of not requiring sharpening makes these types of mechanical pencils very popular.

For drawings produced on detail paper, vellum, or cloth, lead made with a mixture of graphite and kaolin (clay) is used. In order to obtain smudge-free drawings on polyester films, specially formulated plastic-base leads are used extensively. Both lead types are available in wood-cased pencils as well as refills for mechanical pencils in both standard and microthin styles.

Drafting leads are available in 18 different grades of hardness. The lead diameter and the hardness of the lead are inversely related; that is, larger-diameter leads are used to provide strength to the softer grades and the diameters become progressively smaller as the leads become harder. Thus the different diameters produce varying degrees of line widths and varying degrees of line darkness, which is related to the grade of lead hardness. For example, the soft leads, which range in grade from very soft (7B) to soft (2B), will produce bold, wide lines which are very dark, almost black. Lead grades, sizes, hardnesses, and applications are shown in Table 1–2.

The *very soft lead* grades (5B, 6B, and 7B) are seldom used in electronic drafting because they are too easily smeared. The *soft* leads are rarely used for producing schematic diagrams but are used to shade or highlight pictorial diagrams of packaging assemblies.

TABLE 1–2 Grades of Drafting Pencil Leads

Lead Size	Grade*	Grade†	Hardness	Applications
	9H	—	Very hard	Charts, diagrams, graphs: very thin and very light gray lines which are difficult to repro-duce on blueprint machines
	8H	—	Very hard	
	7H	—	Very hard	
	6H	K6H	Hard	For most construction linework on electronic drawings (not for reproduction)
	5H	K5H	Hard	
	4H	K4H	Hard	
	3H	K3H	Medium	For finished linework; pencil tracings to be blueprinted
	2H	K2H	Medium	
	H	KH	Medium	
	F	KF	Medium	For lettering and freehand sketching
	HB	KHB	Medium	
	B	KB	Medium	
	2B	K2B	Soft	For shading work
	3B	K3B	Soft	
	4B	—	Soft	
	5B	—	Very soft	Usually too soft for electronic drafting work; for architec-tural drawings and various types of artwork
	6B	—	Very soft	
	7B	—	Very soft	

*For vellum and cloth.

†For polyester film.

The *medium lead* grades (3H to B) produce thinner lines which are not as dark as those resulting from the use of softer leads. Grades F, HB, and B are used for technical sketching, arrowheads, and let-tering, while grades 3H, 2H, and H are used on drawings that will be reproduced on blueprint copiers.

The *hard leads* (grades 9H to 4H) will produce finer lines with lighter shade. Grades 6H, 5H, and 4H are used for construction lines on engineering drawings that are not intended to be reproduced. Grades 9H, 8H, and 7H are used primarily for work that requires extreme accuracy, such as charts, diagrams, and graphs. The lines resulting from these leads do not reproduce well in blueprint copiers.

As seen in Table 1–2, the lead grades used on polyester film are more limited than those used on detail paper, vellum, or cloth. Plastic-based leads are coded with the letter K (also CF) to distinguish them from standard graphite leads.

1-3

Pencil-Point Sharpening

Maintaining a good point on any type of pencil used is essential to realize good quality and consistent line work. Wood-cased pencils require two sharpening steps. The unmarked end of the pencil is first inserted into a standard hand- or power-operated sharpener which has been equipped with drafting blades. See Fig. 1–3a. This sharpener will remove only the wood casing, leaving about $\frac{3}{8}$ in. of lead

(a) Sharpener with drafting blades

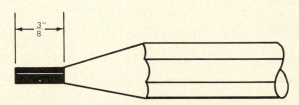

(b) Results of sharpening with drafting blades

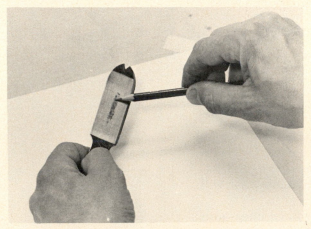

(c) Point shaping on metal file

(d) Removing loose graphite particles

FIGURE 1–3 Sharpening procedures for wood case pencils.

exposed which has not been shaped. This is shown in Fig. 1–3b. The lead is then pointed to the desired conical shape by passing the point back and forth over a metal file while simultaneously rotating the pencil for a uniform shape. This operation is shown in Fig. 1–3c. Care should be taken to avoid sharpening to a needle point which would break off as soon as it is pressed to the paper.

When wood-cased pencils are pointed with the use of a metal file or sandpaper, fine graphite particles will remain on the lead point. These particles must be removed to avoid smearing the drawing. Graphite particles can be removed by wiping with a cloth or depressing the point several times into a block of Styrofoam. This is shown in Fig. 1–3d. Because the sandpaper pad is also laden with graphite dust, care should be taken in handling and storing the pad to avoid smudging clean surfaces.

The standard 2-mm-diameter lead used in mechanical pencil holders can be pointed and shaped using a simple *lead pointer* or *rotary pointer*. The use of a simple lead pointer is shown in Fig. 1–4a. This lead pointer consists of a tubular body which guides the mechanical pencil and a set of carbide blades for sharpening the lead. Approximately ½ in. of lead is first exposed at the end of the holder jaws. The pencil is then inserted into the pointer, gently pressed inward, and rotated clockwise. The graphite particles removed during sharpening simply fall into the bottom receptacle, which is emptied periodically. This sharpener produces a conical point that is ideal for drafting. Similar to the wood-cased pencil, the sharpened point must be cleaned of graphite particles using cloth or Styrofoam.

The rotary pointer shown in Fig. 1–4b is a more sophisticated variation of the simple lead pointer. To use this sharpener, ½ in. of lead is first exposed from the pencil jaws. The pencil is then inserted into the access hole or nose piece, which is on the top cover of the sharpener. Gentle downward pressure is applied to the pencil while the top section of the pointer, together with the pencil, are rotated clockwise several times. For uniform results, the pencil should not be allowed to rotate on its own axis but held and rotated with the pointer cover. As with all sharpening techniques, the lead thus sharpened must be cleaned of graphite dust before use.

As can be seen, lead sharpening is somewhat time consuming in addition to being messy. It is for these reasons that the use of microthin leads with their appropriate holders are finding increasing popularity with draftspeople. Work need not be interrupted for lead sharpening and graphite dust is essentially nonexistent.

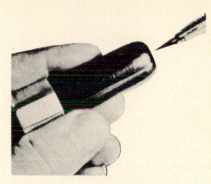

(a) Lead pointer for mechanical pencils

(b) Rotary pointer

FIGURE 1–4 Mechanical lead pointers. [(a) courtesy of Alvin & Co., Inc.; (b) courtesy of Kratos/Keuffel & Esser.]

1-4

Erasing Pencil Marks

Erasers are used to remove pencil marks and smudges from drawings. The three common eraser materials are (1) *soft rubber*, (2) *hard rubber*, and (3) *vinyl*. The soft rubber or "pink" erasers are generally used to clean smudges from paper or vellum. When erasing

Pencil   Motor driven

FIGURE 1–5   Various eraser configurations. [Courtesy of Kratos/Keuffel & Esser.]

smudges, excess eraser pressure should be avoided to prevent damage to the surface of the drawing sheet. For removal of construction or finish lines, a hard rubber eraser is more suitable. With the use of a hard rubber eraser, even more care is necessary to avoid excessive pressure. Both hard and soft erasers are available in cubes or pencil-top forms, as shown in Fig. 1–5. Vinyl erasers are used to remove pencil marks and smudges from drafting film. Even though polyester films are more durable than other types of drafting sheets, excessive eraser pressure should again be avoided.

Electric erasers, also shown in Fig. 1–5, are equipped with a slow-speed motor and a chuck which accepts tubular-shaped erasers. Usually, only soft rubber material is used for these erasers. Because the eraser motion is powered, extremely light pressure is all that is required.

Proper erasing, either manually or electrically, requires that the eraser be free of graphite particles from previous erasures. If there are any visible dark markings on an eraser, its surfaces should be cleaned by rubbing them on a clean piece of scrap paper before use.

After any erasing is done on the drawings, particles of eraser material will be deposited. Those particles should be removed by brushing off with a *dusting brush*, shown in Fig. 1–6. Attempting to remove the eraser particles by wiping with your hand could result in smudging the drawing.

Any pencil line drawn on a drafting sheet will produce some loose graphite dust which remains on the drawing. Tools, equipment, and hands invariably pick up these particles and tend to smear them on the drawing. To remove these minor smears, a *dry cleaning pad*, also shown in Fig. 1–6, is used. A dry cleaning pad consists of a bag made of a loosely woven fabric and containing fine eraser granules of vinyl material. To clean smears from drawings, the dry cleaning pad is gently tapped over the drawing, releasing eraser particles. The pad is then gently rubbed over the smeared area. *Excessive pressure will deteriorate the drawing's line quality and should be avoided.* After

Dusting brush   Dry cleaning pad

FIGURE 1–6   Draftsman's cleaning tools. [Courtesy of Kratos/Keuffel & Esser.]

FIGURE 1–7   Using eraser shield.

the smears are removed, the eraser particles are swept off the drawing with a dusting brush.

When it is necessary to erase an incorrect line or mark from the drawing, it is important that adjacent lines or marks are not disturbed. For this purpose, an *eraser shield,* shown in Fig. 1–7, is extremely helpful. Eraser shields are made of thin sheet metal with a variety of cutout shapes. To use this aid, the shield is held firmly against the drawing with the appropriate cutout exposing just the line or mark to be removed. The eraser is rubbed over this area with the shield in place, preventing the eraser from removing adjacent lines or marks.

To consistently obtain clean drawings, you should make a conscious effort to observe the 10 practices for drafting cleanliness that are listed in Table 1–3.

TABLE 1–3   Practices for Maintaining Drafting Cleanliness

1. Periodically wash hands and use talcum powder to absorb natural skin oils.
2. Frequently remove graphite particles from drafting equipment with a clean, dry cloth.
3. Avoid sharpening pencils over drawing board or drafting equipment.
4. Clean pencil points of graphite particles after each sharpening.
5. Store pencil sharpeners and sandpaper pads away from drawings and equipment.
6. While drawing, do not rest hands or arms directly on penciled areas.
7. Slightly raise the T-square and triangles before moving them across the drawing surface.
8. While drafting, periodically sprinkle a small amount of powdered Artgum onto the work to pick up graphite particles.
9. Remove Artgum particles with graphite from the drawing using only a dusting brush.
10. Cover the drawing with clean paper or cloth if it is to be allowed to stand for a prolonged period of time before completion.

1-5

Ink Drawings

With improved transparent tracing paper and reproduction equipment, pencil drawings are usually sufficient for most purposes. However, ink drawings are considered superior in terms of finer overall drawing appearance and improved line quality and result in better reproductions, including those prepared for microfiche. Ink drawings require a greater degree of skill by the draftsperson. However, where costs are justified, ink drawings are preferred where overall high-quality work and reproductions are of principal concern.

The most widely used pen for producing ink drawings is the *technical fountain pen* shown in Fig. 1–8. This pen is suitable for producing linework and lettering, both freehand and mechanical. (Lettering is discussed in Chapter 2.) The pen point is shouldered so that smudges and ink blobs will not result when used with straightedges and templates.

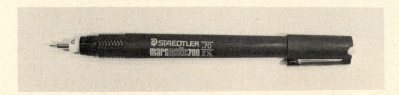

FIGURE 1–8   Technical fountain pen.

Technical fountain pen points are precision ground and can be used with or without edge supports even when held at angles of 80 degrees. In general, however, it is recommended that these pens be held perpendicular to the drawing surface. The points are made of polished tungsten carbide, sapphire jewel, or chrome-hardened stainless steel. The tungsten or sapphire points are preferred for use with drafting film because of their higher abrasive qualities. The stainless steel point, which is more economical, is appropriate for use with vellum. To provide for a variety of line widths, a selection of these pens is necessary. Table 1–4 shows various line widths and the pen point sizes required to produce these widths. Technical pens are also made to be used with a compass. The use of a compass is discussed in Section 1–12.

TABLE 1–4   Most Widely used Technical Fountain Pen Point Sizes

Point size	4 x 0	000	00	0	1	2	2.5	3	4	5	6	7
Millimeters	0.13	0.18	0.25	0.35	0.40	0.50	0.70	0.80	1.00	1.20	1.80	2.00
Inches	0.005	0.007	0.01	0.014	0.016	0.020	0.028	0.030	0.039	0.047	0.071	0.077

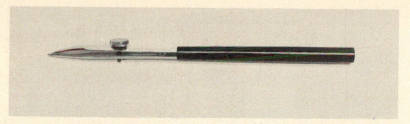

FIGURE 1–9 Straight inked lines can be drawn with this type of ruling pen.

The use of the technical pen requires a minimum of skill and results in high-quality drawings. Its cartridge design prevents ink from drying, thereby eliminating clogging problems. These pens require only occasional filling after a prolonged period of use.

A *ruling pen*, shown in Fig. 1–9, is used to draw straight or curved lines. This pen is used with a guide, such as a T-square, triangle, or French curve. An advantage of a ruling pen is that its tip, which is adjustable, will provide any line width. However, its use requires a greater degree of skill and the tips need constant attention to result in high-quality work. For these reasons, the technical pen has essentially replaced the ruling pen for producing inked drawings. The technical pen saves time and effort and is convenient to use. The resulting drawings are of the highest quality.

To avoid smears and smudges when preparing ink drawings, several precautions need to be observed. First, you must avoid moving the T-square or triangles over freshly inked lines. Second, intersecting previously drawn lines that are still wet will cause fillets at the intersections of those lines. Finally, avoid inking center lines over compass marks. This could cause ink blots on the reverse side of the drawing. Center lines should be inked before the compass is used.

For correcting errors in ink drawings, soft rubber and vinyl erasers or liquid ink remover are used. Because ink is more stubborn to remove than pencil, extra care must be taken to avoid excessive eraser pressure, which would damage the drawing surface. The area to be erased sould be placed directly over a hard surface, such as a plastic triangle, to support the drawing. Liquid ink remover, used with blotter material, is used primarily on vellum and drafting cloth. After its use, light passes with the eraser may be necessary to complete the erasure.

1-6
Drafting Boards and T-squares

Drafting boards are generally made from strips of selected kiln-dried soft wood. These wood strips are glued together to form a solid core having smooth true-plane top and bottom surfaces. Usually, both surfaces of a drafting board may be used for drawing. To form the working edge (or running guides) to accommodate the head of a T-square, the board has either vertical strips of wood fastened to the

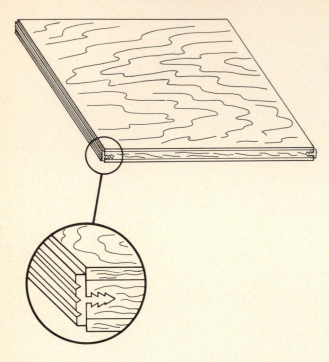

(a) Lightweight metal edge drawing board

(b) Large format drafting table

FIGURE 1–10 [(b) courtesy of Keuffel/Esser & Kratos Company.]

shorter side edges by tongue-and-groove joints and glue or aluminum end cleats deeply embedded and running along these edges. Portable drafting boards are available in a wide variety of sizes ranging from 9×12 in. to 31×42 in.

The board selected for a specific drawing should be several inches longer and wider than the drawing paper to be used. For general student use, a board that will accommodate B- or C-size paper is usually adequate (see Section 1–1). Most industrial applications require the use of a board which is suitable for the use of D-size paper. A typical portable drafting board is shown in Fig. 1–10a. A drafting table, used for large drawings or layouts, is shown in Fig. 1–10b.

Drafting table tops are made with a soft wood work surface similar to those of portable drafting boards. They are also available with kiln-dried hardwood or high-density plastic tops. Drafting tables have the advantage of an adjustable work surface height and tilt angle for more comfortable drafting. Provided as an integral part of the table or as a separate unit are a series of drawers for the convenient storage of drafting equipment and supplies. Drafting tables are available in a variety of sizes, ranging from 23×31 in. to 44×96 in.

Before using the working surface of a drafting board or table, it should be inspected for any surface defects or indentations, such as compass point holes or hard-lead pencil impressions. Because the wood used in the construction of drafting boards is soft, surface damage often results from normal use. For this reason, it is common to cover the surface of the board with vinyl or linoleum which is trimmed

FIGURE 1–11 **Light table for drafting.** [Courtesy of nuArc Company, Inc.]

to the exact size of the board and permanently fastened with double-sided tape, spray adhesive, or a mastic-type bonding agent. In applying this cover, care should be taken that ripples or bumps are not present. The application of these materials provides a slightly harder surface on which to draw, which is less susceptible to damage. The surface is self-sealing to compass point holes and has fast recovery to hard-lead pencil impressions. In addition, these surfaces are easier than wood to clean.

Another common type of drafting table is called a *light table*. This is shown in Fig. 1–11. The work surface is a heavy glass plate which may be provided with a precise ruled grid to aid in accuracy of design. Positioned below the glass is a light diffusion sheet made of translucent white plastic. This sheet is located between the glass surface and the fluorescent lamps. Light tables are especially advantageous for tracing and for laying out printed circuit boards. (Printed circuit board layout is discussed in Chapters 7 through 11.)

The *T-square* is used in conjunction with a drafting board to draw horizontal lines and to support the edges of triangles to draw lines at various angles. The T-square is constructed of two parts, the *head* and the *blade*, which are rigidly fastened together to form a true 90-degree angle. See Fig. 1–12. The inner edge of the head, which runs along the side of the drafting board, and the edges of the blade are called the *working edges*. Both the head and the blade are commonly made of hard wood, although stainless steel and aluminum T-squares are also available. The wooden T-square has clear plastic bonded to each working edge along the entire length of the blade. These are shown in Fig. 1–12a. The purpose of these clear plastic edges is to allow viewing of the lines beneath the working edge of the blade. To minimize ink smearing, the plastic edges are manufactured so as to be slightly elevated above the drawing surface. Wooden

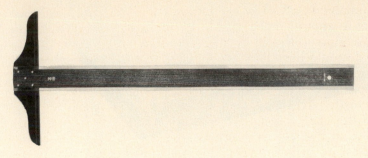

(a) T-square

(b) Testing the 90° angle between the head and blade of a T-square

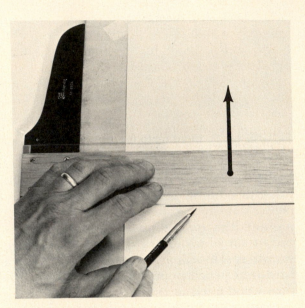

(c) Testing blade edge parallelism

FIGURE 1–12 **Basic drafting T-square.** [(a) courtesy of Kratos/Keuffel & Esser.]

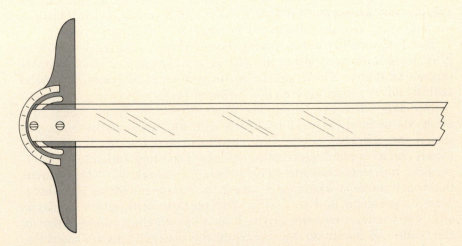

FIGURE 1–13 **T-square with adjustable protractor head.**

blades are available in a variety of lengths, ranging from 15 to 48 in. Longer metal T-squares are available up to 72 in., although these long sizes are more difficult to use. The T-square used with any drafting board should have its blade length 2 to 4 in. longer than the horizontal length of the board.

Prior to use, the T-square should be checked to see that it is square and that the blade edges are parallel. The 90-degree angle between the head and the blade can quickly be verified with the use of a triangle, as shown in Fig. 1–12b. With the triangle positioned as shown, there should be no gaps between the edges of the triangle where they touch along the edges of the blade and head for the T-square to have a true 90-degree blade-to-head angle.

To check that the blade edges are parallel, the head is held firmly against the side of the board and a line is drawn approximately the length of the upper blade edge. The T-square head is then moved up and the lower blade edge is aligned with the drawn line. This is shown in Fig. 1–12c. If there is any space between the line and the lower blade edge, this is an indication that the blade edges are not parallel. If the space is significant, the T-square should be discaded.

In use, the T-square is positioned on the drafting board with its head tight against the left edge of the board. This position is for right-handed draftspersons. If left-handed, the head is positioned against the right edge of the board. In either case, the blade will extend along the entire work surface and its edges will be parallel to the top and bottom edges of the board.

In addition to the standard drafting boards and T-squares, there are also other styles which are very useful, especially for perspective drawings. The T-square shown in Fig. 1–13 is equipped with a protractor head graduated to 180 degrees with an indicator on the arm for setting the desired angle. Thumbscrews secure the head in position once the angle has been set. This T-square has a blade made entirely of clear plastic for total surface visibility.

1-7

The Parallel Ruling Blade and Drafting Machines

The *parallel ruling blade* drafting unit, shown in Fig. 1–14a, is a modification of the drafting board and T-square. The ruling straightedge is similar to the blade portion of a T-square. This straightedge is attached to a drafting board or table through a series of guide wires and wheels which are located at the vertical edges of the board and the ends of the blade. The blade ends ride flush with the edges of the board. This suspension system holds the blade slightly off the board surface, which prevents smudges when it is moved. A gentle push or pull with one hand easily moves the blade vertically across the board surface. The entire length of the blade continually maintains its parallelism with the top and bottom edges of the board at any point in its up-and-down travel. Slight downward pressure on the blade will firmly position it in place. This unit is especially helpful on long drafting boards or tables where long standard T-squares

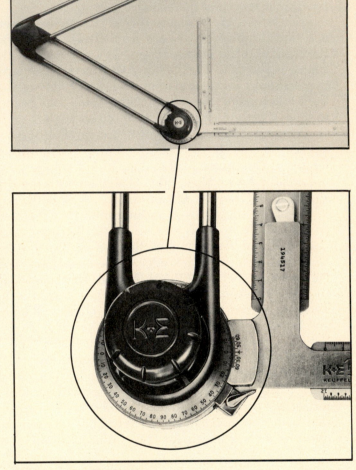

(a) Parallel ruling blade

FIGURE 1–14 Drafting machines. [(a) Courtesy of Kratos/Keuffel & Esser; (b) courtesy of Keuffel/Esser and Kratos Company.]

(b) Arm–type drafting machine

tend to be cumbersome to handle. The blade is made entirely of clear plastic, which provides complete drawing surface visibility.

The parallel ruling blade unit has become more popular than the separate drafting board and T-square because it is more convenient to use. In industry, however, the *drafting machine* has replaced both systems for reasons of increased accuracy, convenience, and speed. A typical arm-type drafting machine is shown in Fig. 1–14b. These units are available for either left- or right-handed operation. It consists of two rigid plexiglass blades, both having precision machine-engraved scales. A description of the graduations and divisions of these scales is provided in Appendix IV. The horizontal blade is usually longer than the vertical blade. These blades meet at a central hub called an *indexing head* with a fixed angle of 90 degrees between them. The blade scales can be moved over the entire work surface of the board to any location or position. The ball joints in the elbow and pivot points allow for smooth movement and for the scales to lie flat on the drawing surface. In addition, the blades can be easily lifted over objects on the drawing table. The indexing head is equipped with a 150- or 360-degree protractor with automatic 15-degree indexing and free angle setting controlled by a thumb release. When depressed, the thumb release button on the hub allows the protractor head to move the fixed blades to any desired angle. This angle is locked into place by releasing the button. The protractor head is provided with markings to 1 degree or with a double vernier scale which can be read to 5 minutes of arc. Adjustable spring tension at the clamp end for counterbalancing and a disk brake in the elbow prevents the scales from slipping out of position when used with inclined boards.

The drafting machine is usually permanently mounted to a large drafting board or table with a clamp. Small portable boards with drafting machines are also available. In additon to the plastic blades, some drafting machines are equipped with metal blades. All of these blades are scaled in either English or metric units.

1-8

Drafting Triangles

Triangles are used in conjunction with the T-square to draw vertical lines as well as angles other than 90 degrees to the horizontal. Typical triangles are shown in Fig. 1–15. They are available in metal, wood, plastic, or professional-quality acrylic plexiglass. Triangles made from the plastic materials are either clear or tinted and are by far

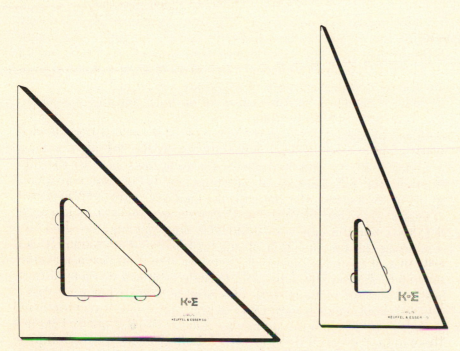

FIGURE 1–15 Standard 45° and 30° x 60° triangles. [Courtesy of Kratos/Keuffel & Esser.]

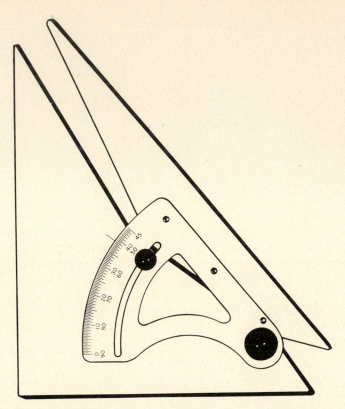

FIGURE 1–16   Triangle adjustable from 0° to 90°. [Courtesy of Kratos/Keuffel & Esser.]

the most popular. The thickness of plastic triangles is approximately 0.10 in. The size of standard drafting triangles is given by the length of the longest leg. The most widely used sizes are 6, 8, 10, and 12 in. The two standard geometric styles of triangles are the 45 degree and the 30×60 degree, both of which are shown in Fig. 1–15. The 45-degree triangle is available up to 16 in. long, and the 30×60-degree triangle, up to 24 in. long.

As will be shown in the next section, these two triangles are used to draw either vertical lines perpendicular to the T-square or lines at angles of 30, 45, and 60 degrees. By using the two styles in combination, the full 360-degree spectrum can be obtained in increments of 15 degrees, starting with the smallest angle of 15 degrees.

Angles other than those in the 15-degree increment steps are often required. For this purpose, *adjustable triangles* may be used. By combining the function of a protractor (discussed in Section 1–10) with that of triangles, any angle between 0 and 90 degrees can be accurately drawn. The adjustable triangle is shown in Fig. 1–16. It has a protractor element with two rows of graduations. The outer row indicates angles from 0 to 45 degrees to be used with one base of the triangle, while the inner row indicates angles of 45 to 90 degrees to be used with the other base. Both rows are graduated in $\frac{1}{2}$-degree increments. The protractor element is attached to the triangle with a metal screw post and knurled-knob locking screw. This arrangement forms a fulcrum point for the protractor. The triangle element is rigidly attached to the protractor element. To adjust the desired angle between the hypotenuse and either base, the knurled-knob screw is loosened and the protractor is moved to align the graduation mark for the angle with the index mark. When the angle is set, the knob screw is tightened. The knurled knob also serves as a convenient handle for ease in operating the unit.

When an adjustable triangle is used with a T-square, lines at angles which fall in any one of the eight half-quadrants between 0 and 360 degrees can be constructed. If a drafting machine is not available, the adjustable triangle offers wide versatility with many of the advantages of the machine.

1-9

Basic Drafting Techniques

Before continuing with the discussion of drafting equipment and supplies, we will consider the proper use of those already presented and the basic techniques involved for the preparation of high-quality drawings.

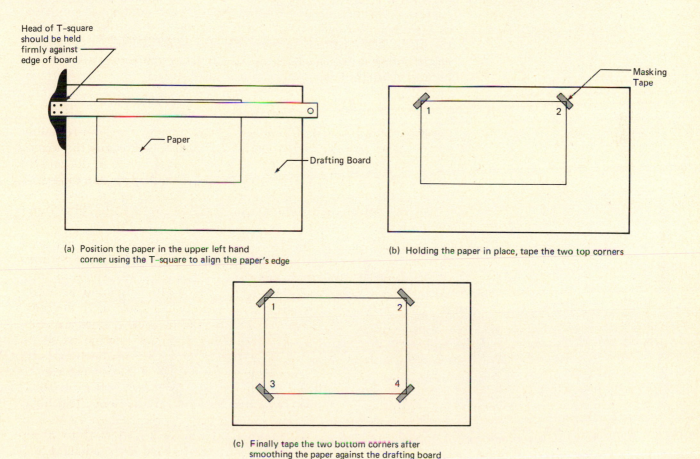

(a) Position the paper in the upper left hand corner using the T-square to align the paper's edge

(b) Holding the paper in place, tape the two top corners

(c) Finally tape the two bottom corners after smoothing the paper against the drafting board

FIGURE 1–17 Taping paper to the drafting board.

The first step in producing a drawing is to position and secure the paper to the drafting board. The paper is placed flat onto the board and positioned in the upper left corner of the work surface. See Fig. 1–17. In this position, your hands will have good support when working over the entire drafting area. (Left-handed people would position the paper in the upper right corner of the board.) Because the paper is close to one edge of the board, it minimizes any error of horizontal lines drawn along the edge of the T-square due to flexing or bending of the blade.

The top edge of the paper is next aligned with the blade of a T-square, the head of which is held firmly against the left edge of the board. This is shown in Fig. 1–17a. With the paper thus aligned, it is firmly held in place and the T-square is moved down out of the way. The two upper corners of the paper are then taped to the board using 2-in. lengths of drafting or masking tape placed diagonally across these corners. See Fig. 1–17b. The paper is then smoothed to the left bottom corner and it is taped on a diagonal. Again, the paper is smoothed from the top edge to the lower right corner and taped. See Fig. 1–17c. The paper is now in a convenient working position. Because the top edge of the paper was initially aligned with a T-square, lines drawn will be both horizontal and perpendicular to the edges of the paper.

Horizontal lines are drawn using the top edge of the T-square blade as a guide and a sharp-pointed drafting pencil. Begin by positioning the blade edge precisely in the position where the line is to be drawn using both hands. The blade is held firmly in this position with the fingers of the left hand (alternate if left-handed). The line is drawn

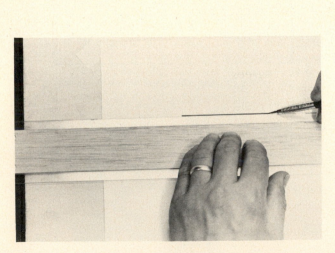

Using top edge of T-square
to draw a horizontal line

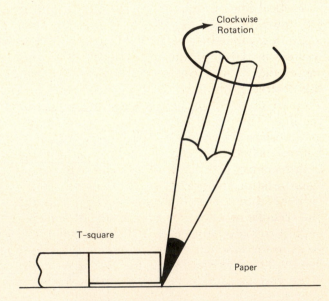

Pencil point should contact the T-square's
edge constantly as the horizontal line is drawn

FIGURE 1–18   Method for drawing a horizontal line.

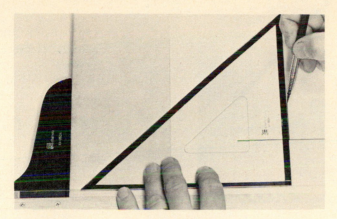

FIGURE 1–19 Constructing perpendicular lines.

from left to right (again, alternate if left-handed). The point of the pencil is held tightly against the edge of the blade and tilted at a 60-degree angle to the paper in the direction of the line to be drawn. With even pressure applied, the pencil is drawn tightly along the blade while constantly rotating the pencil clockwise between your thumb and index finger. This technique is shown in Fig. 1–18. The result of the application of even pressure and rotating the pencil will produce a uniform line width. In addition, rotating the pencil prevents the formation of a flat surface on the lead tip and thus requires less frequent sharpening. The technique of drawing the pencil along while rotating it is not difficult and, with practice, can be mastered. The results obtained are superior to other line-drawing methods.

Vertical lines which are perpendicular to horizontal or base lines are drawn with the use of either a 45-degree or 30 x 60-degree triangle in combination with a T-square. This is shown in Fig. 1–19. A triangle is placed on the top edge of a T-square with one 90-degree angle edge flat against the blade. This combination is then moved to the desired position to draw the vertical line. The triangle and the T-square are held firmly in place with one hand while the other draws the line. The pencil is held at an angle of approximately 60 degrees in the direction of the motion. Vertical lines may be drawn top to bottom or bottom to top along the edge of a triangle. With the point in firm contact with the working edge of the triangle, the line is drawn, again rotating the pencil.

Inclined lines at various angles to the base line can be drawn with one triangle or a combination of triangles. Figure 1–20a shows a 45-degree triangle being used with a T-square to draw a line at 45 degrees to the base line. The positioning of single triangles used to draw lines at 30, 45, 60, and 90 degrees is shown in Fig. 1–20b. Also shown is the various combination of two triangles to draw a variety of other angles.

Lines parallel to each other may be quickly drawn with a T-square and triangles. The hypotenuse of the triangle is first set at the desired angle. The T-square blade is then placed against the base of

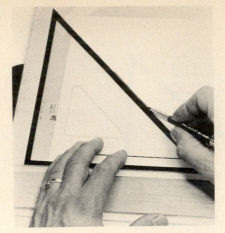

FIGURE 1–20 Using drafting triangles.

(a) Constructing a 45° line

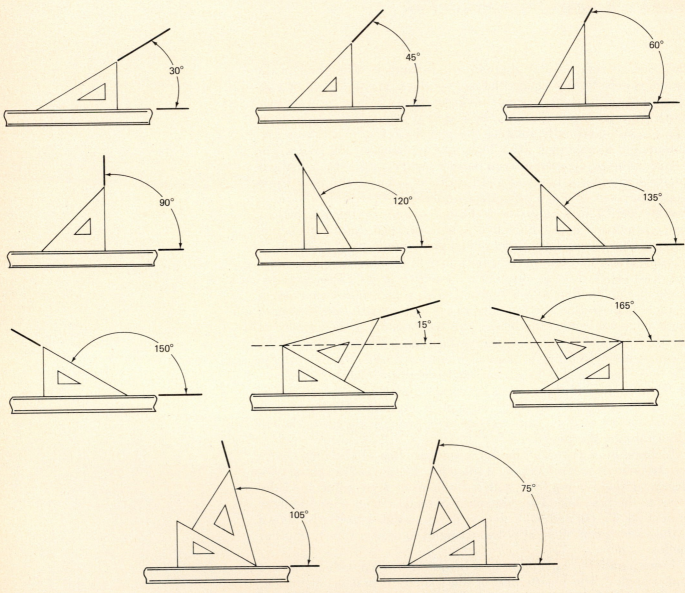

(b) Various angles for drawing inclined lines using triangles individually and in combination

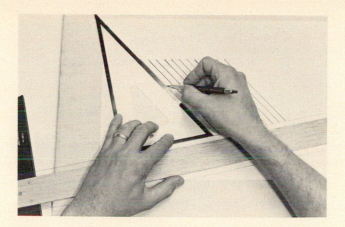

FIGURE 1–21   Constructing parallel lines at odd angle.

the triangle. The first line is drawn along the hypotenuse of the triangle. The triangle is then moved along the blade with the T-square held firmly in place. Any other line thus drawn will be parallel with the initial line. This technique of drawing parallel lines is shown in Fig. 1–21. Angles that are not attainable in single or in any combination of triangles can be drawn with an adjustable triangle (previously discussed) or with a protractor, which is treated in the next section.

The procedure for drawing a perpendicular to an inclined line is shown in Fig. 1–22. After drawing the initial inclined line, one of the sides of the triangle adjacent to the 90-degree angle is positioned along the inclined line. The perpendicular is drawn along the *other* adjacent side.

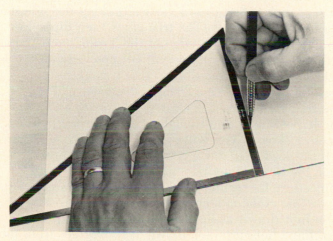

FIGURE 1–22   Constructing a perpendicular line to a previously drawn line.

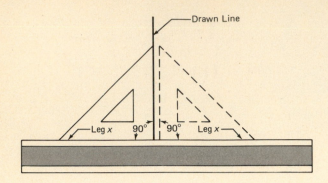

(a) Test for accuracy of 90° angle

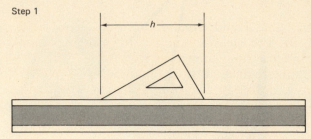

Step 1

A base line equal in length to h is first drawn along the edge of the T-square

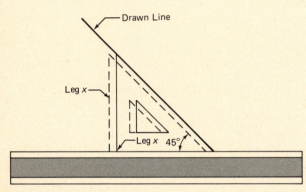

(b) Test for accuracy of 45° angle

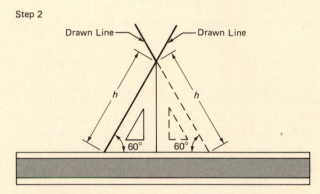

Step 2

(c) The 60° angles are accurate if all drawn lines are equal in length to dimension *h*

FIGURE 1–23 **Testing the triangle's straightness and accuracy of angles.**

To be assured of drawing accuracy, triangles should be tested for straightness before use. This is done in a similar way to the testing of the T-square (Section 1–6). The method for testing triangles is shown in Fig. 1–23. The test for the 90-degree angle is shown in Fig. 1–23a. A line is first drawn vertically along the side that is perpendicular to the T-square. The triangle is then flipped over, keeping the same leg against the edge of the T-square. If the angle is a true 90 degrees, the vertical leg will fall precisely on the line. Any serious deviation is cause to discard that triangle.

The testing of a 45-degree triangle is shown in Fig. 1–23b. One leg of the triangle is placed against the T-square and a line is drawn along the hypotenuse. The triangle is then turned over and rotated 90 degrees clockwise. This will place the opposite leg against the T-square. If the 45-degree angle is true, the hypotenuse of the triangle will fall on the drawn line.

The procedure for testing the 60-degree angle is shown in Fig. 1–23c. A base line equal in length to the largest leg of a 30 x 60-degree triangle is first drawn along the edge of the T-square. Two intersecting inclined lines are then drawn at 60 degrees from the ends of the base line. The resulting triangle will be equilateral if the 60-degree angle is true.

1-10

The Protractor

Semicircular and circular *protractors* are drafting instruments that are used to measure angles formed by line segments. Two styles of protractors are shown in Fig. 1–24. The semicircular protractor measures angles of 0 to 180 degrees. The circular protractor effectively represents two semicircular protractors and can thus measure angles of 0 to 360 degrees.

Protractors are graduated in degrees with minor divisions typically in $\frac{1}{2}$ degrees. The most popular protractors are made of transparent plastic and range in thickness from 0.05 to 0.09 in. The

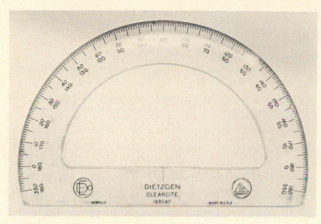

(a) Semi-circular protractor

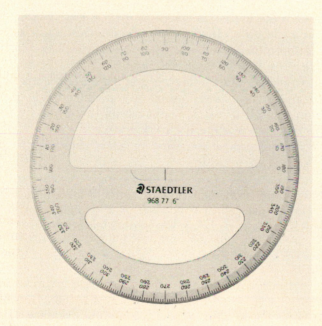

(b) Circular protractor

FIGURE 1–24 Two protractor styles. [(b) courtesy of Staedtler/Mars.]

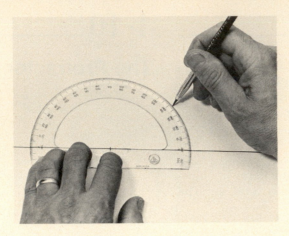 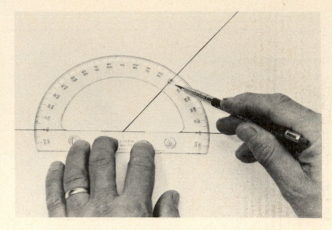

(a) Initial set-up to construct a 38° angle   (b) Measuring a 50° angle

FIGURE 1–25 Using a protractor.

semicircular protractor can also be used in conjunction with a straightedge to construct any angle between 0 and 180 degrees. The construction of a 38-degree angle with the horizontal base line is shown in Fig. 1–25a. To lay out or to measure angles, two scales are provided on the protractor. The outer scale runs clockwise from 0 to 180 degrees and the inner scale runs counterclockwise from 0 to 180 degrees. A point directly below the 90-degree graduation mark is located at the center of the protractor base to serve as the origin of the angle to be read or drawn. The protractor is aligned with a horizontal base line along its inside reference edge, which is the distance between the left and right 0-degree graduation marks. The measuring of a 50-degree angle is shown in Fig. 1–25b. The protractor is first placed so that its reference edge aligns with one leg of the angle to be measured and its center mark placed at the origin of the angle. The line forming the angle can thus be easily read as 50 degrees from the *right* 0-degree graduation mark reading the inner scale counterclockwise. The complementary angle is found by reading from the left 0° mark clockwise on the other scale. Note in Fig. 1–25b that this reading is 130 degrees.

Another style of protractor, with an adjustable arm, is also available. This semicircular protractor is made of nickel silver and is approximately 8 in. in diameter. It is equipped with a vernier scale which allows more accurate measurements of angles to degrees and minutes. The adjustable arm is used as a straightedge for forming angles, which reduces drawing time considerably.

1-11

Drafting Scales

Drafting scales are precisely subdivided measuring instruments which are used to lay out or measure lines in a drawing. They are not intended to be used as straightedges to draw lines. Their use is limited to measurement only and allows the draftsperson to make a variety of scaled drawings, that is, larger, smaller, or full size of the original subject, in most cases without having to make mathematical conversions. Scales are available in English or metric units. The specific scale and the size of the paper to be used are both dictated by the size of the original subject.

Scales range in length from 6 to 24 in. The 12-in. scale is the most commonly used size in electronic drafting. These scales are available in two basic cross-sectional shapes, flat and triangular. These are shown in Fig. 1–26. The beveled edges on the flat scales, shown in Fig. 1–26a–f, allow the scale marks to be positioned close to the drafting surface, which aides in accurate reading or transfer of dimensions. The two-bevel style shown in Fig. 1–26f has two ribs on the bottom surface. One or both of those ribs may be made of rubber for holding the scale more firmly in place.

The triangular-shaped scale, shown in Fig. 1–26g and h, can accommodate as many as 12 different overlapping scales due to its six scale surfaces. Although the triangular scale saves the costs of requiring many flat scales, it is easy to lose track of which scale is being used as the scale is moved from one location to another on the drawing. To overcome this problem, a *scale guard* not only helps to keep track of the scale being used but also serves as a convenient means of handling the scale.

All scales have either *open-divided* or *fully-divided* scale markings along their length. An open-divided scale has only one major unit at the beginning of the scale, subdivided into its minor divisions. The remaining scale length is divided into only the major divisions. The fully-divided scale has the major divisions subdivided by its minor divisions throughout the length of the scale.

Scales are classified as *civil engineer's, decimal, metric, fractional, architect's,* or *mechanical engineer's.* Several different scale configurations and types are shown in Fig. 1–27.

Civil engineer's scales, formerly referred to as *chain scales,* are graduated in decimal units. The major divisions are 1 in. apart and fully divided with subdivisions of 10, 20, 30, 40, 50, or 60 equal parts. For drawings requiring measurement or layout in decimal units, the

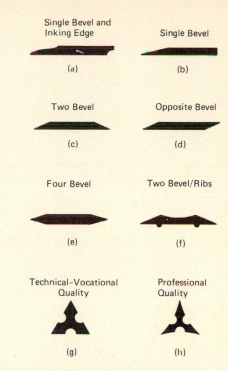

FIGURE 1–26 Cross-sectional shapes of flat and triangular scales.

FIGURE 1–27 Different drafting scale configurations. [Courtesy of Alvin & Co., Inc.]

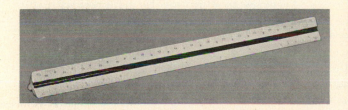

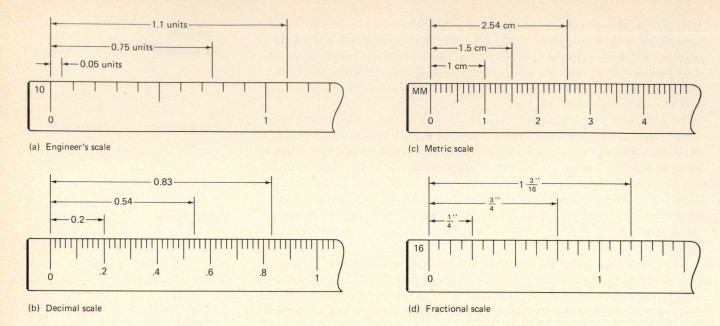

(a) Engineer's scale

(b) Decimal scale

(c) Metric scale

(d) Fractional scale

FIGURE 1–28 Four examples of reading drafting scales.

civil engineer's scale (commonly called the engineer's scale) or the decimal scale may be used. Some examples of readings with the engineer's scale are shown in Fig. 1–28a. These scales are made for preparing reduced drawings with major divisions representing 1 foot or even larger units as required by use of civil engineers.

The *decimal scale* shown in Fig. 1–28b, has major division of 1 in. and subdivisions marked every 0.2 in. (two-tenths of an inch) and minor divisions of 0.02 in. (twenty-thousandths of an inch). Thus, from the 0 index point, the first minor division represents 0.02 in., the second 0.04 in., the third 0.06 in., and so on. Decimal scales are available in full- and half-scale units. A variety of readings are shown in Fig. 1–28b to demonstrate the use of the full-size decimal scale.

Metric scales are graduated in units conforming to the metric system of measurement and are commonly 30 centimeters (cm; $\approx$12 in.) in length. The metric scale shown in Fig. 1–28c is fully divided with major divisions in centimeters and subdivisions in millimeters. Subdivisions in $\frac{1}{2}$ mm are also available.

A *fractional scale*, with some example scale readings, is shown in Fig. 1–28d. The fully-divided scale shown has major divisions of 1 in., subdivided into $\frac{1}{8}$-in. units with minor subdivisions of $\frac{1}{16}$ in. Thus, from the 0 index point, the distance to the first mark is $\frac{1}{16}$ in., to the second is $\frac{1}{8}$ in. (or $\frac{2}{16}$ in.), to the third is $\frac{3}{16}$ in. to the fourth is $\frac{1}{4}$ in. ($\frac{2}{8}$ or $\frac{4}{16}$ in.), and so on. Accuracy to $\frac{1}{32}$ in. can be achieved with this scale by estimating the halfway distance between each $\frac{1}{16}$-inch mark, which represents $\frac{1}{32}$ in.

Appendix V lists the conversion of metric, decimal, or fractional numbers into either the English or metric systems.

Architect's scales are graduated in the English system and are used for making reduced, actual-size, and enlarged drawings. The major divisions on these scales represent 1 foot. For example, one-half scale means that $\frac{1}{2}$ in. is equal to 1 foot. The triangular type has scales that read from left to right with others completely overlapping those to read from right to left. Each of these scales has a 0 index positioned at both ends of each scale edge. Little confusion results from the use of this scale, since each end is clearly marked with its scale size designation near its 0 index. A fully-divided major unit for that scale designation is also provided outward from each 0 index mark. Each minor subdivision represents a specific fractional inch. The measurement in fractions of an inch for the minor division of each scale size is given in Appendix VI for architect's and mechanical engineer's scales. The fully-divided portion of the scale must be used to measure a value larger than an even number of feet. To measure any distance under 1 foot, the subdivided major unit at the end of the scale is used.

When an enlarged drawing of an object is required, the $1\frac{1}{2}$- or 3-in. scales are used to make $1\frac{1}{2}$ or 3-times full-size drawings. For twice actual-size drawings, use the full-size scale and double each of the actual measurements before laying them out on the drawing. For example, to represent $2\frac{1}{2}$ in., lay out 5 in. on the scale.

To make a one-half scale drawing, the full-size scale is used by dividing the actual measurements of the object by 2. For example, to lay out 3 in. to one-half scale, measure and mark $1\frac{1}{2}$ in. on the drawing.

The *mechanical engineer's scales* are very similar to the architect's scales except that the major divisions between 0 index marks represent 1 in. Thus each minor subdivision represents part of an inch. Refer to Appendix VI for the measurement of the minor divisions for each of the scale sizes. When measuring distances under 1 in., the subdivided major unit at the end of the scale should be used.

The mechanical engineer's scale is most useful for making reduced drawings where measurements are in inches and fractions of an inch. For example, for a one-half scale drawing, a distance of $5\frac{9}{16}$ in. can be measured off directly using the one-half scale without involving calculations.

Appendix VII provides a guide for determining the scale reduction when using any of the designated scales on architect's, civil engineer's, mechanical engineer's, and metric scales.

1-12

The Compass

A *compass* is used to draw precise circles and arcs. The three basic styles are the *bow*, *drop*, and *beam* compass. These are shown in Fig. 1–29a. For most electronic drafting applications, the bow compass is commonly used. This instrument consists of two legs that meet at a pivot point. One leg is provided with a needle point to locate the center of a circle. The end of the other leg has a clamp which will accept standard $\frac{1}{16}$-in. drafting lead or an adjustable technical pen adapter. A fulcrum bearing is located at the pivot point, which is

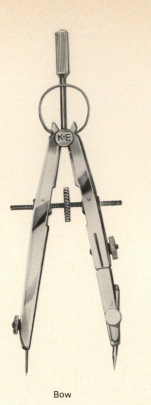 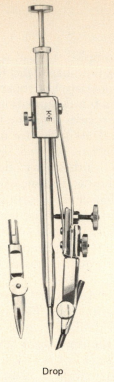 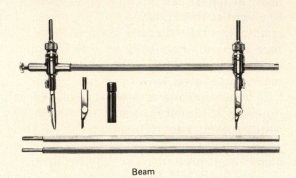

Bow

Drop

Beam

(a) 3 Styles of compasses

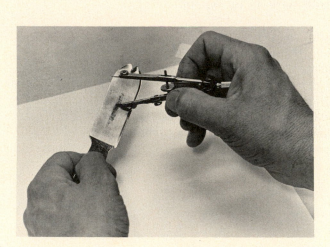

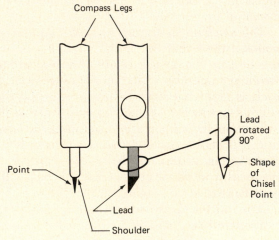

Compass Legs

Point

Lead rotated 90°

Shape of Chisel Point

Lead

Shoulder

(b) Sharpening compass lead

(c) Correct position of compass lead

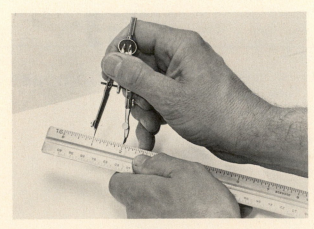

32

(d) Setting compass

(e) Constructing circle

surrounded by a spring steel crown. A knurled handle is attached through the top of the crown to the bearing. Adjustment of the spacing between the legs is made by turning a thumb wheel which is positioned on a spindle threaded into the compass legs.

To draw a circle properly with a compass, the end of the lead must first be sharpened into a beveled configuration by carefully shaping it on a sanding block. This is shown in Fig. 1–29b. (Remember to remove the graphite residue from the lead before use.) The lead is then clamped into the leg of the compass with the flat portion of the lead tip positioned outward from the leg into which it is clamped. The lead is then pushed into the clamp to a distance so that its tip is slightly above the needle point of the other leg, but not higher than its shoulder. See Fig 1–29c. (Needle points that are not provided with a shoulder should not be used on a compass. These were made to be used with dividers and can cause holes that are larger than necessary when used with a compass.) If the lead is properly positioned, it should just touch the surface of the paper when the needle point is about halfway into the paper.

The leg spacing is set by first positioning the needle point over a convenient index mark on the desired scale. With the point held firmly, the thumb wheel is rotated between the thumb and index finger until the point of the lead is directly in line with the scale division that is equal to the radius (one-half the diameter) of the circle to be drawn. See Fig. 1–29d.

Refer to Fig. 1–29e for techniques in drawing a circle. The needle point of the previously set compass is positioned into the center of the circle. The knurled handle is then grasped by the thumb and forefinger and, with the lead resting on the paper, the compass is rotated in a clockwise direction (counterclockwise if left-handed). It is good practice to make the first rotation with a light pressure and then check the diameter of the circle with a scale. It is much easier to erase lightly drawn lines if an error is detected. When satisfied that the circle diameter is correct, additional compass rotations are made with increasing pressure applied until the desired line weight is obtained. While rotating the compass, it should be tilted slightly forward in the direction of the drawing rotation.

Prepared compass leads are available in $1\frac{1}{4}$-in. lengths with sharp chisel points. They come in grades of HB, F, H, 2H, 3H,, 4H, 5H, and 6H. When using the compass to draw construction lines, 4H or 6H leads should be used. If the line weights of circles or arcs are to be black, softer leads should be used. It is not advisable to place heavy downward pressure on the compass to make lines darker. Rather, the lead used in the compass should be one grade softer than that used for the drawn lines.

The bow compass is generally used for drawing circles having diameters of approximately 2 to 12 in. For circles having diameters of about $\frac{1}{32}$ to $2\frac{1}{2}$ in., the drop compass is used. For large circles, a beam bar may be attached to the leg of a bow compass, which normally holds the lead. These bars greatly extend the drawing capacity of the bow compass. Beam compasses are better suited for drawing circles having diameters of 16 in. or larger.

When making drawings that require arcs or circles to be tangent to a straight line, it is recommended that the arc or circle be drawn first. It is easier to draw a straight line tangent to a circle than the other way around.

For many electronic drafting applications, the extreme accuracy of circle size and placement obtainable with a compass is not a rigid requirement. In these cases, drafting templates are used extensively with greater saving of time, yet with no sacrifice of line quality. Drafting templates are discussed in the next section.

1-13

Drafting Templates

Drafting templates are drawing aids that contain a variety of cutouts having different shapes and outlines. Their use greatly reduces drafting time. Templates are generally 0.030 in. thick and are made of clear or tinted plastic. They provide just about any standard drafting symbol, including specialized applications in fields of chemistry, mathematics, architecture, piping, welding, highway mapping, mechanical, fluid power, and landscape.

For applications to electronic drafting, standard graphic templates include most of the common component symbols. *Logic* templates are used to produce circuit schematic diagrams that are basically digital in nature. *Linear* templates are used where the circuit is largely analog (i.e., transistors, op-amps, etc.). For printed circuit board design, scaled component layout templates include many common electronic component outlines. (Printed circuit board layout with the use of templates is discussed in detail in Chapters 7 and 8.)

The most widely used templates are the *circle* template and the lettering guide, which is discussed in Chapter 2. The circle template consists of a series of precision die-punched holes. Different styles of circle templates include various ranges of hole diameters in decimal, fractional, or metric sizes. Combinational templates provide a series of hole sizes that are designated in decimal, fractional, and metric sizes on the same template. For convenience, some circle templates have decimal, inch, or metric scales engraved along the edges.

It is not possible for one circle template to provide the full range of sizes to satisfy every requirement. Appendix VIII compares the range of circle sizes for a number of common template styles. The cutout sizes of the circles are always slightly larger in diameter (by approximately $\frac{1}{32}$ in.) than the size stamped or printed beside each hole. This is to compensate for the thickness of the pencil point as the holes are drawn.

The circle template, as well as a selection of other popular templates widely used in electronic drafting, are shown in Fig. 1–30.

To use the template effectively, construction lines used as guides are required for proper positioning of the symbol. The use of a grid system, such as that on standard graph paper, also may serve as an accurate guide. These guide lines are viewed through the transpar-

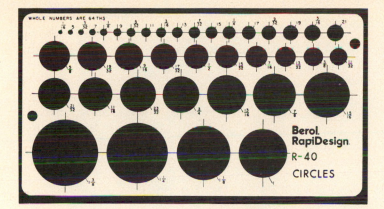

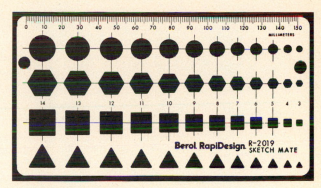

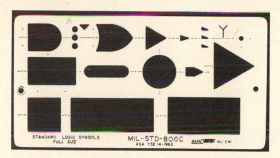

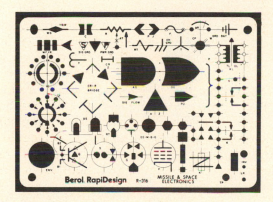

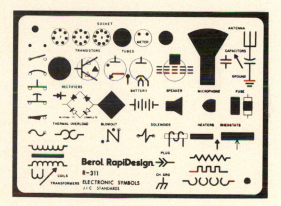

FIGURE 1–30 Popular templates used in electronic drafting. [Courtesy of Berol USA.]

ent templates for aligning the symbol. The positioning of a circle template together with the proper way to hold the pencil are shown in Fig. 1–31. Note that the pencil is held vertically with the point constantly in contact with the edge of the circle where the template contacts the drawing surface. Also note that the four index marks, which are engraved about each hole and 90 degrees apart, are in perfect alignment with the two previously constructed center lines used to locate the position of each hole center. To be useful, these center lines should be drawn dark enough to be sufficiently visible through the template. Of course, these center lines must be longer than the diameter of the circle so that they pass completely beyond the four index marks on the template. Otherwise, it will be difficult, if not impossible, to align the hole properly.

Circle templates are also available with ink button lifts, called *bosses*, which are an integral part of the template. Although this

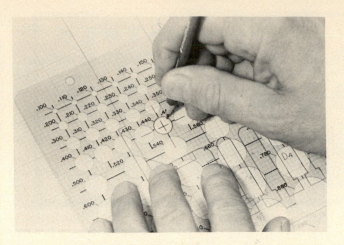

FIGURE 1–31   Template and pencil positioning to construct circle.

template may be used with either pencil or technical pen, the bosses allow more convenient use for drawing with pen. For this application, the bosses are positioned face down, which causes the circle edges to be slightly elevated from the drawing surface. See Fig. 1–32a. This reduces the possibility of ink smearing or smudging the

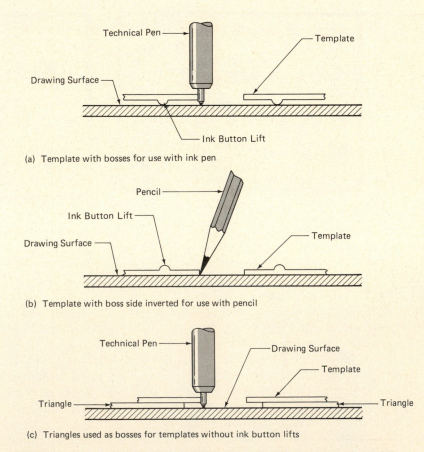

(a) Template with bosses for use with ink pen

(b) Template with boss side inverted for use with pencil

(c) Triangles used as bosses for templates without ink button lifts

FIGURE 1–32   Positioning templates for use with ink pen or pencil.

drawing. When used with pencil, the template is positioned so that the bosses are facing upward, thus allowing the circle edges to lie flat against the drawing surface. See Fig. 1–32b.

Any style of template that is not fitted with bosses may be used with a technical pen by raising it slightly off the drawing surface. This can be effectively accomplished by placing the template over two triangles which have been spaced far enough apart so that their edges will not interfere with the pen tip as the symbol is drawn. See Fig. 1–32c.

1-14

French Curves and Sweeps

When drawing charts or graphs, it is often necessary to produce lines that have irregular shapes or contours that are not possible to draw with a compass or templates. For these purposes, the *French curve* or *adjustable curve* are used. The French curve is available in a variety of forms, sizes, and shapes. Some typical French curves are shown in Fig. 1–33a. They are made from clear or tinted plastic.

FIGURE 1–33 French and adjustable curves. [(a) courtesy of Kratos/Keuffel & Esser; (c) courtesy of Utley Company, Inc.]

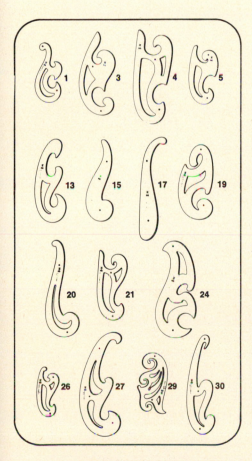

(a) Plastic French curves

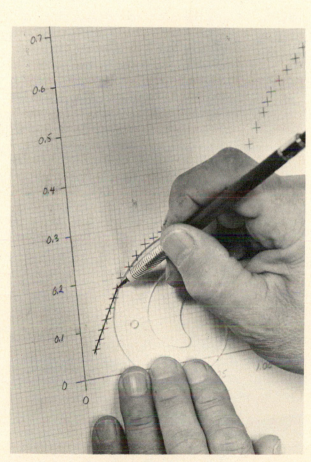

(b) Using a French curve

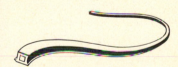

(c) Adjustable curve

Cross sectional view

The curves are designed to aid in drawing contoured lines by matching the shapes of various sections of the intended line shape with those portions of the curved edge that best align with these shapes. This procedure of alternately matching contoured sections and blending line segments with each other is continued until a smooth continuous line is drawn. Use of a French curve to construct an irregularly shaped line is shown in Fig. 1–33b. Note that a section of the French curve is made to align with as many points on the intended line as possible. The segment of the line is drawn and the curve is again repositioned to align with other points. This is repeated as many times as necessary to ensure that a smooth continuous line results.

An *adjustable curve* or *sweep* is shown in Fig. 1–33c. It consists of a square-section lead core encased in a smooth vinyl body. This curve can be shaped into a pattern to align with the required line contour. If it is to be used with ink, the side of the curve with a raised rib or flange should be used.

____ EXERCISES _____

1–1 Perform the following tests to ensure the accuracy of your T-square and drafting triangles:
(a) Test the 90-degree angle on both the 45- and the 30–60-degree triangles.
(b) Test the 45-degree angle on the 45-degree triangle.
(c) Test the 60-degree angle on the 30–60-degree triangle.
(d) Check that the T-square edges are parallel.

(e) Check that the T-square head is at a 90-degree angle to the blade edges.

1–2 Using decimal, fractional and metric scales, measure the sizes of the elements shown in Fig. 1–34 and record all values obtained on a table similar to that provided.

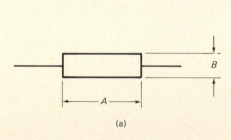

(a)

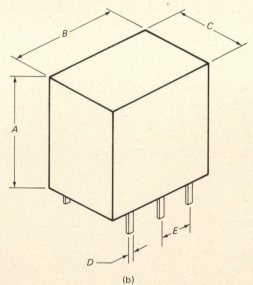

(b)

FIGURE 1–34

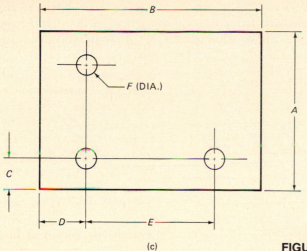

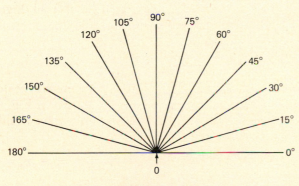

(c)

FIGURE 1–34 (cont.)

Figure	Units	A	B	C	D	E	F
(a)	Decimal			—	—	—	—
	Fractional			—	—	—	—
	Metric			—	—	—	—
(b)	Decimal						—
	Fractional						—
	Metric						—
(c)	Decimal						
	Fractional						
	Metric						

1–3 Using a T-square and triangles, divide 180 degrees into 15-degree increments and draw 2-in. line segments radiating from point 0 as shown in Fig. 1–35.

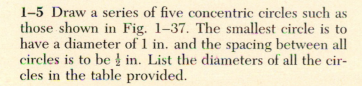

FIGURE 1–35

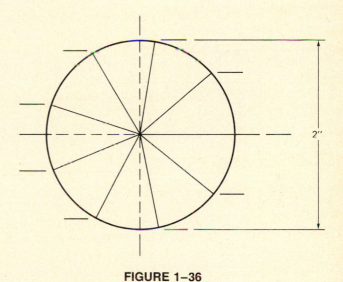

FIGURE 1–36

1–4 Draw a circle having a 2-in. diameter with a template and locate its center. With a protractor, divide the circle into nine equal sections as shown in Fig. 1–36 and label the angles at the points of intersection between each radius and the circumference.

1–5 Draw a series of five concentric circles such as those shown in Fig. 1–37. The smallest circle is to have a diameter of 1 in. and the spacing between all circles is to be $\frac{1}{2}$ in. List the diameters of all the circles in the table provided.

Circle	Diameter
A	
B	
C	
D	
E	

FIGURE 1–37

1–6 Draw a grid system similar to the one shown in Fig. 1–38. The grid is to be four divisions per inch using a 6-in. square. Locate uniform plotted points in the positions shown and draw them using $\frac{3}{16}$-inch line segments. With a French curve, draw a smoothly contoured line through the plotted points.

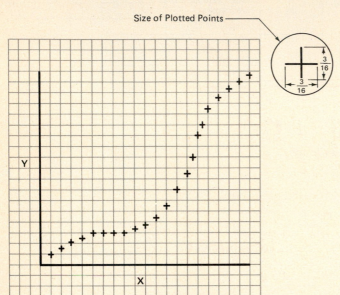

FIGURE 1–38

1–7 Draw the object shown in Fig 1–39 to twice the actual size from the values given in the table provided. Show dimensions in decimal values, fractional values, and metric values.

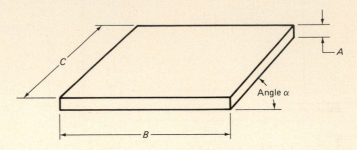

A	B	C	Angle α
0.40″	3.35″	4.12″	30°
$\frac{1}{8}$″	$2\frac{1}{2}$″	$1\frac{5}{8}$″	45°
5 mm	63 mm	78 mm	60°

FIGURE 1–39

2

Linework, Lettering, and Dimensioning

_____ LEARNING OBJECTIVES _____

Upon completion of this chapter on linework, lettering, and dimensioning, the student should be able to:

1. Know the accepted line conventions used in drawings.

2. Letter drawings freehand using the single-stroke Gothic style and the microfont systems.

3. Correctly use lettering triangles, guides, templates, and devices.

4. Properly use dry transfers for lettering.

5. Know the proper use of the conventions for dimension lines, numerical values, notations, and symbols on drawings.

6. Locate dimensions on drawings.

7. Properly use leaders to convey dimensional information.

8. Correctly dimension angles.

9. Accurately locate the position of holes.

10. Understand fractional, decimal, and metric systems of dimensioning.

11. Understand the conventional methods of dimensioning.

2-0

Introduction

Standard conventions for linework, lettering, and dimensioning have been adopted for converting engineering ideas into detailed drawings. It is essential that the draftsperson be familiar with these internationally recognizable conventions so that the pictorial information is presented in a neat, clear, and precise form to minimize any possibility of confusion or misinterpretation.

In electronic drafting, component and device code numbers, letters, and values need to be clearly and correctly formed. For this reason, guidelines to freehand lettering are presented in this chapter. In addition, dry transfer lettering and lettering guides are included.

Electronic equipment is typically packaged in sheet metal enclosures which are fabricated to specific sizes using dimensional drawings. The draftsperson needs to present these drawings in a standard format that is familiar to the fabricating technician. The standard conventions for dimensioning are presented in detail in this chapter.

2-1

Line Conventions for Electronic Drafting

Standard line symbols have been adopted by the American National Standards Institute (ANSI) to aid in conveying detailed pictorial information. These standards are often referred to as the "alphabet of lines." Each type of line on a drawing is drawn in a specific way and has its particular meaning.

All lines are boldly drawn so that they may easily be reproduced in a blueprint machine. The only exception to this is construction lines whose sole purpose is to aid in alignment. The thicknesses of the lines vary from approximately 0.5 to 0.8 mm. In general, the larger the drawing and the more significant the portion being drawn, the thicker it is drawn. When photoreduction of the drawing is not required, three different line weights (widths) may be used to aid in highlighting the features. Prominent portions of the drawing would be drawn with thick lines and hidden lines drawn thinner ($\approx$0.5 mm). Secondary lines, such as those used for leaders, center lines, dimension lines, ditto lines, extension lines, and phantom lines, may be drawn even thinner. Figure 2–1 shows standard line symbols and the recommended range of thicknesses together with a brief description of their application in electronic drafting. In all cases, the selected line width must be consistent throughout the drawing in addition to being uniformly sharp and opaque.

The thickness of *outlines* are drawn thick or very thick since they are used to highlight important or critical circuitry, such as sensitive operational amplifier inputs, a large ground bus, or noise-sensitive circuitry in high-speed digital applications. The thick lines make the outlined areas more prominent to the eye.

Visible lines are used in drawing the majority of device and part edges in a pictorial assembly drawing. On circuit schematic diagrams, these lines are used in the construction of component or device symbols as well as all circuit interconnections, with the exception of those where outlines are used.

Construction lines are the only lines not drawn boldly. They are intended to be used as guides for lettering and numbering, to outline the work before preparing the final drawing, to estimate size, and to center objects. Construction lines are drawn lightly and are usually erased after they have served their purpose. If they are drawn with the harder leads (4H, 5H, or 6H), erasing will not be necessary since these lines are not reproducible on blueprint machines.

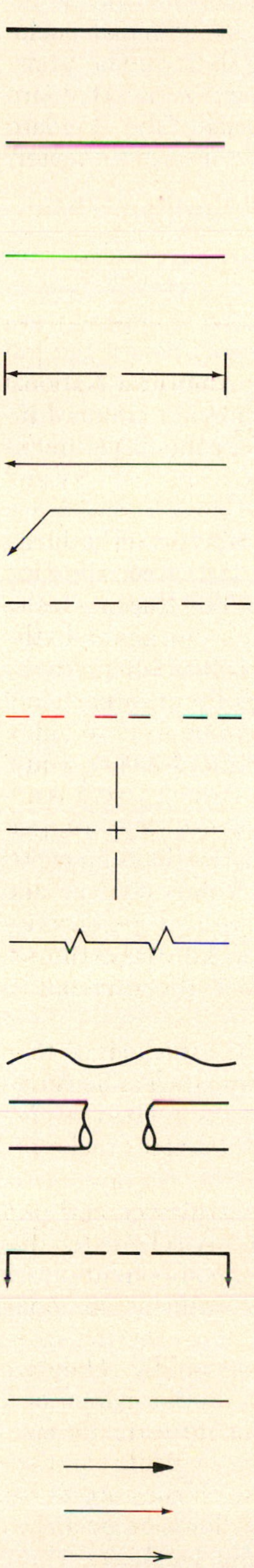

Line Symbol	Line width	Name	Application
————————	Thick to very thick (1.0–1.2 mm)	Outline Line	Used to outline critical circuitry or borders
————————	Thick (0.7–1.0 mm)	Visible Line	Outline parts, symbols and most circuit schematic paths
————————	Thin (0.18–0.35 mm)	Construction/ Guide Line	Typically light lines either erased or which are not reproducible
⊢←——————→⊣	Thin (0.18–0.35 mm)	Extension and Dimension Lines	Extension-lines defining extent of dimension. Dimension-used for detailing size with linear measurements. Value inserted between arrows
←————	Thin (0.18–0.35 mm)	Leaders	For parts or dimensional information
– – – – – – –	Medium (0.35–0.50 mm)	Hidden/Bend Lines	For metal fabrication, hidden views, future add-ons, etc.
– — – — – —	Medium (0.35–0.50 mm)	Ditto Lines	Repeated detail
─┼─	Thin (0.18–0.35 mm)	Center Lines	Part or hole center
∿∿─────	Thin (0.18–0.35 mm)	Long-break Lines	Reduce the actual or scaled length of a repetitive information
∿∿∿	Thick (0.7–1.0 mm)	Short-break Lines	Same as above
)( )(	Thick (0.7–1.0 mm)	Cylindrical-break Lines	Same as above for drawing with round cross-sections
↓┌─ ─ ─┐↓	Thick (0.7–1.0 mm)	Cutting Plane Line	Designates position of cutting plane and direction of view for sectional drawings
— — – — —	Thin (0.18–0.35 mm)	Phantom Lines	Used to indicate parts either adjacent or removed. Also lines of motion or alternate position of moving parts.
——▶	Approximately 2 mm wide by 6 mm long	Arrowheads	Used to indicate either the end of a dimension line or direction of a sectional view or points to a specific item.

FIGURE 2–1 Standard line symbols used in electronic drafting.

Extension and *dimension lines* are drawn thin but boldly in order to be easily reproducible. The dimension line is broken in the middle between the two extension lines to allow space for the designated distance (with tolerance if required). The dimension line terminates at each end with an arrowhead. See Fig. 2–1. The width of the arrowhead should be one-third the length with the point just touching the extension line. Depending on the scale of the drawing, the length of the extension lines should terminate between $\frac{1}{8}$ and $\frac{1}{16}$ in. below the arrowhead. The origin of the extension line should be brought up to, but not touch, the line being dimensioned.

Leader lines are drawn thin with medium-width arrowheads and are used as informational pointers. For example, in crowded schematic diagrams, leaders can be used to identify parts or indicate values by "pointing" at the part and placing the information at the tail end (i.e., opposite of the arrowhead). Leaders are also used for dimensioning. For example, they may be used to show the dimensions of small holes in a chassis, the sizes of which are too small to dimension conveniently by other means.

Hidden or *bend lines* are medium width with the line sections drawn longer than the spacing between each line. The lengths of both the lines and the spacings should be uniform across the entire distance of the hidden line. These lengths are dependent on the drawing scale. In general, $\frac{1}{4}$-in. line segments separated by $\frac{1}{8}$-in. spacings are suitable for most drawings. Hidden or bend lines are used for (1) showing the hidden edges of an object and (2) marking the position along which a bend is to be made in a sheet metal drawing.

Ditto lines are also medium width in a series of two identical line segments with a small spacing between each line and a larger spacing between each group of two. Ditto lines indicate repeat work such as showing the second through the seventh channels of an eight-channel shift register. These lines make it unnecessary to draw the complete circuit detail.

Center lines are drawn thin and are used to locate the center of holes, such as drill or punch holes in a chassis. The center line is drawn with alternating long and short segments. At the center point of the hole, the two short line segments of both center lines, drawn at 90 degrees to one another, should cross. Center lines should always terminate in a long line segment which extends beyond the perimeter of the hole.

When a portion of an object is to be reduced on the drawing, *break lines* are used. For example, if it is inconvenient to show the full length of a potentiometer control shaft because of space restrictions, a cylindrical break line may be used. These lines are drawn thick since they are usually extensions of lines used to outline parts. *Long* and *short* break lines are employed for purposes of conserving space on a drawing where the length of a part can be reduced or repetitive detail eliminated. Short break lines are drawn thick, while long break lines are thin straight lines interrupted at regular intervals by freehand zigzags.

Cut-plane lines indicate the plane in which a part of an object has been removed. These lines are drawn thick and include an arrowhead at each end and at 90 degrees to the split line segments. These arrowheads indicate the direction of the sectional view.

Finally, thin *phantom lines* are used to outline the orientation of detailed parts to other sections of an assembly without having to draw the detail of these sections.

The application of the line symbols just described will be demonstrated in the next several chapters.

2-2

Freehand Lettering

The style of lettering recommended by ANSI is the *single-stroke Gothic* style, which is the most common form used in electronic drafting. Gothic-style letters and numbers may be drawn either vertically or inclined at approximately 68 degrees to the right. The vertical lettering is more popular since it is easier to read. Gothic letters are formed in either uppercase or lowercase style. On a specific drawing, consistency in the use of vertical or inclined lettering is essential. Combining both forms on a drawing is not acceptable practice. Of course, lower- and uppercase letters may be used on an individual drawing.

Lettering is best accomplished with a medium-grade lead such as F or HB, as specified in Table 1–2. All freehand lettering should be evenly proportioned and spaced with uniformity of height, slant, and alignment. The height of lettering is related to the size of the drawing, but is generally $\frac{1}{8}$ to $\frac{1}{4}$ in. for most electronic drafting applications. Proper lettering skills require practice for developing neatness and speed. The beginner should approach this practice with patience and not be discouraged at the first attempt. From one to four types of strokes are required for the proper formation of Gothic numbers and letters. These strokes and the characters to which they are associated are shown in Fig. 2–2.

The complete alphabet of uppercase vertical Gothic letters and numbers is shown in Fig. 2–3. In order to demonstrate the relative height and width of each character as well as its shape, they are drawn on a 6×6 square grid. In addition, the numbered arrowheaded line

FIGURE 2–2 Common strokes and associated lettering.

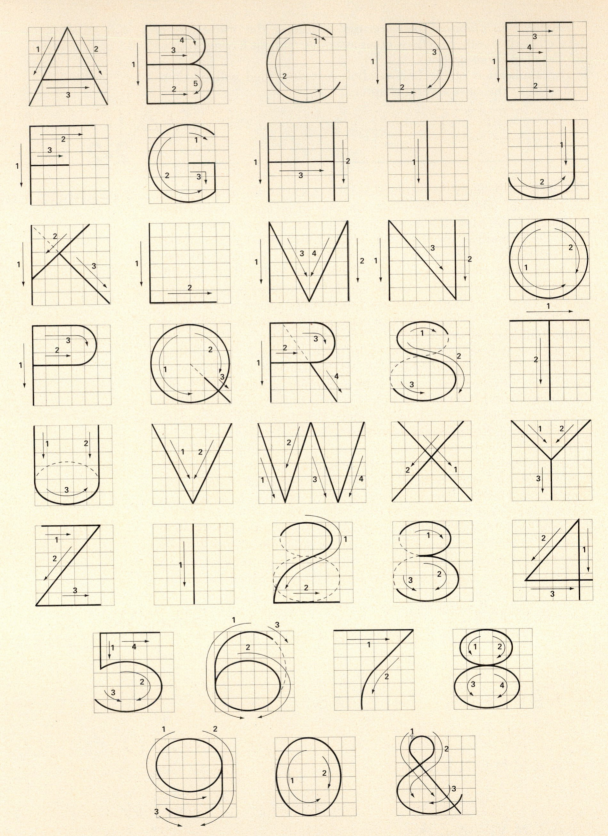

FIGURE 2–3 Gothic style upper-case vertical letters, numbers, and symbols.

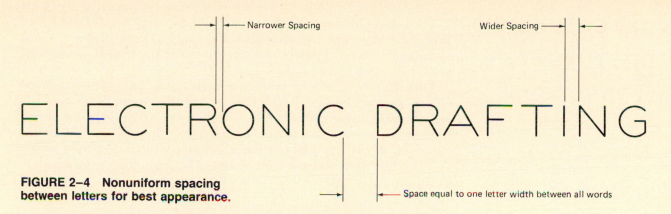

Narrower Spacing

Wider Spacing

ELECTRONIC DRAFTING

FIGURE 2-4 Nonuniform spacing between letters for best appearance.

← Space equal to one letter width between all words

segments illustrate the recommended order and direction of stroke for properly constructing each of the letters and numbers. Note that the letter W is 8 squares wide. All of the others fall within the 6-square grid. Note also that only the letters A, M, O, Q, T, V, X, Y, and Z are as wide as they are high, that is, 6 squares. All other characters are five-sixths as wide, with the exception, of course, of the number 1. Finally, the letter O is a full 6 squares wide, while the number zero (0) is only 5 squares wide.

The spacing between individual letters and between words is very important in order that the information be easily followed with the eye. Spacing between words is normally fixed at a distance which will accept one letter, usually the equivalent of the letter O. The spacing between letters and numbers will vary and will depend on their appearance being even. See Fig. 2-4. The spacing between letters having straight edges, such as the letters E, N, H, I, M, and so on, should be greater than for letters having slanting or rounded edges, such as the letters W, A, O, C, and so on.

The recommended distance between lines of lettering is $\frac{1}{2}$ to 1 times the height of the letters. The distance between lines of unrelated technical information should be at least 2 times the height of the letters.

abcdefghijklmnopqrstuvwxyz

(a) Vertical lower-case Gothic letters

ABCDEFGHIJKLMNOPQRST
UVWXYZ   1234567890&

(b) Slanted upper-case Gothic letters and numbers

abcdefghijklmnopqrstuvwxyz

(c) Slanted lower-case Gothic letters

FIGURE 2-5 Several Gothic style letters and numbers.

ABCDEFGHIJKLMNO

PQRSTUVWXYZ

1234567890

FIGURE 2–6 **Microfont lettering and numbering.**

The other Gothic styles, lowercase vertical, uppercase slanted, and lowercase slanted, are shown in Fig. 2–5. These styles are not as commonly used in electronic drafting as is the uppercase vertical style.

Microfilming of large drawings has become common in recent years. When 35-mm film is enlarged back to full size, there is a loss of letter resolution which results in a reduction of clarity. This loss of resolution is caused by the use of Gothic-style letters. To overcome this problem, ANSI has adopted the *microfont* lettering system, which is reproducible without loss of clarity. The microfont lettering style is shown in Fig. 2–6.

2-3

Guide Lines for Freehand Lettering

Freehand lettering begins by first drawing a series of very light guide lines with a hard pencil lead, such as 4H. These lines will not be reproduced in a blueprint machine. Two parallel horizontal guide lines are first drawn so that the distance between them will be equal to the desired letter height. The upper line will limit the top of the uppercase letters and the lower line will limit the bottom of these letters. Each capital letter drawn should touch both of these lines but not extend beyond them. A third horizontal guide line centered between the two previously drawn is extemely helpful in forming the uppercase letters B, E, F, H, P, R, and S and the numbers 3 and 8.

In addition to the horizontal guide lines previously drawn, a series of randomly spaced vertical guide lines are drawn which extend through the horizontal lines and at 90 degrees to them. These are shown in Fig. 2–7a and are used as an aid in alignment. Where large and small capital letters are to be used, as shown in Fig. 2–7b, a guide line is drawn between the upper and lower lines previously drawn so that the height of the small capital letters will be three-fifths to two-thirds that of the large letters.

For drawing lowercase vertical lettering together with uppercase lettering, four horizontal guide lines are necessary. These are shown in Fig. 2–7c. Starting at the uppermost guide line and moving in descending order, these lines are termed *cap line, waist line, base line,* and *drop line.* Lowercase letters are also three-fifths to two-thirds of the height of the capital letters. Therefore, the distance between the base line and the waist line are drawn to this ratio. The distance between the base line and the drop line is made the same as that between the waist line and the cap line. To serve as an example of these distances, if the height of the uppercase letters is to be $\frac{3}{16}$ in., the waist line will be $\frac{3}{16} \times \frac{2}{3} = \frac{6}{48}$ or $\frac{1}{8}$ in. above the base line. The distances between the waist and cap lines and the drop and base lines will then be $\frac{3}{16} - \frac{1}{8} = \frac{3}{16} - \frac{2}{16} = \frac{1}{16}$ in. each. The drop line is often omitted since the only letters extending below the base line are g, j, p, q, and y.

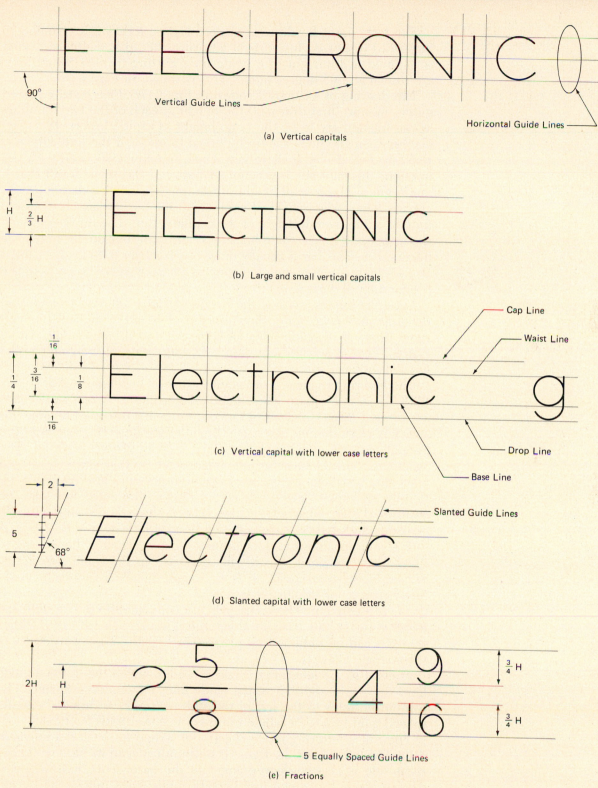

(a) Vertical capitals

(b) Large and small vertical capitals

(c) Vertical capital with lower case letters

(d) Slanted capital with lower case letters

(e) Fractions

FIGURE 2–7 Guidelines for freehand lettering.

The techniques for drawing slanted Gothic lettering are similar to those just discussed for vertical lettering. Because slanted lettering is inclined to the right, the guide lines cutting the horizontal lines are drawn at a 68-degree angle. This is shown in Fig. 2–7d.

The drawing of fractions is shown in Fig. 2–7e. Five equally spaced guide lines are first drawn. If the height of each of the two numbers in the fraction are to be the same height as capital letters, the spacing between the lines should be equal to one-half the letter height. Refer to Fig. 2–7e. If there is a whole number associated with the fraction, this number is positioned between the second and fourth guide lines. The overall height of the fraction is twice that of an uppercase letter. The individual numbers in the fraction are about three-fourths of the overall height of the capital letter. The numbers touch the upper and lower guide lines (first and fifth) but do not touch the fraction bar. This bar is usually drawn horizontally but may be slanted at a 45-degree angle when lettering in a narrow space, such as in the preparation of parts lists.

Lettering guides, such as the *Braddock/Rowe lettering triangle* and the *Ames lettering guide*, shown in Fig. 2–8, aid in the accurate and rapid drawing of guide lines for lettering and dimensioning. These guides serve as a faster and more efficient means of drawing parallel lines at uniform spacings.

The lettering triangle is designed so that its hypotenuse slides along the edge of a T-square when standard spacings between guide lines are to be drawn. Columns of tapered holes are provided to allow for a variety of guide line spacings. The holes are tapered to prevent the breaking of the pencil point and are used in groups of three. The spacing of each of these groups is arranged so that properly spaced lines may be drawn for the use of both upper- and lowercase letters. If only uppercase letters are to be drawn, the waist line may be omitted. Using adjacent vertical sets of three holes automatically provides the correct space between lines of lettering.

The number under each column of holes designates the spacing in $\frac{1}{32}$ in. between holes which will serve as cap lines and base lines. The numbers used are 3, 4, 5, 6, 7, and 8, which represent guide line spacings for uppercase letters of $\frac{3}{32}$, $\frac{4}{32}$ ($\frac{1}{8}$), $\frac{5}{32}$, $\frac{6}{32}$ ($\frac{3}{16}$), $\frac{7}{32}$, and $\frac{8}{32}$ ($\frac{1}{4}$) in. respectively. These spacings are conveniently drawn with the hypotenuse of the triangle along the T-square. A variety of other spacings can be obtained by placing each of the other two edges of the triangle along the T-square.

The slot in the lettering triangle is used to obtain the inclined guide lines drawn through horizontal guide lines for slanted lettering. The column of holes to the left of the slot are used for drawing equally spaced guide lines for whole numbers together with fractions.

The Ames lettering guide consists of a rigid frame and a movable disk. The left side of the frame is perpendicular to its base while the right side is at a 68-degree angle to the base. This drafting aid is used to draw horizontal and vertical guide lines quickly and accurately, in addition to 68-degree inclined lines for letters $\frac{1}{16}$ to 2 in. in height. The frame has one column of equally spaced holes $\frac{1}{8}$ in. apart over a range of 0 to 2 in. while the disk has four columns of

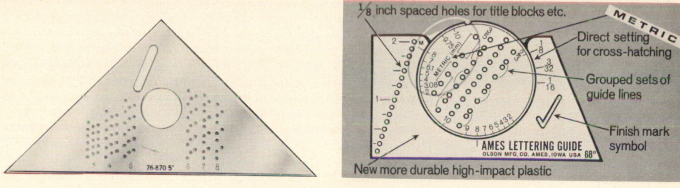

(a) Lettering triangle and Ames lettering guide

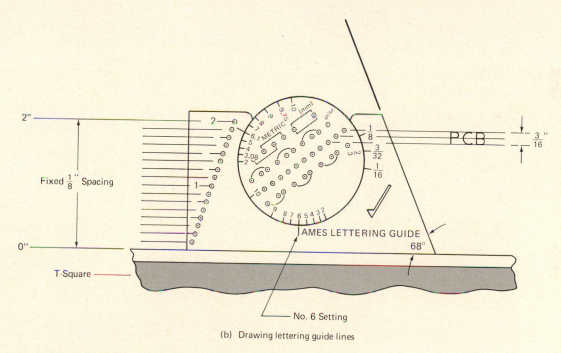

(b) Drawing lettering guide lines

FIGURE 2–8 Lettering triangle and Ames lettering guide. [(a) courtesy of Utley Co., Inc. and Kratos/Keuffel & Esser.]

holes. From left to right, the first column is for drawing guide lines to metric dimensions. When a disk number to the left of this column is aligned with the metric index mark on the frame, one of a selection of 10 guide line spacings are available, ranging from 2 to 10 mm apart. The next column of holes is in groups of three for the drawing of guide lines for which the height of lowercase letters is to be three-fifths that of uppercase letter. The third column is of equally spaced holes and the fourth column has its holes again in groups of three but for drawing guide lines for which the height of lowercase letters will be two-thirds that of uppercase letters.

The selection of guide line spacings for the last three columns is made by aligning the appropriate disk number with the index mark which is located near the bottom of the frame. These numbers range from 2 to 10 and represent spacings in 32nds of an inch, thus providing guide line spacings from $\frac{1}{16}$ $\left(\frac{2}{32}\right)$ to $\frac{5}{16}$ $\left(\frac{10}{32}\right)$ in.

The drawing of guide lines with this versatile aid is done much the same as with the lettering triangle. After the desired spacing has been set, the pencil point is placed into the appropriate tapered hole and the base of the frame is smoothly drawn along the edge of a T-square. This is shown in Fig. 2–8b.

The Ames lettering guide can also be used to draw vertical guide lines by sliding the base of the frame along the vertical side of a triangle which is resting on a T-square.

2-4

Lettering Guides, Templates, and Devices

Other aids which are available to give the draftsperson alternatives to freehand lettering are *plastic lettering guides*, *lettering templates*, and *lettering devices*. Lettering guides are shown in Fig. 2–9a. The lettering guide has a plastic body containing cutouts which form letters and numbers. These are available for drawing both vertical and slanted styles for numbers and upper- and lowercase Gothic letters.

For lettering with the guide, it is supported along the edge of a T-square which serves as a base to maintain the position of the letters. The first letter cutout is positioned in the desired location and the guide and T-square are held firmly in place with the left hand. The letter is then drawn by holding the pencil perpendicular to the paper and tracing the outline of the letter. See Fig. 2-9b. To form the next letter, the guide is moved so that the desired letter is positioned with appropriate space allowed between letters. Because the lettering guide is transparent, previously drawn letters can be easily seen, thus making the task of spacing rather simple.

Forming letters with the guide can also be easily done with a technical pen. For this purpose, a specially designed lettering joint is available. The barrel of the pen is unthreaded from the tubular point and rethreaded onto the end of the movable joint. The pen is clamped into the vertical position while allowing the barrel to be held at the normal writing angle.

FIGURE 2–9 Lettering guides make the task of lettering easier. [(a) courtesy of Berol USA.]

(a) Plastic lettering guide   (b) Using the lettering guide

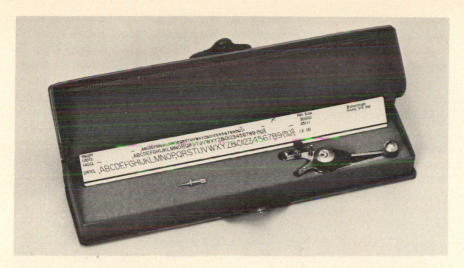

FIGURE 2–10 Lettering template with adjustable scriber and technical pen. [Courtesy Staedtler/Mars.]

Lettering guides are undercut so that ink or graphite smears are avoided when sliding the guide from one character to another. However, when medium lead is used to draw letters, graphite particles will form on the edges of the guide, necessitating periodic cleaning.

A *lettering template* with an adjustable scriber and a technical pen is shown in Fig. 2–10. Vertical or slanted letters may be formed as the guide pin accurately follows the grooved letters in the template. The tail pin of the scriber runs along a base line which is always the same distance from the bottom of all letters on each template. This allows the interchanging of templates or the turning over of a template to form the characters on the reverse side without having to move the T-square against which the template is placed.

The *Varigraph* lettering device, shown in Fig. 2–11a, produces both different size and different style letters. The alphabet and num-

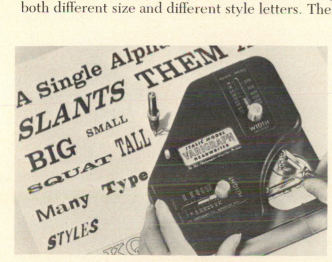

(a) Varigraph lettering guide

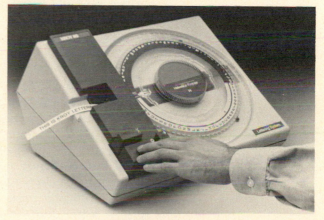

(b) Kroy lettering machine

FIGURE 2–11 Lettering machines. [(a) courtesy of Varigraph, Inc.; (b) courtesy of Kroy, Inc.]

bers of each style are engraved in durable metal matrices. A wide variety of different letter sizes can be obtained from just one matrix since there are adjustment controls for height, width, and slant of letters. This elaborate device has a pointed metal stylus which traces the engraved character on the matrix, resulting in that character being automatically reproduced in ink as single-stroke or "built-up" (multiple-line) lettering.

The *Kroy lettering machine*, shown in Fig. 2–11b, provides for a wide variety of type sizes (8 to 36 point) in popular type styles to be reproduced on tape. This method of lettering is rapid and requires little skill.

The *lettering typewriter*, by its unique design, allows the characters to be typed directly onto the drawing without having to remove it from the drafting board. The typewriter moves along an indexing rail that is placed against a T-square. As each key is struck, the typewriter will automatically index one space. The keys are arranged in alphabetical order and little skill is required once the technique of registering the typed lines has been learned.

2-5

Dry Transfer Lettering

Freehand lettering requires a great deal of practice for mastering the proper techniques. Another form of lettering which is less demanding is with the use of *dry transfer* or *pressure-sensitive* lettering sheets. These polystyrene film sheets are 5 mils thick and are typically 4×7, 5×9, or $8\frac{1}{2} \times 11$ in. in size. They support a variety of letters, numbers, and symbols which are composed of opaque black plastic film. The sheets are provided with a waxed backing sheet which protects the characters but will not adhere to them. This type of lettering can be applied to almost any smooth surface, such as acetate, paper, drafting film, glass, metal, wood, and so on. It will last indefinitely provided that the lettered surface does not come in contact with an abrasive material. The polystyrene sheet is dimensionally stable and will not stretch or distort as the letters are transferred.

To use dry transfer lettering, the protective backing sheet is first removed. The lettered side of the polystyrene sheet is then placed against the surface to be lettered and the desired character is aligned into position. Guide lines, such as those used for freehand lettering, are helpful for positioning dry transfer characters. Once aligned, the letter to be transferred is lightly, but firmly, rubbed with a burnishing tool or simply with a hard-lead pencil (at least 6H) or a ball-point pen. A series of overlapping vertical and then horizontal strokes are applied over the entire character. This is shown in Fig. 2–12. After the character has been burnished, the support sheet is gently raised from the work surface. The sheet is then moved to position the next letter and the rubbing procedure is repeated. Once a word or line of information is completed, the waxed backing sheet is placed over the lettered section and the entire area is gently burnished to complete the work. The resulting lettering can be done in a fraction of

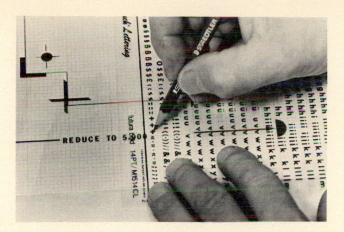

FIGURE 2–12   Application of dry transfer lettering.

the time required by freehand lettering and is superior in uniformity, crispness, and clarity.

Because dry transfer characters consist of plastic film inks, they can crack if the drawing is rolled tightly or folded. This fact needs to be considered when using this method of lettering. Dry transfer lettering is not widely used in general electronic drawings but is invaluable for printed circuit board artworks which require literal and numerical information as well as electronic symbols to be transferred.

2-6

Dimension Line Format

Before the fabrication of any object can be initiated, a working drawing showing its complete detail must first be made. Included in this drawing will be complete instructions for fabrication to meet preestablished specifications. One form of these instructions is *dimensioning,* which is a conventional system of lines, numerical values, notations, and symbols which have been developed by ANSI (American National Standards Institute) and ISO (International Organization for Standardization). If the basic rules and standards of dimensioning are adhered to when making a finished drawing, the information for the fabrication of a piece of equipment will be clear and complete.

Dimensioning indicates the shape, overall size, and relative position of all parts that make up a piece of equipment. It is essential, therefore, that the draftsperson become familiar with the form, meaning, and proper use of dimensioning symbols and techniques. When preparing a finished drawing, two questions must be in the forefront of the draftsperson's mind. These are: (1) is the dimensional information complete and clear, and (2) can the unit be fabricated completely with the information provided? Constant scrutiny of the drawing as it is being developed is needed so that the answer to both of these questions is in the affirmative.

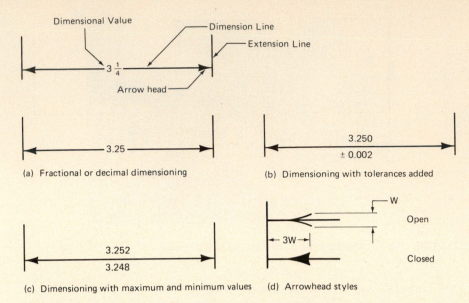

FIGURE 2–13   Conventional dimensioning format.

Dimension lines are thin solid lines terminating in arrowheads which show both the direction and length of the dimension. See Fig. 2–13. The point of the arrowhead touches but does not extend beyond the *extension line* which is used to extend a visible line of the object on the drawing. The dimension line is broken to allow space for the numerical value, either in fractional or decimal form, to be placed. If the dimensional information requires a tolerance to be included, the dimension line is not broken. The nominal dimension is placed above the line and the tolerance value below the line. This is shown in Fig. 2–13b. Dimensional tolerance may also be indicated by showing the maximum and minimum numerical values. In this case, the largest value of the dimension is placed above the line and the smallest value below the line. See Fig. 2–13c.

The arrowheads at the ends of each dimension line show the precise limit of the dimension which is set at the tip of the arrowhead. Arrowheads may be drawn as either *open* or *closed*. Both of these styles are shown in Fig. 2–13d. Regardless of the style, all arrowheads should be drawn with their length equal to three times their width. See Fig. 2–13d. In addition, all arrowheads on a drawing should be the same size and style.

2-7

Location of Dimensions

Dimensions should be located, whenever possible, outside the body of the object. This is not always possible with some complex drawings, however. In any case, the dimensional information should be easily and clearly read when the drawing is held horizontally. Vertical dimensions may be shown with the numerical value either hor-

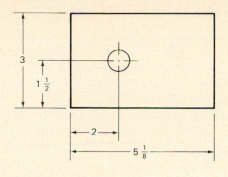

(a) Vertical dimensions shown with horizontal values

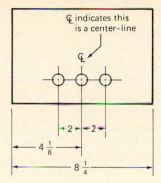

(c) Dimensioning to a center line

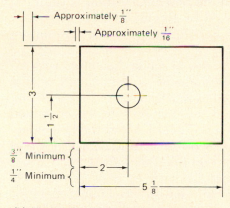

(b) Vertical dimensions are read by rotating this drawing 90° clockwise

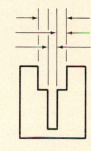

(d) Tight dimensioning may require that gaps be provided in the extension lines

FIGURE 2–14 **Basic use of extension and dimension lines.**

izontally or vertically. These are shown in Fig. 2–14a and b, respectively. By convention, vertically numbered dimensions are drawn so that they can be horizontally viewed by rotating the drawing 90 degrees in a *clockwise* direction. These dimensions are *never* positioned in a direction which requires the drawing to be turned counterclockwise for horizontal viewing.

As shown in Fig. 2–14b, the extension lines begin approximately $\frac{1}{16}$ in. from the visible object line and extend about $\frac{1}{8}$ in. beyond the last dimension line arrowhead. Dimensions should be arranged so that the smallest dimension is closest, but not less than $\frac{3}{8}$ in., to the body of the object. Progressively larger dimensions are then added, terminating with the longest overall dimension placed farthest away from the object. The minimum distance between these dimension lines should not be less than $\frac{1}{4}$ in. As shown in Fig. 2–14a and b, the dimensional values should be staggered so that they are not in line with each other. This allows a clearer presentation and eliminates the possibility of misinterpretation. Where dimensions are referred to a central extension line, such as shown in Fig. 2–14c, they may be placed on the same horizontal line, which presents a clear and more easily read appearance. For extremely tight dimensioning, it may be necessary to provide a gap in an extension line so that it will not cross through an arrowhead. This is shown in Fig. 2–14d.

2-8

Leaders

A *leader* is a thin line drawn at an angle and terminating with an arrowhead or a dot. Its purpose is to convey dimensional information or a note to an object. The angle is large (usually 45 or 60 degrees to the horizontal) and has a horizontal line segment (shoulder) which is approximately $\frac{1}{4}$ in. long. This shoulder is drawn to the center height of the dimension or note being conveyed. The proper use of leaders is shown in Fig. 2–15. Several methods of dimensioning large and small holes are shown in Fig. 2–15a, b, and c. In Fig. 2–15a, the arrowhead of the leader touches the edge of the hole and points directly at its center. The numerical information is centered on the $\frac{1}{4}$-in. shoulder and is followed by the abbreviated note *DIA* (diameter) so that there will be no confusion as to the meaning of this dimension.

In Fig. 2–15b, the leaders are drawn as circle diameters with arrowheads at both ends which touch the edge of the hole. Note that the abbreviation DIA is again added to this leader. As shown in Fig. 2–15b, the leader and dimensional information may be contained within the circle or extended with a shoulder beyond the hole edge. The dimension of the hole may also be described with the use of standard dimension lines and extension lines, as shown.

When defining small holes to be machined by drilling or punching, the leader provides the hole size and the machining operation to be employed. Examples of leaders used for this application are shown in Fig. 2–15c.

When an arc is to be dimensioned, a leader is used in a similar way to that shown in Fig. 2–15b, with the difference that the di-

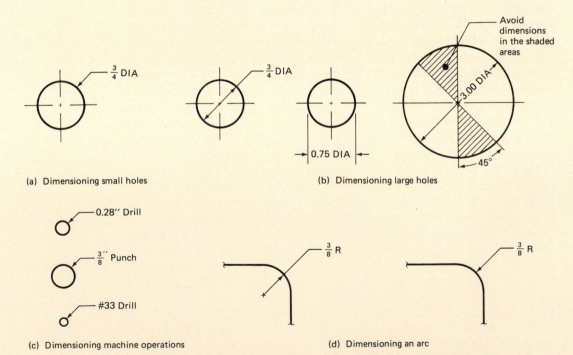

(a) Dimensioning small holes

(b) Dimensioning large holes

(c) Dimensioning machine operations

(d) Dimensioning an arc

FIGURE 2–15 Several methods of dimensioning circular detail.

mension given is the *radius*. Two methods of dimensioning an arc are shown in Fig. 2–15d. In the left-hand figure, the leader extends from a point shown with a cross mark, which defines the center of the arc and terminates at the arc with an arrowhead. Thus the leader length from cross mark to arrowhead is the radius of the arc. The leader then continues beyond the arc and a shoulder is used followed by the numerical dimension and the letter R to specify that it is the radius of the arc that is being dimensioned. If the radius is of sufficient length, the dimension may be inserted inside the arc in a break in the leader. Otherwise, either method shown in Fig. 2–15d may be used.

When a drawing requires the use of many leaders, the following practices should be adhered to for improved appearance:

1. Closely positioned leaders should run parallel to each other.
2. Leaders should never cross each other.
3. Leaders should cross as few lines as possible.
4. Leaders should not be excessively long.

2-9

Dimensioning Angles

Angles are dimensioned by arcs or curved segments with the size of the angle located either inside or outside the visible outline of the angle. Both methods are shown in Fig. 2–16. If the angle is large and the complete information can be placed within the outline of the angle, two arc segments with arrowheads point outward from the interior and touch the sides of the angle. Sufficient space between arc segments is allowed to provide the angle size. See Fig. 2–16a. If the angle is small, resulting in it being inappropriate to use the method just described, *outside* arc segments which point inward and touch the angle outline are used. The angle size may then be located inside or outside the angle outline, depending on the interior size of the angle. See Fig. 2–16b and c. Crowding of dimensional information can also be avoided when working with small angles by the use of extension lines as shown in Fig. 2–16d. A solid arc is drawn within the extension lines and the angle size positioned immediately to the right of the arc.

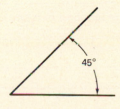

(a) Large angles allow use of inside arc segments

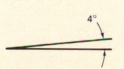

(b) Smaller angles may require outside arc segments

(c) Very small angles will require the value of the angle to be placed on the outside

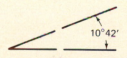

(d) To minimize crowding, extension lines can be used for angle dimensioning

FIGURE 2–16 Methods of dimensioning angles.

The most suitable method of dimensioning angles is most often dictated by the height of the lettering used and the requirement that crowding of information be avoided.

2-10

Locating Holes

The standard method for showing the position of the center of a circle or hole on a drawing is by *location dimensioning*. This gives the horizontal and vertical distances of the hole center from a reference point or reference sides of the object which are at right angles to each other. See Fig. 2–17. (Dimensioning to the edge of a hole is unacceptable practice.) Perpendicular center lines which intersect at the center of the circle are first drawn. They are then extended beyond the outline of the object and form two of the four required extension lines. Unlike standard extension lines as discussed in Section 2–6, which are spaced $\frac{1}{16}$ in. from the outline of the object, these center lines continue unbroken through the outline. The two remaining extension lines are drawn at right angles at the reference point as shown in Fig. 2–17. The horizontal and vertical dimensions are then placed between pairs of extension lines to locate accurately the center of the hole.

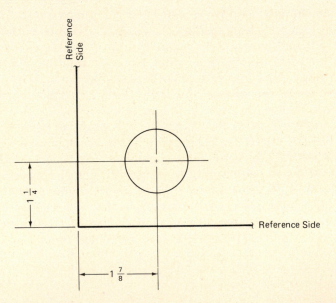

FIGURE 2–17   Standard method of dimensioning the location of a circle.

2-11

Fractional, Decimal, and Metric Dimensions

Straight-line or *linear* measurements used in electronic drawings use either the *English* system or the international system of units (SI), also referred to as the *metric* system. The two standard units of measure in the English system are the *foot* (abbreviated *ft* and its symbol ') and the *inch* (abbreviated *in.* and its symbol "). The foot is divided into 12 in. and the inch is further subdivided into fractions, typically, $\frac{1}{2}$, $\frac{1}{4}$, $\frac{1}{8}$, $\frac{1}{16}$, $\frac{1}{32}$, and the smallest subdivision, $\frac{1}{64}$ of an inch. Fractional dimensions are normally used when the accuracy required does not exceed $\pm\frac{1}{64}$ in. For greater accuracy, decimal subdivisions are used. The inch unit can be decimally subdivided into $\frac{1}{10}$ (0.1 in.), $\frac{1}{100}$ (0.01 in.), $\frac{1}{1000}$ (0.001 in.), $\frac{1}{10000}$ (0.0001 in.), $\frac{1}{100000}$ (0.00001 in.), and $\frac{1}{1000000}$ (0.000001 in.). Dimensioning a drawing with a mixture of decimal and fractional values is not good practice. The exception to this rule is when all dimensioning is done in decimals, but nominal sizes for machine hardware such as bolts, nuts, and screw threads are defined in fractions.

The standard unit of measure in the metric system is the *meter*, which is subdivided in multiples of 10, that is, 10 decimeters (dm), 100 centimeters (cm), and 1000 millimeters (mm), which is the smallest subdivision. Two dimensioning standards are used in the metric system, one using a decimal point and the other using a comma to separate the whole number from its fractional portion. For example, assume that an object has a measurement of 14.35 mm. This can also be expressed as 14,35 mm. Both designations mean that the object is 14 mm long with an additional length of $\frac{35}{100}$ of a millimeter.

Appendix V provides a table to convert common fractional and decimal inches to millimeters, and vice versa. These conversions are based on the equivalencies of 1 meter (m) = 39.37 in. and 1 in. = 2.54 cm (25.4 mm).

2-12

Methods of Dimensioning

Although there are a number of dimensioning methods available for use by the draftsperson, the following four methods are employed almost exclusively in electronic drafting: (1) *base-line* or *datum-line* dimensioning; (2) *coordinate* dimensioning; (3) *center-line* dimensioning; and (4) *continuous* dimensioning. Each of these is shown in Fig. 2–18 on page 62.

Base-line or *datum-line* dimensioning, shown in Fig. 2–18a, uses one horizontal line, typically the bottom or base line of the object, and one line perpendicular to it, usually the left edge of the object, as *datum reference lines*. All dimensions are made from these lines. The x and y axes (horizontal and vertical lines) should always be at the extreme bottom and the extreme left of the object, respectively. This places their intercept point or origin in the left bottom corner of the drawing. Thus all dimensions are made vertically upward and horizontally to the right of this point, resulting in all positive numbers for dimensional values. The selection of any other datum-line positions would result in negative numbers and possible confusion in measurement.

Another form of datum-line dimensioning is shown in Fig. 2–18b. In this method, the dimension line terminates at the extension line. The use of only a single arrowhead is required in this form of dimensioning. Note that the longest dimension is shown separately. This method of dimensioning is preferred for work requiring close

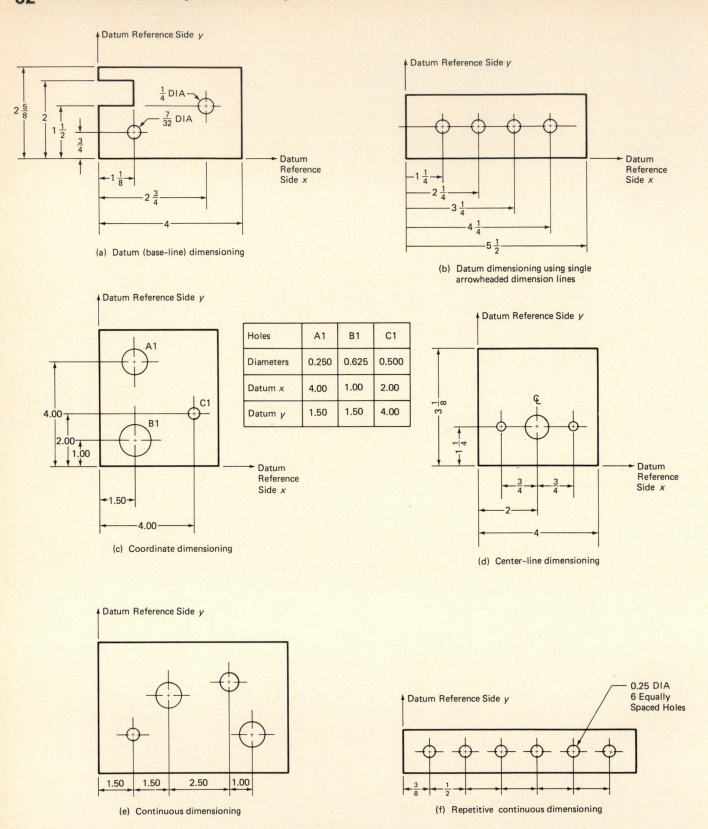

FIGURE 2–18 Several dimensioning techniques all using datum referencing.

tolerances since each significant location point is dimensioned individually from the datum line. This prevents accumulating a series of tolerances, which can easily lead to measurement error.

The *coordinate* system of dimensioning, shown in Fig. 2–18c, is a form of datum-line dimensioning. It is preferred when a large number of locations are to be dimensioned and the use of the datum-line system, requiring many dimension lines and numerical values, could lead to confusion. The coordinate system is commonly used in chassis design, printed circuit board layout, and other applications where a large number of hole locations are to be dimensioned. Coordinate dimensioning, similar to datum-line dimensioning, requires the establishment of two datum lines, from which all measurements are referred. Again, the bottom datum line is defined as x and the perpendicular datum line at the extreme left is labeled y. All holes are drawn with center lines and designated for individual identification by a system of letters or numbers. A table is then provided on the drawing which lists (1) the hole identification designation, (2) the diameter of the hole, (3) the dimension in the x direction that the hole's center is located from the datum line, and (4) the dimension in the y direction that the hole is located from the datum line. See Fig. 2–18c. This system not only eliminates the complicated and confusing appearance that results from the use of an inordinate number of dimension lines, but when combined with the standard base-line system, it is readily adaptable to computer-aided manufacturing (CAM). This is so because the dimensional information is provided with x-y coordinates which refer to a common reference point.

Center-line dimensioning is often combined with base-line dimensioning when critical placement of holes associated with a common part is required. Refer to Fig. 2–18d. Base-line dimensioning is used to locate the center of the large hole, showing its x and y position relative to the datum point. The two holes on either side of the large hole are then dimensioned from the extensions of their center lines to those of the large hole. It can be seen that center-line dimensioning is used when the requirement of critical placement of holes relative to *one another* is more important than locating their positions relative to the datum point.

Continuous dimensioning, shown in Fig. 2–18e, combines dimensions in a successive string. Only the first hole, labeled x in the figure, is dimensioned from the datum point. All of the other holes are then dimensioned from each other's center line, left to right. Where repetitive continuous dimensioning is specified, a simplified version of continuous dimensioning, shown in Fig. 2–18f, may be used. Although continuous dimensioning reduces the space required for individual dimension lines, it results in an accumulation of tolerances which must be accounted for in layout work. Therefore, this system of dimensioning should be used only when a high degree of accuracy in layout work is not a requirement.

The standards established for electronic drafting have been developed by professional organizations, manufacturers, and users who are concerned with all aspects of engineering. The purpose of these standards is to provide industry with methods and techniques which have been widely approved and adopted as being good practice.

Conforming to these standards promotes minimal confusion, controversy, and misinterpretation. Because these standards are revised periodically, latest issues should be obtained when making references. Where design of equipment for the armed services is involved, the draftsperson should become familiar with the use of military standards.

EXERCISES

2–1 Draw the symbols listed below using the correct pencil grade. The symbols are to be a total length of 3 in. and spaced 1 in. apart. Refer to Fig. 2–1.

(a) Leader   **(d)** Hidden line
(b) Dimension and   **(e)** Center line
     extension lines   **(f)** Short break line
(c) Visible line   **(g)** Cut plane

2–2 Construct the six sets of guide lines as shown in Fig. 2–19. Using correct freehand techniques for vertical Gothic-style lettering (see Fig. 2–7a and c), letter the information required using all capital letters in parts a through c and lowercase letters in parts d through f.

2–3 Using the vertical Gothic style, draw freehand the sequence of numbers and fractions starting with $\frac{1}{16}$ and progressing to $1\frac{1}{2}$ in $\frac{1}{16}$ increments. Refer to Fig. 2–7e for the construction of guide lines for drawing fractions. Use the value of $\frac{1}{4}$ in. for dimension H of Fig. 2–7e.

2–4 Reconstruct the portion of the printed circuit board shown in Fig. 2–20. Determine only the hor-

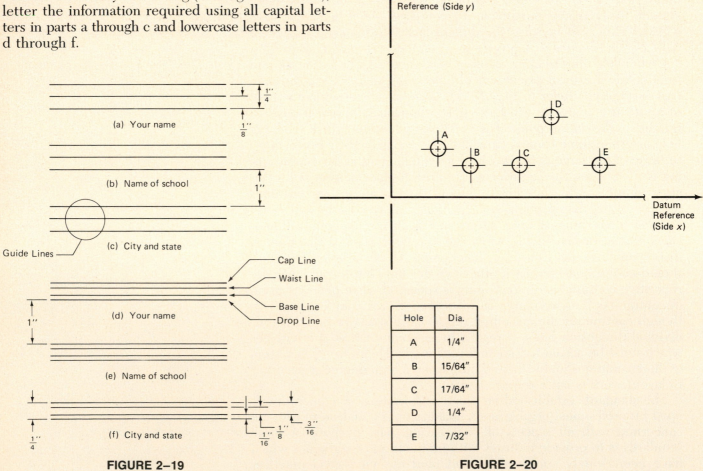

FIGURE 2–19

Hole	Dia.
A	1/4"
B	15/64"
C	17/64"
D	1/4"
E	7/32"

FIGURE 2–20

izontal dimensions for all hole locations from the y datum by measuring directly on Fig. 2–20 with a decimal scale. Use open arrowheads on all dimension lines. All dimensions are to be in decimal form. With the information provided in the table of Fig. 2–20, replace each of the letters shown beside each hole with the correct dimension.

2–5 Construct a table similar to the one shown in Fig. 2–21. With the information listed in Fig. 2–21 and with reference to Appendix V, convert the dimensions given in millimeters (mm) to decimals and record these values in your table.

2–6 Redraw the plate shown in Fig. 2–21. Using base-line dimensioning, redimension the figure with the decimal values recorded in your table of Exercise 2–5. Show all hole sizes with decimal values using leaders for their dimensioning.

2–7 Draw a table similar to the one shown in Fig. 2–21. With the information provided and with reference to Appendix V, convert the dimensions given in millimeters (mm) to fractions and record these values in your table. Redraw the plate of Fig. 2–21 and dimension it with fractional values by the continuous dimensioning method.

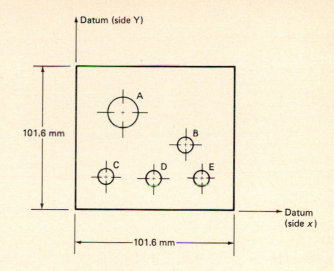

Hole	Dia. (mm)	x	y
A	12.7	25.4	76.2
B	9.53	76.2	38.1
C	6.35	19.1	19.1
D	6.35	50.8	19.1
E	6.35	82.6	19.1

FIGURE 2–21

3

Electronic Circuit Graphical Symbols

Upon completion of this chapter on electronic circuit graphical symbols, the student should be able to:

1. Know the basic techniques of drawing electronic circuit symbols.

2. Graphically represent the following:
   a. Circuit path connections.
   b. Common or ground points.
   c. Different types of wire and cable.
   d. Common connectors.
   e. Sources of electric power.
   f. Circuit protection devices.
   g. Indicating lamps.
   h. Switches and relays.
   i. Resistors, capacitors, and inductors.
   j. Transformers.
   k. Semiconductor diodes, transistors, and devices.
   l. Integrated circuits.
   m. Logic circuits.

3. Properly use electronic symbol drafting templates.

3-0

Introduction

Electronic circuit schematic diagrams are drawings that provide all pertinent information relative to a circuit using internationally recognizable graphical symbols. Each unique symbol represents a specific electrical, electromagnetic, or electronic component or device. Additionally, each symbol is provided with a *reference designator*, which is the abbreviation of the component name or letter designation and is positioned close to the symbol. This coding system not only provides an efficient means for complete identification of each component symbol but also serves to correlate information appearing on other drawings associated with the schematic diagram, such as *assembly drawings*, *wiring diagrams*, certain mechanical drawings, and the system's *parts list*. Finally, together with each graphical symbol and the reference designator, is the value of the component or the identifying number of the device, which completes the part's description.

This chapter deals with the drawing of common symbols used in electronic circuit diagrams. These symbols are presented, whenever possible, in groups of components which are common to a general classification. After the discussion of various common symbols in a group, a drawing of one or more of those symbols on a grid system is presented to illustrate the proper drawing proportions. The use of a grid system ensures uniformity of proportions and sizes of symbols.

All of the symbols presented are provided with the conventional reference designator as well as typical values and number types for the various components and devices. Finally, the use of electronic drafting templates to aid in preparing electronic circuit diagrams is presented.

The techniques for drawing circuit schematic diagrams with the use of the symbols discussed in this chapter are presented in Chapter 4.

3-1

Drawing Electronic Graphical Symbols

Graphical symbols are drawn as visible (single-stroke) lines, the same as those used in lettering (see Chapter 2). The line width of the symbols should be uniform throughout the drawing. Best results are obtained with the use of H, 2H, or 3H leads, discussed in Chapter 1

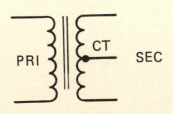

(a) Graphical symbol for a transformer

(b) Transformer

FIGURE 3–1 Equivalent symbol for component.

and Section 2–1. The overall size of the symbols is not a primary consideration. However, they should not be so small that difficulty in their interpretation results, nor so large as to dominate any area of the drawing when compared to other symbols. Relative size consistency in any drawing should be maintained. The size of the symbols is normally dictated by the overall size of the drawing and the available space.

Most graphical symbols used in electronics can be drawn in any orientation without altering the meaning of the symbol. For convenience, they may be drawn as shown in this chapter, drawn upside-down, or rotated clockwise or counterclockwise.

It may be well to point out here that graphical symbols of electronic components represent their functions in a circuit and have no relationship to the physical appearance of the parts. This fact may cause initial confusion until the association between the symbol and the appearance of the part it represents is established. To illustrate this difference, the graphical symbol for a transformer is shown in Fig. 3–1a and a photograph of an actual transformer is shown in Fig. 3–1b.

3-2

Representing Circuit Path Connections

Components and devices, which are represented by graphical symbols on a circuit diagram, are connected (wired) together to form electrical circuit paths, resulting in a functional system. The connecting paths are made with lengths of wire, through conductive metal paths on a printed circuit board, through wired connections to the metal frame of a chassis, or by other wiring methods. Regardless of the wiring method used, circuit paths between components are drawn as solid single lines on the circuit diagram by one of the two systems shown in Fig. 3–2. These are termed the *dot* and the *no-dot* methods for representing electrical path interconnections.

In the dot connection system, shown in Fig. 3–2a, each conductive path is drawn as a line segment and the electrical connections between conductor paths is represented by a small, round, solid dot. Points at which electrical conductors cross each other without an electrical connection are called *crossovers* and are drawn without the dot. See point a in Fig. 3–2a.

In the no-dot system, shown in Fig. 3–2b, perpendicular lines terminating at other lines are considered to be electrically connected to each other. The crossover, shown at point b, again represents two wires which cross each other without an electrical connection. A connection between two paths at a point requires that two separate lines, each representing a circuit path, be drawn with a slight space between the intersecting points. These are shown as points c and d in Fig. 3–2b.

Whichever connection system is used, all wiring paths are drawn as horizontal and vertical lines. Connecting lines and crossovers should be drawn at right angles to each other. Diagonal lines should be avoided except in special cases such as the drawing of bridge and multivibrator circuits.

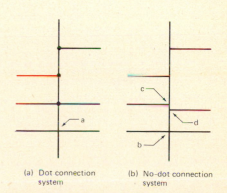

(a) Dot connection system   (b) No-dot connection system

FIGURE 3–2 The dot and no-dot connection system for wiring interconnection.

Many common electronic symbols are discussed in the following sections of this chapter. These and other useful symbols are provided in Appendix IX.

3-3

Common Points of Electrical Connections

In the construction of electronic equipment, it is common practice to have some connections made to specific points of electrical reference. These connections are made to the metal chassis or frame, a common conductor, or to ground (earth). The symbols for each of these reference points are shown in Fig. 3–3.

(a) Earth or ground   (b) Chassis or frame   (c) Common reference point

FIGURE 3–3 Connection symbols to indicate circuit return paths.

When an electrical connection is to be made to the earth, typically through a water pipe buried in the ground, it is symbolized as shown in Fig. 3–3a. These ground connections are usually considered as 0 volts potential. An example of this is the contact in a wall receptacle which accepts the U-prong of a standard three-prong plug. This U-prong is connected to ground by means of interconnecting wiring to a water pipe.

Another common reference point for electrical connections is the metal chassis or frame of the equipment. Because these are commonly made of highly conductive metals, they may be used as part of the circuit wiring by making connections to their surfaces at convenient points. The connections made to these metal surfaces may be at a higher potential than those made to earth or ground. The symbol to represent connections made to chassis or frames is shown in Fig. 3–3b.

To distinguish common electrical points of reference at the same potential which are not connected to the chassis or to ground, the symbol shown in Fig. 3–3c is used. These connections are typically made to a circuit ground bus wire and are called *common returns* or *circuit returns*. An example of this type of reference point is the common terminal of a split-voltage power supply. All circuit connections returned to this point of reference are shown by the common reference point triangular symbol. In hybrid circuits (i.e., circuits containing both analog and digital elements) the analog and digital reference points may both be shown with the common reference triangle, provided that an appropriate designation is included. These designations may be simply adding the term *analog* or *digital* in parentheses beside each common reference point symbol. The proper proportions for drawing the three symbols used for common electrical points are shown in Fig. 3–4.

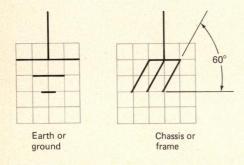

Earth or ground   Chassis or frame

60°

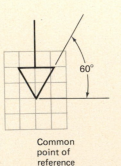

60°

Common point of reference

FIGURE 3–4 Suitable proportions for drawing symbols used to represent common points of electrical connections.

As will be seen in Chapter 4, the use of these symbols eliminates the necessity of having to draw long lines to represent circuit returns, thus presenting a less complicated drawing.

3-4

Wire and Cable

A variety of drafting techniques are used to illustrate and distinguish the different types of wire or cable used to form electrical interconnections in a circuit diagram. Depending on the circuit, a conductor may be one wire, a group of wires (cable), or any other element, such as the chassis, which is capable of providing a path for current to flow. In this section we consider only the wires being used as the conducting media.

Various wire and cable symbols used in electronic drafting are shown in Fig. 3–5. When two or more wires are combined in a group or sheath, they are called a *cable assembly*. The wires are drawn as solid parallel lines, each line representing a conductor in the group. These lines are surrounded by a perpendicular solid loop, as shown in Fig. 3–5a, to indicate that they are grouped into a cable.

When two wires are twisted together, as is done to minimize adverse electromagnetic field effects, they are again drawn as solid parallel lines, each passing through a solid circle. These circles are drawn similar to the number *eight*, as shown in Fig. 3–5b.

Symbols that represent *shielded conductors* are shown in Fig. 3–5c and d. Shielded conductors find application in low-frequency circuits where the impedance of the conductor is not critical. These conductors have an insulated solid inner signal conductor which is encased in a tightly woven pattern of fine-diameter wire which comprises the shield. Shielded conductors are commonly available with or without an insulated covering or jacket. The symbol for shielded cable is a broken loop encompassing the conductors. See Fig. 3–5c and d. If the shield is to be electrically connected to another metal element serving as the reference point, the appropriate symbol is added to the broken loop. The symbol in Fig. 3–5d indicates that the shield is electrically connected to the chassis.

The symbol for *coaxial cable* is shown in Fig. 3–5e. This cable is a shielded conductor which is used for transmission of radio-frequency (RF) energy with low loss. The shield is usually connected to the chassis or ground with the use of a special connector. Unlike most wire, coaxial cable is available with a specific characteristic impedance, such as 50, 75, or 95 ohms.

A method of showing a large number of conductors grouped together to form a cable is shown in Fig. 3–5f. It is implied that all of the wires in the cable are electrically insulated from one another. The cable trunk, drawn on the left side of the figure, contains all eight conductors and is drawn as a single solid line. Even though the individual conductors branch off, as shown in Fig. 3–5f, and the number of conductors remaining in the trunk successively diminish, the trunk continues to be drawn as a single line of uniform width. Cables are designated with the letter W on electronic drawings.

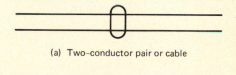

(a) Two-conductor pair or cable

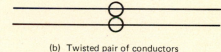

(b) Twisted pair of conductors

(c) Single conductor shielded

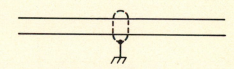

(d) Two-conductor pair shield connected to chassis

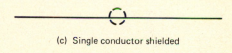

(e) Coaxial cable

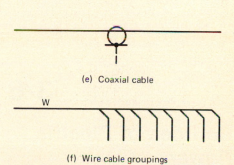

(f) Wire cable groupings

FIGURE 3–5   Various wire and cable symbols.

3-5

Connectors

An extremely large variety of connectors are available to make convenient electrical connections between electronic circuits. Due to the vast number of connectors, a selection of common types will be presented here. These are shown in Fig. 3–6. All connectors consist of two major elements which form a two-part electrical system when connected together. One element is the *male* or *plug* assembly and is intended to mate with the second element, the *female* or *jack* assembly, to form the electrical connection.

The simplest method of illustrating a connector pair for a single conductor connection, such as a banana plug and jack, is shown in Fig. 3–6a and b. The male plug is drawn as an arrow, in line with and facing the symbol for the female jack. Note the reversal of the arrowhead on the jack symbol. The reference designator for the plug is P and for the jack, the letter J.

Figure 3–6c and d show the symbol for a multiple plug and receptacle connector for multiconductor terminations. Typically, the plug and jack sections are shown mated together. The plug symbol is distinguished from that of the jack by rounding its corners. The individual conductors terminating on either side of the connector are commonly identified on the bodies of the plug and jack sections. A series of sequential numbers or letters define the individual conductors in the group.

The symbols for a connector that is usually associated with household wiring are shown in Fig. 3–6e and f. The plug portion (Fig. 3–6e) is drawn with a circle in which two blackened rectangles represent the *hot* and *neutral* leads or prongs. The extended circle with the flattened bottom represents the third prong, which is the ground (service or equipment) lead of the connector. The jack or receptacle portion (Fig. 3–6f) is drawn identical to the plug except that the outlines representing the contacts which receive the prongs are not blackened.

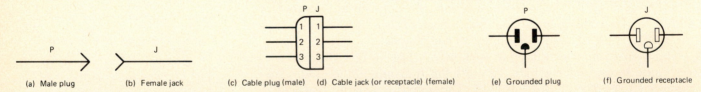

(a) Male plug   (b) Female jack   (c) Cable plug (male)   (d) Cable jack (or receptacle) (female)   (e) Grounded plug   (f) Grounded receptacle

FIGURE 3–6 Various symbols used for connectors.

3-6

Sources of Electric Power

One source of dc (direct current) power is the single primary cell commonly referred to as a *battery*. A single cell is usually defined as having a nominal value of 1.5 volts dc (Vdc). Cells have two electrodes connected to external terminals, one designated as *positive* (+) and the other as *negative* (−). The symbol for the positive terminal is represented by a long line drawn perpendicular to the conductor line. The negative terminal is represented by a short line which is about one-half the length of the positive terminal line. The symbol for a single cell is shown in Fig. 3–7a. Although the polarity of the cell is implied by the length of the lines, each terminal should be identified with a + and − sign close to the appropriate line to eliminate the possibility of confusion. See Fig. 3–7a.

When a group of rechargeable (secondary) cells are combined into a multicell source, such as that in an automobile storage battery, the symbol is a series of single-cell symbols, as shown in Fig. 3–7b. To be exact, a single-cell symbol should be drawn for each 1.5 volts of the battery value. However, a multicell battery drawn with *three* equally spaced single cell symbols with the value of the battery placed beside the symbol is accepted practice. See Fig. 3–7b. The reference designator for both single-cell and multicell batteries is the letters BT.

A nonspecific type of power source (i.e., no designated value of voltage), either dc or ac (alternating current), is shown as a circle with the appropriate number of conductors. See Fig. 3–7c. If it is a dc power supply, the reference designator is A or PS and the positive terminal is marked with a + sign positioned close to the appropriate point of the symbol.

Sources of ac power are represented by a circle with the shape of one complete ac sine wave centered in the circle. A sine-wave generator or oscillator of fixed-frequency output is shown in Fig. 3–7d. If the wave shape was triangular, the symbol for one cycle of a triangular wave would be placed in the center of the circle. To complete the identification of the ac source and to avoid the possibility of confusion, the type of supply should be shown beside the symbol (e.g., 12 Vac, 1000-Hz sine wave).

If the frequency of the generator or oscillator is adjustable, an arrow is drawn through the symbol, as shown in Fig. 3–7e. Sources of ac power are designated by the letter Y.

Suitable proportions for drawing common symbols of power sources are shown in Fig. 3–8.

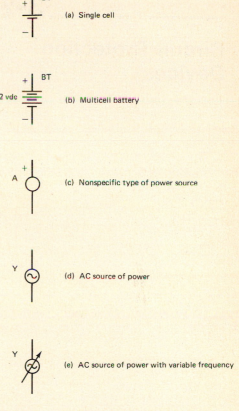

(a) Single cell

(b) Multicell battery

(c) Nonspecific type of power source

(d) AC source of power

(e) AC source of power with variable frequency

FIGURE 3–7 Symbols used to represent sources of power on circuit schematics.

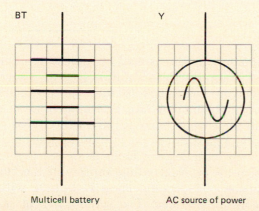

Multicell battery

AC source of power

FIGURE 3–8 Suitable proportions for drawing symbols used to represent sources of power.

3-7

Circuit Protection Devices

Circuits are protected from current overload or fault conditions by either *fuses* or *circuit breakers*. Fuses are wired in the conductor path and their symbol is drawn as a cycle of a sine wave or as a rectangle, both of which are shown in Fig. 3–9a. The reference designator for a fuse is the letter F. When the rated current or voltage of a fuse is exceeded as the result of an overload or fault condition, the heat generated will melt the metallic element, resulting in an open circuit and an interruption in the current path. Fuses are a one-time protection device and therefore must be replaced when "blown."

Circuit breakers are designed to "trip" or disconnect a protected circuit from the source of power whenever a specified overload condition occurs. Unlike fuses, circuit breakers need not be replaced but simply reset after the fault has been corrected. The symbol for a *single-pole circuit breaker*, which is used for protecting a single-conductor path, is shown in Fig. 3–9b. Figure 3–9c shows the symbol for a *double-pole circuit breaker* which protects two conductors simultaneously. The dashed line between each half-circle in the symbol represents a mechanical linkage (ganging) between the movable contacts. Even if an overload condition exists in only one conductor path, both conductors will be disconnected when the breaker is tripped.

Circuit breakers are designated by the letters CB on schematic diagrams and are rated by the amount of current which causes them to trip.

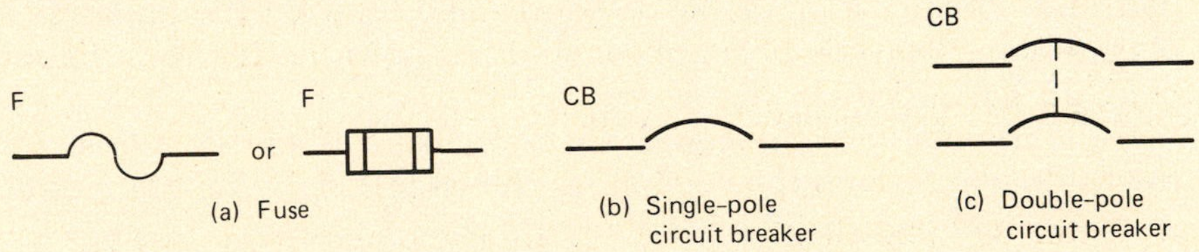

FIGURE 3–9   Devices used to protect circuitry during fault conditions.

3-8

Indicating Devices

Lamps are commonly used as signal or indicating devices to (1) signal when power is applied to a circuit or system, (2) monitor the status of existing current conditions, (3) indicate the sequence of events when specific circuit functions are activated, and so on. *Incandescent* lamps emit light when current passes through a metal filament which is sealed in a glass envelope containing inert gas or a vacuum. The symbol for an incandescent lamp is shown in Fig. 3–10a. The loop drawn within the external outline represents the filament. A more general symbol for an indicating lamp is a circle with a letter that indicates the color of the lens. The symbol in Fig. 3–10b includes

(a) Incandescent indicating lamp

(b) Pilot lamp (amber)

* Abbreviation	Color
A	amber
B	blue
C	clear
G	green
O	orange
OP	opalescent
P	purple
R	red
W	white
Y	yellow

* To avoid confusion with meter or relay symbols the letter should be placed outside the lamp outline with the filament drawn.

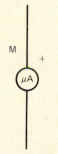

(c) Meter in microampere units

Abbreviation	Function
A	Ammeter
MA	milliammeter
µA UA	microammeter
V	voltmeter
OHM	ohmmeter
W	wattmeter
F	frequency meter
t°	temperature meter

FIGURE 3–10   Various indicators used on circuit schematics.

the letter A, indicating an *amber* lens. The table shown in Fig. 3–10b lists 10 lens colors and their letter designations. As seen in both Fig. 3–10a and b, the reference designator for indicating lamps, or *pilot lights*, is the letters DS.

In electronics, many types of meters are used to indicate the magnitude of a variety of parameters, such as current and voltage. The symbol for a meter is also a circle, which is further identified by the abbreviations of the units that it is to measure. See Fig. 3–10c. The table in the figure lists some common meter applications with their letter designations, which are placed within the meter symbol to indicate the parameter to be measured.

If the meter is intended to measure dc values, a plus (+) sign is included as part of the symbol and is placed close to the positive terminal. Figure 3–10c shows the symbol for a meter used to indicate dc values of current in microampere units. The reference designator for meters is the letter M.

The need for clear symbol identification is apparent when lamp and meter symbols (both circles) appear in the same drawing. Reference designators eliminate any possibility of misinterpretation.

3-9

Switches

Mechanical *switches* are used in electronics to allow circuit paths to be manually connected or disconnected between one circuit element and another. To connect circuit elements, a switch is said to *make* (make contact) or be *closed*. For disconnecting elements, a switch is said to *break* (break contact) or be *open*. All switches consist of one or more fixed *contacts* and one or more movable parts called *poles*.

Figure 3–11a shows the same type of switch in two possible states. If the normal contact position (before energizing) is *open* (i.e., pole contact not made), the switch is designated as NO, which represents *normally open*. A switch whose normal contacts are *closed* (i.e., pole contact made) is a *normally closed* type of switch and designated NC. Both switches shown in Fig. 3–11a are the *single-pole single-throw* (SPST) type. Each has one movable pole and one fixed contact. The fixed contact shown as a darkened triangle in Fig. 3–11a indicates that the switch is a nonlocking device such as a momentary on/off pushbutton switch.

In Fig. 3–11b, the fixed pole as well as the pivot point of the movable pole of both switches are shown as circles. These are the symbols for locking-type switches which do not automatically return to their initial contact position when released.

A *single-pole double-throw* (SPDT) switch consists of two fixed poles and one movable contact. The symbols for both the nonlocking (circular fixed contact) and the locking (darkened triangle fixed contact)

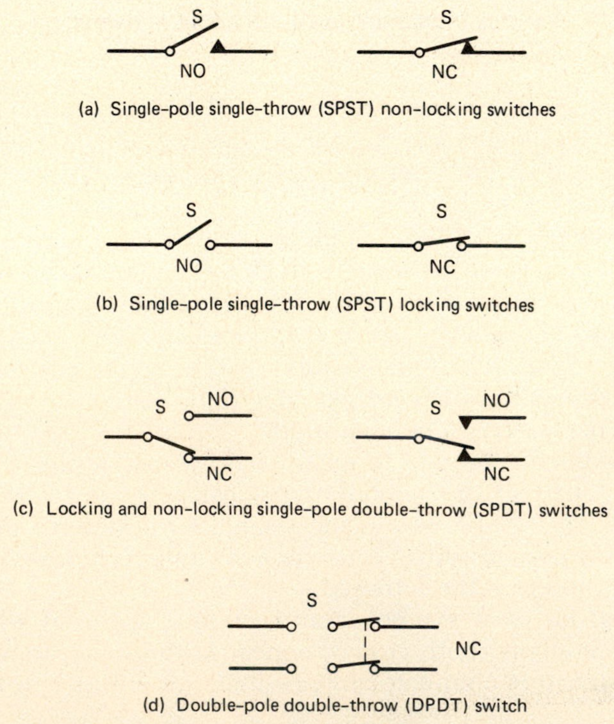

(a) Single-pole single-throw (SPST) non-locking switches

(b) Single-pole single-throw (SPST) locking switches

(c) Locking and non-locking single-pole double-throw (SPDT) switches

(d) Double-pole double-throw (DPDT) switch

FIGURE 3–11 Some common switch symbols.

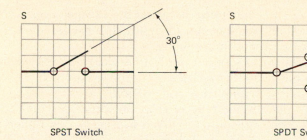

SPST Switch   SPDT Switch

FIGURE 3–12 Suitable proportions for drawing symbols used to represent switches.

types are shown in Fig. 3–11c. The contact where the connection is made is designated NC and the other contact, which is not making contact, is designated NO.

The symbol for a *double-pole double-throw* (DPDT) switch is shown in Fig. 3–11d. It has two movable poles and two fixed contacts. The dashed lines shown between the poles indicate that there is a mechanical linkage between both poles. This linkage causes both poles to make or break simultaneously with both common sets of fixed contacts.

The reference designator for switches is the letter S. Suitable proportions for drawing common switch symbols are shown in Fig. 3–12.

3-10

Relays

Relays are electromechanical switching devices that are used to open or close electrical contacts by remote control. The basic parts of a relay are a coil of wire, an iron core, a movable armature spring, and a set of one or more groups of poles and contacts. The symbol for the coil is a rectangle with two lines representing the coil leads. This is shown in Fig. 3–13a. The complete symbol for a relay is formed by combining that for a coil together with a set of switch contact symbols. See Fig. 3–13b. When energy is applied to the coil, the core moves, causing the pole position of the switch to change. This movement breaks the connection between the normally closed (NC)

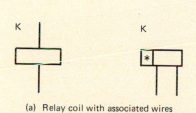

(a) Relay coil with associated wires

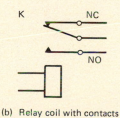

(b) Relay coil with contacts

*Letter Symbol	Relay Type
AC	Alternating Current
EP	Electrically Polarized
FO	Fast Operate
FR	Fast Release
SO	Slow Operate
SR	Slow Release

(c) Various relay types

FIGURE 3–13 Relay symbols and types.

contacts and makes the connection between the normally open (NO) contacts. Relay contact positions are designated in the nonenergized state (i.e., no power applied to the coil).

Even though relays may operate with a variety of contact combinations, they are designated primarily by the *way* in which they operate. Letter symbols within the coil symbol, as shown in Fig. 3–13a, are used to identify the type of relay operation. Some of these letter symbols and their meaning are listed in Fig. 3–13c. Relays are given the letter K as a reference designator.

3–11

Resistors

Resistors are circuit components which are used to establish current, voltage, or power levels in electrical and electronic circuits. The basic unit for resistance is the *ohm*, which is represented by the Greek capital letter omega (Ω). Although rectangles are sometimes used as the symbol for a resistor, the zigzag symbol, shown in Fig. 3–14a, is by far more widely accepted. This symbol represents a resistor having a fixed value with a cylindrical body and axial leads. The zigzag shape represents the component body and the lines at the ends symbolize the leads.

The general symbol for a variable two-terminal resistor, called a *rheostat*, is shown in Fig. 3–14b. This component's resistance is varied by a movable wiper which slides across the resistive material. The slanted arrow in the figure represents the wiper.

When a variable resistor has three terminals which are connected into the circuit, it is defined as a *potentiometer*. This is shown in Fig. 3–14c. Potentiometers are classified as having either *linear* or *audio* (logarithmic) taper. Linear potentiometers change resistance value in direct proportion to the amount of rotation of the control shaft. For example, if the potentiometer shaft is rotated 25% of its total possible rotation, the resistance value between a fixed end and the wiper connection changes by 25%.

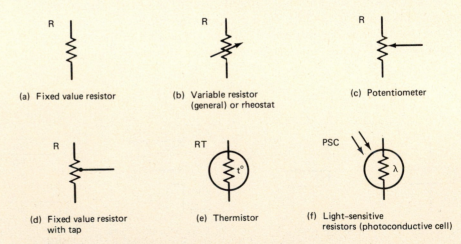

(a) Fixed value resistor

(b) Variable resistor (general) or rheostat

(c) Potentiometer

(d) Fixed value resistor with tap

(e) Thermistor

(f) Light-sensitive resistors (photoconductive cell)

FIGURE 3–14   Several common resistor symbols.

Audio taper potentiometers change resistance logarithmically. That is, the resistance change between a fixed end and the wiper is not linear over the full rotation of the control shaft.

Fixed three-terminal resistors whose third terminal is not adjustable are termed *tapped* resistors. As seen in Fig. 3–14d, the tap is drawn at the end of one of the zigzag points. If the dot system of representing connections is employed (see Section 3–2), a dot is added, as shown, to represent the electrical connection. If the no-dot system is used, the dot is omitted. The reference designator for the standard resistors shown in Fig. 3–14a–d is the letter R. Information on standard resistor values and the resistor color-code identification system is given in Appendices X and XI, respectively.

Special resistor types are available whose resistance change is a function of other parameters, such as voltage, temperature, light and so on. Two such types are the *thermistor* and *photoconductive cell* or *light-sensitive resistor* (LSR), whose symbols are shown in Fig. 3–14e and f, respectively. The thermistor is a transducer whose resistance increases as the ambient temperature decreases. They find wide application in temperature-sensing and control circuitry. The symbol for a thermistor, shown in Fig. 3–14e, is a standard zigzag pattern accompanied by t° to indicate that it is sensitive to temperature. The circle surrounding the resistor symbol in Fig. 3–14e is optional and may be omitted. Thermistors are available in a variety of temperature operating ranges, but they all work on the same principle in that their resistance changes in a nonlinear manner as temperature increases. The reference designator for a thermistor is RT.

The photoconductive cell or light-sensitive resistor (LSR), whose symbol is shown in Fig. 3–14f, is a transducer whose resistance decreases with an increase in light intensity. They find wide application in optoelectronic circuits, such as for on/off and control functions, and in light meters. The LSR is graphically represented by a standard resistor symbol together with the Greek lowercase letter lambda (λ) and two arrows directed toward the zigzag symbol as shown in Fig. 3–14f. Again, the circle shown in the figure may be omitted. Light-sensitive resistors have the letters PSC as their reference designator. Suitable proportions for drawing fixed and variable resistors are shown in Fig. 3–15.

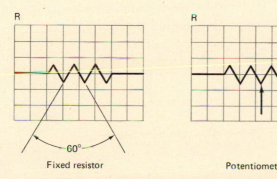

Fixed resistor   Potentiometer

FIGURE 3–15 Suitable proportions for drawing symbols used to represent fixed and variable resistors.

3-12

Capacitors

Capacitors are electrical energy storage elements used extensively in electronic circuits. They oppose a change in dc voltage and block the flow of direct current while permitting the flow of alternating current. Their degree of opposition to the flow of ac current, termed *reactance* (Xc), is inversely proportional to the value of capacitance and the frequency. The basic unit of capacitance is the *farad* and for its reactance, it is the ohm (Ω). The more common unit for capacitance is the *microfarad* (μF), which represents 10^{-6} farad, and the *picofarad* (pF or $\mu\mu$F), which represents 10^{-12} farad. Common values of typical capacitors are given in Appendix XII.

The capacitor basically consists of two conducting metal surfaces which are separated by an insulating material called the *dielectric*. Leads are connected to the metal surfaces to form the capacitor. The value of the capacitor is dependent on its geometry and the type of dielectric used. Common dielectric materials are air, mica, and paper. Capacitors are designated by the letter C on circuit schematic diagrams.

The symbol for a capacitor, shown in Fig. 3–16, is a straight line drawn perpendicular to one lead and an arc drawn perpendicular to the other lead. A *nonpolarized* capacitor, shown in Fig. 3–16a, is one whose lead orientation in the circuit is not critical. If the capacitor's connections in the circuit must be made to a specific polarity, this is a *polarized* capacitor, and a plus (+) sign is added to the symbol close to the straight-line segment that represents the more positive terminal. See Fig. 3–16b.

Variable capacitors (trimmers) consist of a fixed plate or plates (stators) and a movable plate or plates (rotors). The arrow drawn through the capacitor symbol in Fig. 3–16c indicates that it is variable.

In addition to individual capacitors, two or more are available which are combined in a single unit. Their symbol is the dual capacitor shown in Fig. 3–16d. A common connection to both capacitors is brought out as a single lead (usually black) and the remaining connections brought out as separate leads which are identified by a color code, typically red and blue.

To simplify the drawing of capacitors, some electronic industries have discontinued the use of the arc to represent one plate of the capacitor. Rather, they have replaced the arc with a straight line of equal length to the line representing the other plate.

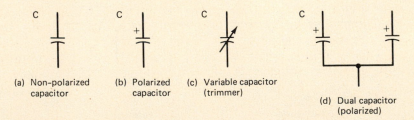

(a) Non-polarized capacitor   (b) Polarized capacitor   (c) Variable capacitor (trimmer)   (d) Dual capacitor (polarized)

FIGURE 3–16 Capacitor symbols.

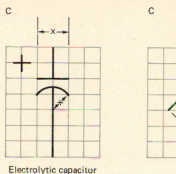
Electrolytic capacitor

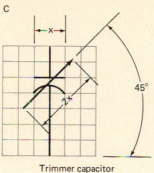
Trimmer capacitor

FIGURE 3–17 Suitable proportions for drawing symbols used to represent fixed and adjustable capacitors.

Suitable proportions for drawing common capacitor symbols are shown in Fig. 3–17.

3-13

Inductors

Inductors are composed primarily of a winding or coil of wire surrounding a core which may be of air or a metallic material. Inductors store energy in the form of a magnetic field that develops around the coil when current flows through the wire. Inductors used in dc circuits as relays, where the magnetic field is used to cause mechanical movement, in addition to low-frequency ac inductors, require soft-iron cores. For applications at moderate or intermediate frequencies (IF), powdered iron or ferrite cores are preferred. Inductors used in radio-frequency (RF) applications are commonly referred to as *chokes* and operate more efficiently with an air core.

Due to their unique electrical characteristics, inductors are used to control both dc and ac current. They find application in filter and tuning circuits since they oppose a change in dc current and are used to restrict ac current by their property defined as inductive reactance (X_L) given in units of ohms (Ω). The reactance offered by the inductor is directly proportional to the value of the inductor and the frequency.

The basic unit of inductance is the henry (H), with the smaller units of *millihenries* (mH) and microhenries (μH) more common to electronic circuit applications. The inductive value is dependent on its geometric construction and its magnetic properties. Inductors are designated with the reference letter L.

An air-core inductor is graphically represented by either one of the two symbols shown in Fig. 3–18a. Although the helix form is extensively used, the group of tangent half-circles are easier to draw and are an acceptable convention.

As shown in Fig. 3–18b, inductor cores composed of magnetic material (soft iron) are represented by two parallel lines drawn ad-

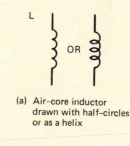

(a) Air-core inductor drawn with half-circles or as a helix

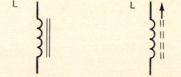

(b) Magnetic core   (c) Adjustable ferrite core

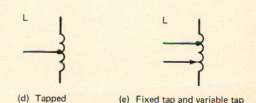

(d) Tapped   (e) Fixed tap and variable tap

FIGURE 3–18 Inductor symbols.

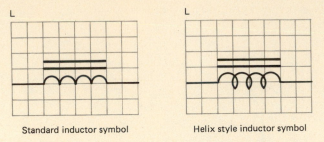

Standard inductor symbol   Helix style inductor symbol

FIGURE 3–19 Suitable proportions for drawing symbols used to represent inductors.

jacent to the basic inductor symbol. For powdered iron or ferrite cores, the symbol is two parallel dashed lines drawn as shown in Fig. 3–18c. Note also the arrow drawn in this figure, which denotes that the core is movable, indicating that the value of the inductor symbolized is adjustable or variable. Figure 3–18d and e shows the symbols for inductors with fixed taps and variable taps. Suitable proportions for drawing the two different styles of inductor symbols are shown in Fig. 3–19.

3-14

Transformers

Transformers are components that are made up of two or more coils that are mutually coupled by magnetic lines of force when current flows through one of the windings. They are used to transfer electrical energy from one winding to another with high efficiency.

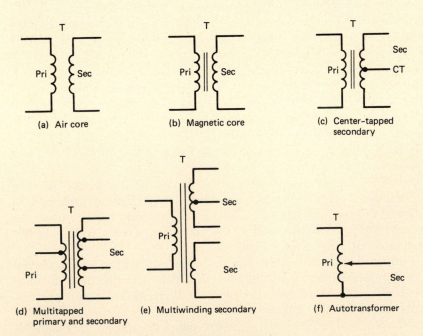

(a) Air core   (b) Magnetic core   (c) Center-tapped secondary

(d) Multitapped primary and secondary   (e) Multiwinding secondary   (f) Autotransformer

FIGURE 3–20 Various transformer symbols.

The transformer symbol is drawn the same as two adjacent inductor symbols as shown in Fig. 3–20. Although the half-circle symbol is shown in the figure, the helix form, as shown in Fig. 3–18, is also extensively used.

Transformers primarily consist of two or more windings on a single core. If it is an air core, no parallel lines are drawn between them, as shown in Fig. 3–20a. Recall that this is similar to an air-core inductor. When the core contains a magnetic material the symbol shown in Fig. 3–20b is used.

The windings of a transformer are identified as the *primary* (pri) and the *secondary* (sec). Normally, power is applied to the primary winding, causing a magnetic field to be generated which is then coupled to the secondary winding. This causes an induced voltage in the secondary which provides current to a load when placed across the secondary winding. When an ac voltage is applied to the primary winding, the voltage induced in the secondary winding is proportional to the ratio of the number of turns of wire in the primary with respect to the number of turns in the secondary. This ratio is called the *transformer turns ratio,* which is stated in the following equation:

$$\frac{Vp}{Vs} = \frac{Np}{Ns}$$

where

V_p = primary voltage
V_s = secondary voltage
N_p = number of turns in the primary
N_s = number of turns in the secondary

For a transformer provided with a tapped secondary winding, the symbol show in Fig. 3–20c is used. If the tap symbol is accompanied by the letters CT, this indicates that it is a *center tap,* which means that there are an equal number of turns of wire to either side of the tap connection. The use of a center-tapped transformer allows the circuit designer more output voltage options.

The symbol for a *multitapped* transformer with both primary and secondary taps is shown in Fig. 3–20d. The use of this type of transformer further increases the circuit designer's options in terms of input and output voltages.

The symbol for *multiwinding* transformers, which have more than one secondary winding, it shown in Fig. 3–20e. Depending on the application, the primary and the secondary windings may be tapped or provided without taps.

A transformer having an adjustable secondary voltage is symbolized in Fig. 3–20f. This type is called an *autotransformer,* whose secondary voltage is a function of the position of the wiper arm, which is illustrated by the arrow pointing into the single winding.

All of the transformers discussed, with the exception of the autotransformer, provide electrical isolation between the source of power and the circuit. This is due to the fact that there is no electrical connection between the primary and secondary windings. They are linked

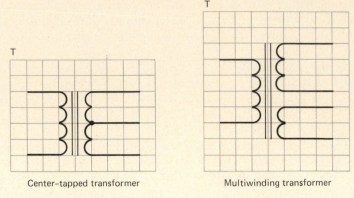

Center-tapped transformer   Multiwinding transformer

FIGURE 3–21 **Suitable proportions for drawing symbols used to represent transformers.**

by a magnetic field only. Although the autotransformer does not provide circuit isolation, it does have the advantage of providing a variable voltage.

The reference designator for transformers is the letter T. Suitable proportions for drawing transformer symbols are shown in Fig. 3–21.

3-15

Semiconductor Diodes

Semiconductor devices are fabricated primarily with two types of silicon material, designated *n-type* and *p-type*. These materials are produced by a process called *doping* in which controlled amounts of specially formulated impurities are added to pure or *intrinsic* silicon. N-type material is formed when the impurities added create *donor atoms* that cause electrons to become the majority current carriers. P-type material consists of silicon which has been doped with impurity materials that create *acceptor atoms* causing *holes* to become the majority current carriers.

A semiconductor *diode* is a two-terminal device fabricated with both n-type and p-type materials. The symbol for a diode is shown in Fig. 3–22a. It consists of an *anode*, represented by the arrowhead, which is formed of p-type material. The *cathode* is represented by a straight line drawn perpendicular to the apex of the arrowhead and is made of n-type material. The diode symbol may be drawn with or without the circle enclosure.

When a voltage of sufficient value is applied across the diode, allowing the anode to be at a more positive potential than the cathode, the diode is considered to be in its *forward bias* mode, which allows current to flow in the direction indicated by the arrowhead. This direction is called *conventional* current flow. (The amount of voltage necessary to cause this current flow is approximately 0.7 V for silicon diodes.)

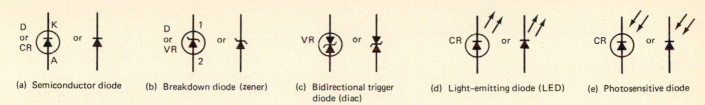

| (a) Semiconductor diode | (b) Breakdown diode (zener) | (c) Bidirectional trigger diode (diac) | (d) Light-emitting diode (LED) | (e) Photosensitive diode |

FIGURE 3–22 Two-terminal devices.

If the polarity of the applied voltage is reversed, placing the cathode at a higher positive potential than the anode, the diode is considered to be *reverse biased* and only a negligible leakage current will flow in a direction opposite to that of the arrowhead. In effect, the diode acts like an electronic switch, finding applications in rectification, detection, switching, and many other circuit functions. The reference designator for a diode is D or CR.

Another type of semiconductor diode is shown in symbol form in Fig. 3–22b. This is the *breakdown diode*, more commonly referred to as the *zener* diode. Note that the symbol is similar to that of a standard diode, with the addition of two short bars extending perpendicular from the ends of the cathode line. The zener diode is used as a *controlled-voltage* device or *regulator* and is made to operate in the reverse bias mode. Its reference designator is D or VR.

The symbol for a *bidirectional trigger diode*, or *diac*, is shown in Fig. 3–22c. It consists of three layers of semiconductor material with the p-type sandwiched between two layers of n-type materials. The two arrowheads pointing in opposite directions indicate that the diac can conduct current in either direction. The reference designator for the diac is also VR.

The *light-emitting diode* (LED) is a semiconductor device which emits light in the visible-light spectrum when properly forward biased. Its symbol, shown in Fig. 3–22d, is that of a standard diode with the addition of two arrows directed away from the symbol, indicating emission of light from the device. LEDs are among the common logic-level indicators used in digital circuits and in digital displays on calculators, clocks, and so on.

The symbol for a *photosensitive diode* (photodiode) is shown in Fig 3–22e. For normal operation, the photodiode is biased in the reverse direction. In darkness, it conducts a very small amount of leakage current but when subjected to light energy, this leakage current increases dramatically. This device finds application as an optical link, with an infrared emitting diode serving as the input to the link. This arrangement allows input voltage to be isolated from output current.

Note that the symbol for the photodiode is similar to that for the LED, the exception being that the arrows are pointing *toward* the device symbol to indicate that light energy is required to activate the device. Both the LED and the photodiode are designated by the letters CR. Suitable proportions for drawing the diode symbol are shown in Fig. 3–23.

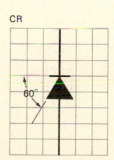

FIGURE 3–23 Suitable proportions for drawing the symbol used to represent a diode.

3-16

Transistors

Transistors are three terminal devices composed of an inner layer of either p-type or n-type material sandwiched between two outer layers of the opposite-type material. Each of these layers is connected to an external lead. The two basic transistor types are the *npn* and the *pnp*, whose symbols are shown in Fig. 3–24a and b, respectively. The three external leads are identified as the *base* (B), which is connected to the inner layer, the *emitter* (E), which is connected to one of the outer material layers, and the *collector* (C), which is connected to the other outside layer. The arrowhead on the emitter symbol indicates the direction of conventional dc current flow. Notice in Fig. 3–24a and b that in the symbol for an npn transistor, the arrow points away from the base, and for a pnp transistor, it points toward the base.

Transistors are used primarily as amplifiers in analog circuits and as switches in digital circuits. The reference designator for transistors is the letter Q.

The graphic symbol for an equivalent npn *Darlington transistor pair* is shown in Fig. 3–24c. This device consists of two individual transistors packaged in a single case with the equivalent three terminals (base, emitter, and collector) connected externally. The Dar-

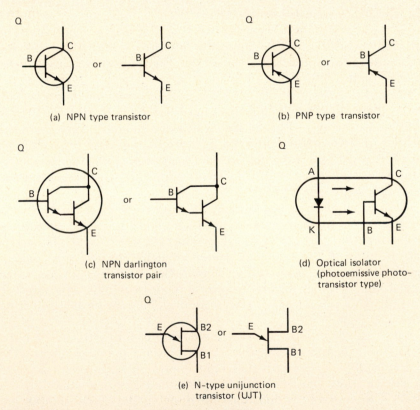

(a) NPN type transistor

(b) PNP type transistor

(c) NPN darlington transistor pair

(d) Optical isolator (photoemissive photo-transistor type)

(e) N-type unijunction transistor (UJT)

FIGURE 3–24 Common transistor symbols.

lington pair is also available as an equivalent pnp type. Darlington pair devices have much larger amplification factors than single transistors. They are widely used in medium- to large-power circuit applications.

The *optical isolator*, whose symbol is shown in Fig. 3–24d, is a special-purpose device. Unlike the optical link arrangement using the light-emitting diode and the photodiode previously discussed, this device is housed in a single case. It consists of an infrared light-emitting diode and a phototransistor which provides the advantage of current gain. When power is applied to the diode, light is emitted and sensed by the phototransistor. Since there is no electrical connection beween the input LED and the output phototransistor, total isolation between these two circuits is attained without the use of expensive and bulky relays or isolation transformers. Whereas the circle enclosing transistor symbols is optional, the oval shown in Fig. 3–24d should always be included to indicate that this is an integral unit with the light-emitting diode (symbolized by two arrows directed away from the diode symbol) contained in the same case.

Some applications of the optical isolator do not require a base connection. Thus these devices are available with either four or five external leads.

The symbol for a *unijunction transistor* (UJT) is shown in Fig 3–24e. This device consists of a section of n-type material with a base lead extending from each end, labeled B1 and B2. The emitter lead is embedded into a small section of p-type material that has been diffused into the n-type material at about 70% of the length from the B1 end, forming a pn junction. When sufficient voltage is applied to the emitter terminal, the resistance between terminals B1 and B2 decreases suddenly and allows current to flow through the device, thus acting like a closed switch.

The UJT is used in oscillator circuits, phase detectors, pulse generators, and timing and trigger circuits. A n-type UJT is shown in Fig. 3–24e.

Suitable proportions for drawing transistor symbols are shown in Fig. 3–25.

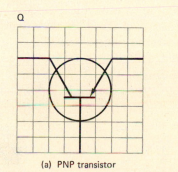
(a) PNP transistor

(b) Unijunction transistor

FIGURE 3–25 Suitable proportions for drawing symbols used to represent transistors.

3-17

Other Common Semiconductor Devices

We shall now look at several other types of semiconductor devices which are commonly employed in electronic circuits.

The symbol for the n-channel *junction gate field-effect transistor* (JFET) is shown in Fig. 3–26a. This is a three-terminal device, each terminal brought out as an external lead. These leads are identified as the *gate* (G), the *drain* (D), and the *source* (S). Whereas the typical bipolar transistor is a current-controlled device, the JFET is a voltage-controlled transistor which finds application in chopper, mixing, and switching circuits. The p-channel JFET is represented by reversing the direction of the arrowhead from that shown in Fig. 3–26a.

Figure 3–26b shows one of several symbols used for various types of *metal-oxide silicon field-effect transistors* (MOSFETs). The symbol shown is for an *insulated gate n-channel depletion-mode* MOSFET. The terminals on this device include the gate, drain, and source, with a fourth terminal to the *substrate* (U). MOSFETs are employed extensively in digital computer switching circuits because of their extremely high input resistance in addition to their being the least temperature-dependent semiconductors.

Another four-layer device is the *silicon-controlled rectifier* (SCR), whose symbol is shown in Fig. 3–26c. Another name for this is a *unidirectional reverse blocking thyristor*. The arrowhead in the symbol shows the direction of conventional current flow between the *anode* (A) and the *cathode* (K) switch terminals. The *control* terminal gate (G) activates the switch terminals into a *closed* state when sufficient signal voltage is applied across the gate-to-cathode terminals. Once an SCR is switched into the conducting mode, it can be turned off only by reducing the anode-to-cathode current below its minimum holding-current value. Thus the SCR allows a small current to control the flow of a large current in another circuit. Unlike a relay, the two circuits in the SCR are not isolated. These devices are used for motor-speed control and in dimmer, timer, and power switch circuits.

The symbol for the *programmable unijunction transistor* (PUT) is shown in Fig. 3–26d. This is a regenerative switch device similar to the UJT but includes a programmable (adjustable) trigger voltage. The PUT is replacing the UJT because of its (1) higher breakdown voltage, (2) lower cost, and (3) adjustable trigger voltage.

A five-layer semiconductor device, called a *triac*, has three external terminals, as shown in the symbol of Fig. 3–26e. These terminals are the *gate* (G) and two switch terminals labeled T1 and T2.

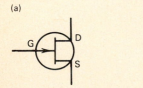
(a)

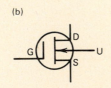
(b)

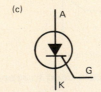
(c)

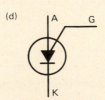
(d)

(e)

FIGURE 3–26 Various semiconductor symbols.

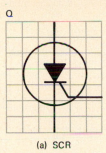
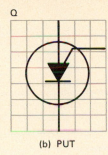
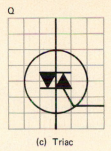

(a) SCR   (b) PUT   (c) Triac

FIGURE 3–27 Suitable proportions for drawing symbols used for various semiconductor devices.

This device is bidirectional, similar to the diac, in that it is able to conduct current through its switch terminals in either direction. The triac is an electronic switch capable of controlling large power levels in 60-Hz ac systems such as motor-speed control, static switching, and lamp dimmers.

The reference designator for all of the devices discussed is the letter Q. Suitable proportions for drawing various semiconductor devices are shown in Fig. 3–27.

The types of devices presented in this section by no means include all of those available, but do represent those commonly employed in electronic systems.

3-18

Integrated Circuits

Integrated circuits are complex devices fabricated from semiconductor material (silicon), as are the devices presented in the preceding several sections. Unlike the other devices, however, an integrated circuit can contain upwards of tens of thousands of transistors together with other circuit elements, such as diodes, resistors, capacitors, and even inductors, all included in one package. Integrated circuits, then, are electronic systems designed to perform specific functions. Since it is not only unnecessary, but highly impractical, to draw the entire circuit contained within the device, integrated circuits are symbolized by *functional blocks.* As shown in Fig 3–28, two types of symbols are used to represent integrated circuits. If the device is basically an amplifier, the triangular symbol, shown in Fig. 3–28a is used. The input or inputs are usually drawn to the left along the vertical edge of the symbol, with the output or outputs at the point facing to the right. The device identification number is commonly included within the body of the symbol such as the 741 operational amplifier (op-amp) identification shown in Fig. 3–28a. The device terminals are identified by numbers corresponding to the pins of the package and may or may not be identified by their individual function on the drawing. It is often the case that the pin arrangement of an integrated circuit shown on a drawing will differ from that of the actual device. This is done to avoid excessive conductor path crossovers on the drawing so as to result in a neater and more readable presentation. The common reference designator for inte-

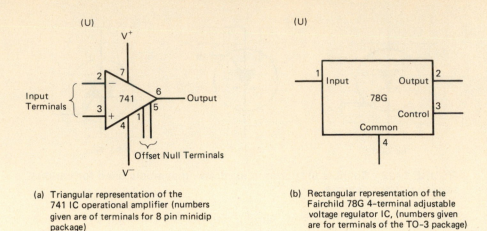

(a) Triangular representation of the
741 IC operational amplifier (numbers
given are of terminals for 8 pin minidip
package)

(b) Rectangular representation of the
Fairchild 78G 4–terminal adjustable
voltage regulator IC, (numbers given
are for terminals of the TO–3 package)

FIGURE 3–28 Two types of symbols used for integrated circuits.

grated circuits is the letter U, although the letters IC are often employed.

Not all integrated circuits serve as amplifiers. They may be regulators, timers, filters, and a variety of other functional circuits. These types of integrated circuits are represented by a simple rectangle as shown in Fig. 3–28b. Both an identification number (usually within the rectangle) and numbered terminals are included on the symbol. Descriptive information of the terminal functions may or may not be included. To serve as an aid in drawing integrated circuits, the template shown in Fig. 3–31c is commonly used.

3-19

Digital Logic Circuits

Digital circuits and digital computers dominate the electronics industry. From the hand-held calculator, to electronic games, to the largest computers, digital circuits influence our daily lives. The electronic draftsperson must become familiar with the basic symbols used to represent the component parts of digital logic circuits.

Principally, all digital circuits work in one of two active states. These are the 1 state, often referred to as the *Hi* or *On* state, and the 0 state, known also as *Lo* or *Off* state. Digital systems are designed to perform a specific task, such as in hand-held calculators, by combining basic circuits to function as a single unit.

The symbols used to represent digital circuits can be divided into two groups, the basic *gate* circuits and the *functional blocks*. Digital circuits are fabricated in standard integrated circuit package configurations and are classified as either SSI (small-scale integration), MSI (medium-scale integration), LSI (large-scale integration), or VLSI (very large-scale integration) packages. These designations refer to the complexity of the device circuitry. If one or a small group of simple gates are packaged together, this would be considered an SSI package, and a microprocessor chip with 40 or more accessible terminals is an example of a VLSI package.

Logic Symbols	Truth Table	Type	

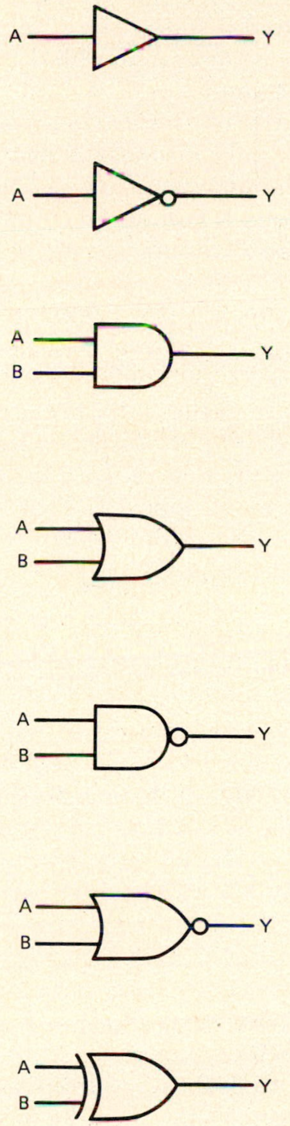

A	Y
0	0
1	1

Amplifier (Buffer) (a)

A	Y
0	1
1	0

Inverter (b)

A	B	Y
0	0	0
0	1	0
1	0	0
1	1	1

AND (c)

A	B	Y
0	0	0
0	1	1
1	0	1
1	1	1

OR (d)

A	B	Y
0	0	1
0	1	1
1	0	1
1	1	0

NAND (e)

A	B	Y
0	0	1
0	1	0
1	0	0
1	1	0

NOR (f)

A	B	Y
0	0	0
0	1	1
1	0	1
1	1	0

Exclusive OR (g)

FIGURE 3–29 Basic logic symbols.

Basic digital logic symbols, together with their *truth tables*, are shown in Fig. 3–29. The truth table shows all possible input levels and the resulting output levels. Note that all inputs are identified by letters, starting with A for a single input, A and B for two inputs, and continuing in alphabetical order for multi-input gates (not shown). The output terminal is identified with the letter Y.

In the truth table for an *amplifier* (current buffer), shown in Fig. 3–29a, both input and output levels are identical. If the input is in a logic 1 state, so also is the output.

The *inverter* gate, as the name implies, is an amplifier with an output which is inverted from its input. If a logic 1 state is applied to the input, the output becomes inverted to a 0 state. The symbol for the inverter, shown in Fig. 3–29b, is the standard amplifier triangle with a circle at the output end of the triangle.

The *AND* gate, whose symbol is shown in Fig. 3–29c, is a two-input device. A 1 state in the output results only if inputs A *and* B are both in a 1 state. On the other hand, the *OR* gate (Fig. 3–29d) has a 1-state output if either input A *or* B are in a 1 state. Of course, a 1-state output will also result if both A and B inputs are in the 1 state.

The complements to the AND and OR gates are the *NAND* (not AND) and the *NOR* (not OR) gates. See Fig. 3–29e and f, respectively. The symbols for these gates are the same as their respective AND and OR gates with the addition of a circle at the output of each to indicate an inversion between input and output levels. When comparing the truth table of the AND gate with that of the NAND gate, it is seen that the output levels are the exact opposite. That is, the output of the NAND gate is the complement of the AND gate. The same is true when comparing the output of the OR gate to that of the NOR gate.

The gate symbolized in Fig. 3–29g is the exclusive OR gate. Refer to the truth table. It is shown that an output 1 state will exist only if input A or input B are in a 1 state, *but not both.*

The basic gates shown in Fig. 3–29 are also available with more than two inputs. In addition, the inputs may also be shown with input inverters (i.e., circles tangent to the gate body on the input side). These gates differ in both name and function from those presented in this section. Because of the large variety of digital circuits available other than the basic gates shown, they are represented by a functional block. A selection of common digital circuits that can be represented by functional blocks is shown in Fig. 3–30.

The *Schmitt trigger* symbol is drawn as a rectangle with one input on the left and two outputs on the right, labeled Q and $\overline{Q}$. See Fig. 3–30a. When the input signal exceeds a specified value, the output terminal Q immediately alters its state (i.e., from 0 to 1 or from 1 to 0). The complement output terminal $\overline{Q}$ is always in the opposite logic state from terminal Q. The Schmitt trigger is designated by a "hysteresis"-shaped symbol drawn within the rectangle as shown in Fig. 3–30a.

The *RS latch,* also known as a *flip-flop* circuit, is designated with the letters FF. See Fig. 3–30b. When a logic level 1 is applied to the Set (S) input terminal and a 0 level to the Reset (R) terminal, the Q output terminal is "set" into the 1 state. Conversely, when the Reset terminal is at the 1 level and the Set terminal at the 0 level, the Q output is "cleared" or reset to 0.

Another common type of digital circuit is the *J-K* flip-flop. See Fig. 3–30c. Even though logic states of 0's and 1's are applied to the J and K input terminals, the output will change state only if a control pulse is applied to the *toggle* (T) terminal. Flip-flops find application in computers as storage elements (memories), counters, and shift registers.

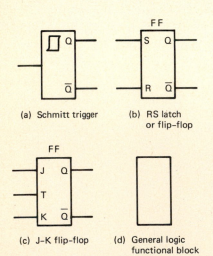

(a) Schmitt trigger (b) RS latch or flip-flop

(c) J-K flip-flop (d) General logic functional block

FIGURE 3–30 Functional block symbols for a few digital devices.

Regardless of the complexity of a digital circuit, it can be represented by the general rectangular block shown in Fig. 3-30d. Each rectangular symbol used must be identified by its specific reference designator or by its descriptive name. In addition, each terminal should be identified by its appropriate pin number (for wiring purposes) as well as its literal designation (i.e., R, S, Q, etc.).

3-20

Electronic Symbol Drafting Templates

As was shown in this chapter, electronic symbols can be drawn uniformly and in correct proportions using conventional drafting equipment. This, however, is extremely time consuming, especially for drawing large, complex circuits. A much more efficient method for producing electronic circuit diagrams is with the use of a wide variety of electronic drafting templates, a selection of which is shown in Fig. 3-31. Each of these templates is commonly available by basic function (i.e., general purpose, discrete analog, or digital logic),

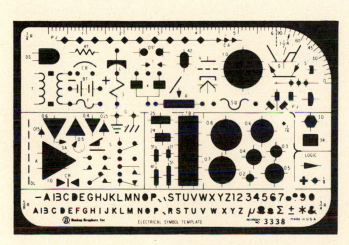

Electrical

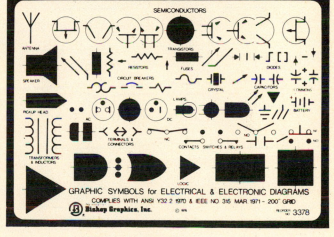

Electrical/Electronic

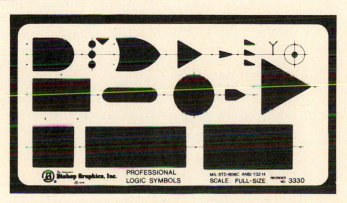

Logic

FIGURE 3-31 Electronic symbol drafting templates. [Provided courtesy of Bishop Graphics, Inc., Westlake Village, California 91359]

Templates should always be used in conjunction with a T-square for proper alignment and should be cleaned regularly. For certain symbols, it is recommended that the template be slightly raised from the surface of the drawing. This allows the conical shape of the pencil lead to fill the cutouts and prevents a wobbly line from resulting. Triangles placed flat beneath the template and spaced to either side of the symbol being drawn serve this purpose. The degree of flex of the template will determine how far apart to place the triangles.

Not only can electronic circuits be drawn much faster with the use of templates, but consistent size and shape of like symbols are assured, resulting in better overall appearance of the drawing.

EXERCISES

3–1 Print the name of each symbol shown in Fig. 3–32.

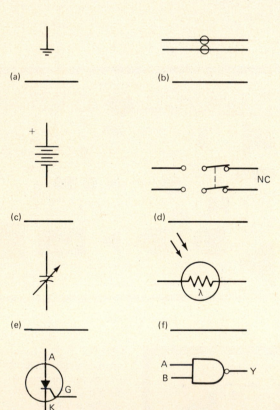

FIGURE 3–32

3–2 Print the reference designator for each of the components and devices listed below.
(a) Incandescent lamp
(b) Circuit breaker
(c) Relay

(d) Resistor
(e) Inductor
(f) Rectifying diode
(g) Programmable unijunction transistor
(h) Integrated circuit

3–3 On a sheet of A-size graph paper having 10 divisions to the inch, draw the symbols for the components and devices listed below. Use standard drafting equipment only. As a guide for the appropriate size, use a $\frac{1}{2}$-in.-diameter circle for the semiconductor device symbols. Draw all other symbols in correct proportion to this guide and to one another.
(a) Zener diode
(b) Light-emitting diode
(c) Npn transistor
(d) Photoemissive transistor-type optical isolator
(e) Potentiometer
(f) Fuse
(g) Ac source
(h) Single-pole single-throw switch
(i) Multiwinding magnetic-core transformer

3–4 On a sheet of drafting vellum and with the use of an electronic symbol template wherever possible, draw the symbols for each of the items listed below.
(a) Polarized electrolytic capacitor
(b) Light-sensitive resistor
(c) Ac-operated relay coil with single-pole double-throw normally closed locking contacts
(d) Dc microammeter
(e) Chassis ground
(f) P-channel JFET
(g) Operational amplifier
(h) Diac
(i) Triac

3–5 On a sheet of drafting vellum and with the use of a logic template, draw each of the following gates with four inputs.

(a) AND   **(c)** NAND

(b) OR   **(d)** NOR

3–6 Using the technique shown in Fig. 3–33, repeat Exercise 3–5 for gates with expanded inputs to accommodate eight connections.

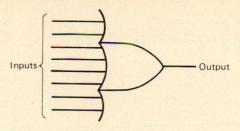

FIGURE 3–33

4

Circuit Schematic Diagrams

LEARNING OBJECTIVES

Upon completion of this chapter on drawing circuit schematic diagrams, the student should be able to:

1. Draw simple analog circuit schematics.

2. Properly show circuit interconnections.

3. Draw complex analog circuit schematics.

4. Draw digital circuit schematics.

5. Draw freehand engineering sketches.

4-0

Introduction

A circuit schematic diagram is a symbolic drawing which provides specific information on (1) the value and type of each component and device, (2) their interconnections between one another, (3) signal and power flow, (4) operating controls, and (5) all other pertinent information necessary for a complete understanding of the circuit and its function. To the engineer and technician, the circuit schematic is the principal drawing required for the design, analysis, fabrication, and troubleshooting of an electronic system. It serves as the primary working document for determining parts placement and, in the case of printed circuit boards, for component and conductor pattern layout. It may be well to point out here that the circuit schematic does not show the physical relationships of components. It is not intended to be a pictorial representation of the packaged system but rather to show how the circuit functions electronically.

It can be seen from the above that the circuit schematic drawing is of the highest order of priority in the electronics industry for initiating the packaging of a system. As such, the schematic must be exact, complete, and neatly presented so there will be no room for misinterpretation. It is, therefore, time well spent in carefully and accurately drawing the schematic in logical and readable form and in providing complete information.

This chapter will show (1) how to read and understand basic circuit schematic diagrams as they appear in finished form, and (2) how to convert rough sketches to a finished circuit schematic.

4-1

Simple Analog Circuit Schematic

The transistor amplifier circuit shown in Fig. 4–1 illustrates a simple analog circuit schematic. Note that standard symbols, as discussed in Chapter 3, are used throughout. The components and devices used in this circuit consist of four resistors, two capacitors, and one transistor. Each of the components and the device is labeled with a reference designator (with an identification number) and with its value. For example, the resistors are labeled as R1 to R4 reading from left to right at the top (R1 and R2) and again from left to right at the bottom (R3 and R4) of the schematic. This is the conventional method of identification since it conforms to the way in which we read and write.

The value of each component is included together with the reference designator. In the case of resistors, the ohm symbol (Ω) is usually omitted from the schematic since it is implied. Thus, resistor R1 is labeled 47K, which denotes a 47,000-Ω resistor. Normally, the power rating and the tolerance of each resistor is not placed next to the component but rather is provided as a note below the schematic. This not only assures clarity by avoiding any crowding of literal and numerical information, but also eliminates much redundancy. In Fig. 4–1, all of the resistors are to have the standard $\frac{1}{2}$-W (watt) rating with tolerance values of ±10%.

Capacitors are identified on the schematic in a similar form as that for resistors. The capacitor labeled C1 is at the left of the schematic and therefore identified first. As with resistors, the value of each capacitor is included next to the component and the units are described in a note below the schematic. Thus the capacitor C1 in Fig. 4–1 has a value of 1.0 microfarad (μF) and C2 has a value of 15 μF. Note that both capacitors are polarized, shown by the plus (+) sign placed next to the appropriate terminals.

The single active device in Fig. 4–1 is an npn type transistor which is identified by the reference designator Q1 (see Section 3–16 for a discussion of transistors). In addition, the transistor is further identified by the code 2N3904, which is the type of transistor to be used. Q1 is therefore uniquely identified, as are the other components in the schematic. Note also that each terminal of Q1 is identified as B (base), C (collector), and E (emitter). This labeling is not commonly shown, however, since the schematic symbol for a transistor implies these terminals and is well understood by engineers and techni-

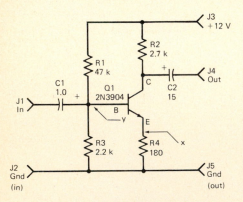

All resistors are 1/2 W ± 10% unless otherwise specified. All capacitor values are in microfarads unless otherwise specified.

FIGURE 4–1 Schematic of a self-biased common emitter transistor amplifier using an NPN device.

cians. They are shown in Fig. 4–1 primarily to allow references made to this circuit in Section 4–2 to be followed more easily.

Reference designators and component values may appear above, below, or to either side of the component. They should always be placed in close proximity to the component, however, to avoid confusion. In addition, the labeling of two closely placed components should be sufficiently separated so as not to cause uncertainty. Most important of all, individual reference designators and values should be assigned to each component and device on the schematic drawing.

External connections to the circuit of Fig. 4–1 are provided through female jacks labeled J1 through J5. Since no specific type of jack symbol is shown on the schematic, each jack should be labeled as to its function. Common practice is to draw the circuit so that the signal flow is from left to right. Thus the input terminal J1 is labeled as IN and is on the far left of the drawing. The signal would normally be applied between terminals J1 and J2 (i.e., input to ground). The signal is then operated on by the circuit (in this circuit, it is amplified) and the processed signal is extracted at the output jacks J4 and J5 (i.e., output to ground). Note that the output jack, labeled OUT, is at the far right of the schematic.

All electronic circuits require dc power to make them functional. In the amplifier circuit of Fig. 4–1, jack J3 is used to connect a +12 Vdc source of power, with the negative side of the source connected to jack J5 (ground).

4-2

Interconnections

Each component and device is provided with external leads which are used to make electrical interconnections between itself and other components and/or devices. An electrical connection is made when two current-carrying elements are connected together in such a fashion as to allow current to flow through each of them.

In its simplest form, an electrical connection is made when two uninsulated (bare) wires are twisted tightly together. This, however, is not considered a reliable connection and the electronics industry exclusively employs soldered or welded connections for long-term reliability. The leads of two elements are connected together electrically and then soldered or welded.

Which leads are to be electrically connected is part of the information provided on a circuit schematic. For example, refer to point x in Fig. 4–1. The interconnecting line between the emitter (E) of Q1 and one end of the resistor R4 are to be connected together electrically. This connection may be made by joining the component and device leads themselves, by adding additional lengths of wire if the distance between these two parts is excessive, or by using the conductive copper foil paths of a printed circuit board. Regardless of the connecting method used, note that point x is a singular connection with no other connections made to any other component or conductive path at this point.

At point *y* in Fig. 4–1, an electrical connection is to be made between the base (B) of Q1, the plus (+) side of C1, one end of R1, and one end of R3. It can be seen that the circuit schematic provides specific and detailed information on the electrical interconnections between all components and devices in order for the circuit to function properly.

When laying out a schematic diagram, there should be uniform density, or balance, of graphical symbols; that is, they should not be crowded in one area while other symbols are spaced apart, leaving large open spaces. To aid in achieving this balance, the following guidelines should be observed:

1. When components are connected between parallel leads, their symbols should be uniformly spaced and their centers aligned.
2. When components are connected to a common lead, their symbols should also be equally spaced and aligned wherever possible.
3. Components connected in series should be shown so that their symbols are equally spaced, in line, and maintain the same size proportions.
4. The line representing component leads should be centered with their symbols. See Fig. 4–1.

4-3

Complex Analog Circuit Schematics

When the circuit schematic becomes more complex, there are several methods used to minimize confusion. Refer to the *negative-acting one-shot multivibrator* circuit shown in Fig. 4–2. Note that the entire circuit is surrounded with a series of broken-line segments. These are called *enclosure lines* and their purpose is to distinguish one specific circuit from other circuits in the system. Thus the circuit of Fig. 4–2 is isolated from other circuits at the perimeter of the enclosure lines, but is intended to be electrically connected at points 1, 2, 3, 4, and 5. These points may or may not be jacks or plugs. Each of these numbered points would correspond to a similar number representing a connection on an associated circuit. The input (point 1, labeled IN) is at the far left and the output (point 2, labeled OUT) is at the far right.

The following describes the functioning of the circuit of Fig. 4–2. Negative signals are applied to point 1, where their amplitude is adjusted by potentiometer R101 and then applied to the inverting input terminals (pin 2) of U1, an IC amplifier. U1 amplifies and inverts the signals and applies them to the noninverting input terminal (pin 3) of U2. The signals are shaped into pulses and only their positive values are passed by diode CR4. That portion of the signal that appears across potentiometer R102 also exists at the output (point 2). The magnitude of the output is adjusted by R102.

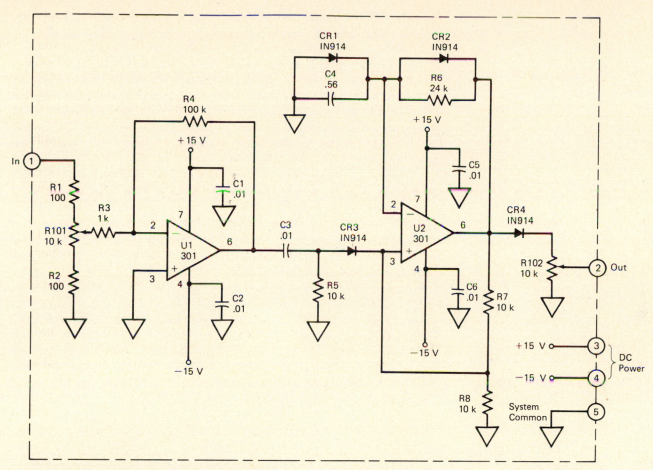

All resistors are 1/2 watt ± 10% unless otherwise specified.
All capacitors are in microfarads unless otherwise specified.

FIGURE 4-2 Negative acting one-shot multivibrator.

To minimize confusion, some shorthand methods of schematic drawing have been employed in the circuit of Fig. 4–2. Note that each pin 7 of both U1 and U2 are to be electrically connected to +15 Vdc. One method of showing this connection would be to draw a connecting line between each pin 7 and then extend that line to the enclosure line, terminating it with a circle. However, this would require crossing over several existing connections and possibly having to crowd components. The method employed in Fig. 4–2 implies the same connections but avoids long lines and crowding. In effect, it states that anything labeled +15 V on the schematic (pin 7 of both U1 and U2) are connected to point 3. Similarly, pin 4 of both ICs are shown to be connected to point 4, seen at the right side of the enclosure lines.

This method of schematic shorthand is extended to include all common connections, which are symbolically represented by triangles. With 10 separate electrical connections represented by triangles in Fig. 4–2, it is implied that they all must be electrically connected together and then connected to the system common (point 5) seen at the right side of the enclosure lines. This method greatly simplifies the drawing and makes it much easier to read. However, care must be taken to draw common symbols clearly so that all such connections are readily seen and understood.

Note in Fig. 4–2 that the guidelines for a well-balanced drawing previously discussed have been followed. There is no crowding of component or device symbols. Sufficient space has been provided around each part for the placement of reference designators as well as component values and device identification numbers. The layout is balanced and the total available space has been used with no isolated crowding.

A properly organized circuit schematic has, wherever possible, all interconnecting paths between parts drawn as either vertical or horizontal lines. Generally, the main axis of component symbols should also be drawn either vertically or horizontally. In the placement of transistors and devices, it is common practice to align them across the schematic from left to right. This results in the best arrangement for obtaining a well-balanced layout.

A circuit schematic diagram should also be attractive to the eye, with no individual part dominating another. All lines drawn for either connections or symbols should have the same thick line weight (0.7 to 1.0 mm; see Section 2–1). The exception to this is the use of very heavy line widths to highlight critical circuitry which may require special attention, such as requiring short lead length or specific component orientation.

4-4

Digital Circuit Schematics

Digital logic circuit schematics can be broadly classified as (1) those comprised mainly of gate symbols, as shown in Fig. 4–3, or (2) those made up largely of functional block symbols, as shown in Fig. 4–4. There is also a third group of circuits which are primarily digital in nature, which combine both gate and functional block symbols as well as analog circuitry.

The schematic diagram of a typical logic circuit consisting of all basic gate symbols is shown in Fig. 4–3. The circuit is shown with enclosure lines (previously discussed), with inputs 1 through 10 at the left and outputs 12 through 15 at the right. The circuit consists of three types of gates. The small triangular-shaped symbols labeled 04 are called inverters. The 04 designation is a shorthand notation to represent the 7404 type TTL (transistor-transistor logic). Also included are the 7402 NOR gates (labeled 02) and the 7400 NAND gates (labeled 00). The shorthand notation of labeling this type of schematic is common practice but not preferred since no standard reference designator is included. A more acceptable method of labeling is that shown for the group of 7400 NAND gates to the right side of Fig. 4–3 which is enclosed in dashed lines and accompanied by the reference designator U3. The dashed lines enclosing all four gates indicate, for this portion of the circuit, one integrated circuit which contains four two-input NAND gates in a single DIP (dual-in-

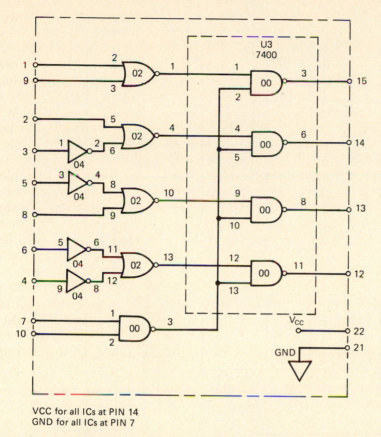

VCC for all ICs at PIN 14
GND for all ICs at PIN 7

FIGURE 4–3 A typical logic circuit consisting of 7400 series TTL gates.

line package). The numbers besides each input and output terminal lead represent the appropriate pin numbers for those terminals on the actual device. It can be seen from the pin numbers that 12 terminals of the DIP package are used. Normally, digital logic schematics for gate circuits do not show the complete wiring paths for the connections to power and ground as part of the circuit. These connections are conventionally understood to be a necessary part of the circuit design. The pin numbers for these connections are generally stated in a note below the schematic, as shown in Fig. 4–3. This note states that pin 14 of each IC must be electrically connected together and also connected to the external terminal labeled Vcc and numbered 22. In like manner, pin 7 of each IC package must also be electrically connected together and also to the external terminal labeled GND and numbered 21.

The 7404 IC package for the four inverters, labeled 04 on the schematic, actually contains six inverters, resulting in having two spares available. Having spare devices is good practice, especially if a fabricated system is to be modified.

The four NOR gates, labeled 02, are contained in a single 7402 package which allows for no available spares. The remaining single 7400 gate, labeled 00, will be obtained from a 7400 IC package, thus having three two-input NAND gates that will be available as spares.

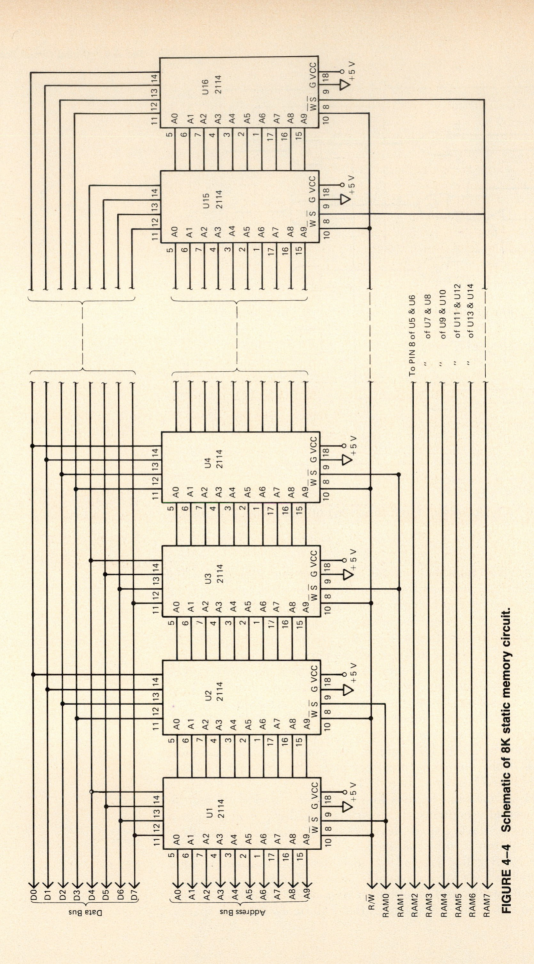

FIGURE 4-4 Schematic of 8K static memory circuit.

The circuit of Fig. 4–3 will thus require four 14-pin DIP IC packages and, when fabricated, will leave two inverters and three NAND gates as spares.

Common difficulties encountered when reading and interpreting gate-dominated digital circuit schematics are: (1) power and ground connections are often implied and not specified clearly; (2) labeling of the various gates frequently is done in shorthand form without the use of standard reference designators; and (3) gates are not always grouped by common function, making it often difficult to determine which gate belongs to which IC package.

Another type of digital circuit schematic is that which employs large numbers of similar functional block symbols, such as the circuit shown in Fig. 4–4. This is the memory section of a microprocessor circuit used in bare-board testing of printed circuits. Because of the extensive duplication of identical wiring of various pin connections, one method of simplifying the drawing is to show only a *partial* schematic. This is done by drawing only the first four and the last two ICs of an actual series of 16. The dashed lines placed horizontally between U4 and U15 indicate a continuation of an identical wiring pattern that exists among all ICs which are located between these two devices. A *broken-line* symbol is also used at each horizontal interconnecting line to further emphasize this identical wiring pattern.

4-5

Freehand Engineering Sketches

A typical form of circuit schematic which may be provided to the draftsperson by a design engineer or technician is shown in Fig. 4–5. It must be assumed that the electrical connections are correct and all information is provided. It is apparent that this schematic is a rough sketch. It has been drawn freehand and is unbalanced, is crowded in some areas, and nonstandard symbols have been used in addition to other deficiencies. Before the finished drawing is at-

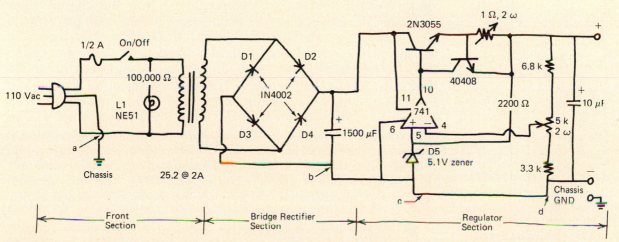

FIGURE 4–5 Rough engineering sketch of regulated power supply circuit.

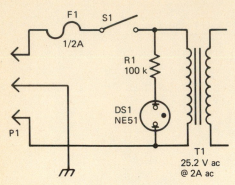

FIGURE 4–6   Front end section of power supply sketch corrected.

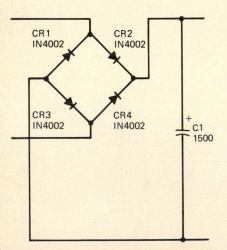

FIGURE 4–7   Corrected bridge rectifier section of power supply schematic.

tempted from this sketch, each component and circuit section should be examined for errors or deficiencies. These need to be corrected immediately. To aid in this task, note that the circuit schematic has been divided into three main sections. The first, or *front*, section (at the left) of a regulated power supply circuit includes everything from the ac plug through the step-down transformer. The second section is the bridge rectifier circuit, including the filter capacitor C1. The third section, to the right of the schematic, is the IC regulator and its associated components.

We will take each of these sections and examine all of its parts for correctness and completeness. In the front section, note that a nonstandard symbol for the 110-Vac plug is shown. Also, the loop in the conductor path at point *a* is incorrectly used, regardless of whether the dot or no-dot system of drawing is employed. The symbol for the chassis connection from the ac plug is also incorrect. The on/off switch symbol needs to be replaced since the intent is to activate the switch which applies continuous power to the system. A latching switch symbol is therefore required. No symbol is shown for the 100,000-Ω resistor. The designation 100K should be placed next to the symbol on the finished drawing with the ohm symbol omitted. The lamp symbol and its reference designator are incorrect. If it is to be a neon lamp, it should be appropriately drawn and labeled DS. Finally, note that both windings of the transformer are drawn reversed.

In general, the front section is well balanced without component crowding, but is missing reference designators F1 for the fuse, S1 for the on/off switch, and T1 for the transformer.

The front section of the power supply circuit with errors corrected and components properly labeled is shown in Fig. 4–6. Note that standard symbols are used, each component is supplied with a reference designator as well as its value, and the dot system of showing circuit path interconnections is employed.

The second section of the power supply circuit of Fig. 4–5 consists of four diodes followed by a filter capacitor. One apparent error is in the drawing of a loop at the lower left side of the rectifier circuit. Also, the connection at point *b* should not be in the position shown along the vertical path coming down from the 1500-μF filter capacitor. Whenever possible, connections should be drawn so as to form straight continuous paths along a given plane. This section of the schematic also uses nonstandard symbols for both the diodes and the electrolytic capacitor. In addition, each component or device should be labeled individually. The method shown in Fig. 4–5 of identifying one of the diodes by its type (i.e., 1N4002) and then drawing arrows to the other diodes from this number is not acceptable practice. Finally, the filter capacitor has no reference designator and is shown with a value of 1500 μF. On the finished drawing, the reference designator C1 should accompany the symbol and only the magnitude of its value is required (i.e., 1500). The units are usually omitted and handled as a note on the drawing, as shown in Fig. 4–9.

All of the deficiencies and errors just discussed for the bridge rectifier section of the power supply sketch have been corrected and a properly drawn schematic is shown in Fig. 4–7. Since the dot method

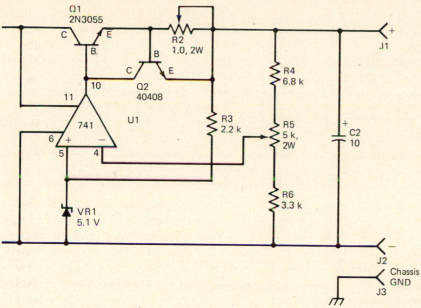

FIGURE 4–8 **Improved regulator section of power supply schematic.**

of showing conductor path connections was adopted for the schematic of the front section of the power supply, it is used throughout the circuit drawing.

The regulator section of Fig. 4–5 will now be examined. The first improvement that can be made for the sake of appearance is to align the vertical paths leading to pins 11 and 6 of the IC so that they are in the same plane. The reference designator for the 741 integrated circuit should be U1. Again, the reference designator for the 5.1-V zener diode could be changed to read VR1. In addition, the other parts (i.e., resistors, capacitor, transistors, and output terminals) need their appropriate reference designators added. The interconnecting wire paths shown at points c and d are stepped unnecessarily. The nonstandard loop appearing to the left of the 5K-ohm resistor needs to be corrected.

Space should be provided to accommodate the total height of the regulator section with the neutral path, labeled minus (−), drawn as a single continuous line. As with the front section of the power supply circuit, the ground symbol used below the negative output terminal is drawn incorrectly since chassis ground is specified. Note that the resistor symbol is missing at the intended position of a 2200-Ω resistor. Also, the symbol for the 3.3K-ohm resistor is drawn incorrectly since the circuit path extends through the symbol. The properly drawn regulator section, including all of the corrections and modifications discussed, is shown in Fig. 4–8.

The corrected sections are combined to form the drawing for the complete power supply circuit. This is shown in Fig. 4–9. Note that standard symbols are used throughout, all parts have their appropriate reference designators, and that there is no crowding of components or wasted space. In addition, the drawing shows a good balance of the parts from left to right and from top to bottom. It can

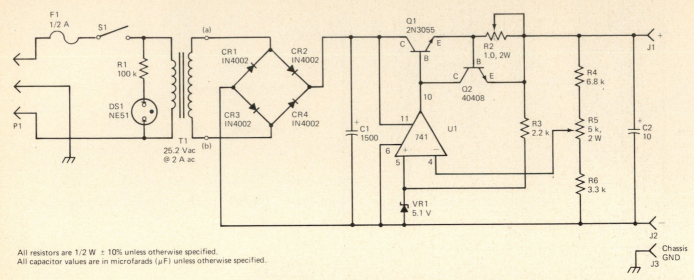

All resistors are 1/2 W ± 10% unless otherwise specified.
All capacitor values are in microfarads (μF) unless otherwise specified.

FIGURE 4–9 Completely corrected power supply schematic from sketch presented in Figure 4–5.

be seen that the most complex section, having the largest number of components, is the regulator circuit. In general, the placement, spacing, and alignment of the components in the most complex section of a circuit governs that of the simpler sections.

EXERCISES

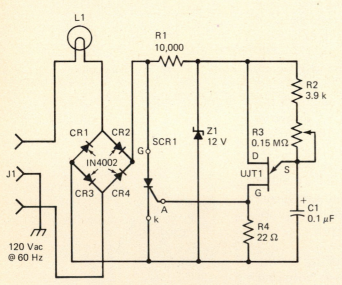

All resistors are 1/2 W ± 10% unless otherwise specified.
All capacitor values are in microfarads (μF) unless otherwise specified.

FIGURE 4–10

4–1 Redraw the full-wave power control circuit shown in Fig. 4–10, improving the positioning of components where necessary to result in a more balanced layout. Correct all errors in labeling, including reference designators, component values, and terminal designations.

4–2 Draw the full-wave bridge rectifier together with a 7805 +5-Vdc regulator shown in Fig. 4–11. Identify the missing components and draw their symbols in the correct positions. Include all values and reference designators as listed in Fig. 4–11.

4–3 Draw a well-proportioned schematic of the Wein bridge sine-wave oscillator shown in Fig. 4–12. Include all component values, reference designators, and footnotes as shown.

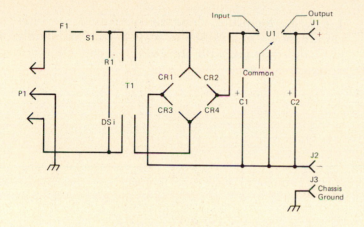

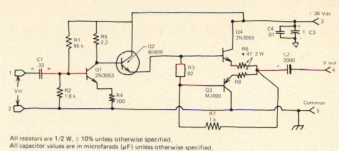

FIGURE 4–13

F1 – Fuse (400 mA Slow-Blo)
S1 – SPST On/Off Switch (Locking Contact)
R1 – 100 KΩ, 1/2 W ± 10% Resistor
DS1 – NE51 Neon Lamp
T1 – 12 Vac @ 1.2 Amps Transformer (Magnetic Core)
CR1, CR2, CR3, CR4, – 1N4002 Diodes
C1 – 1500 μF Electrolytic Capacitor
C2 – 25 μF Electrolytic Capacitor
U1 – 7805 IC Regulator Chip

FIGURE 4–11

4–5 With the use of a logic template, redraw the gate circuit shown in Fig. 4–14. Improve the symmetry and balance of the schematic by repositioning components. Completely label the drawing with the correct terminal and reference designators.

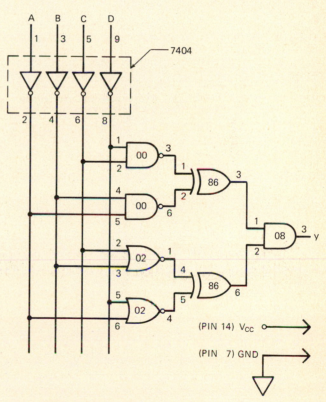

All resistors are 1/2 W ± 10% unless otherwise specified.
All capacitor values are in microfarads (μF) unless otherwise specified.

FIGURE 4–12

4–4 Using the sketch provided in Fig. 4–13, draw a corrected schematic of the audio power amplifier shown. Improve the balance, symmetry, and position of all components. Use standard symbols for all devices and components and label them with the appropriate reference designators and values.

Note: All logic 7400 series TTL

FIGURE 4–14

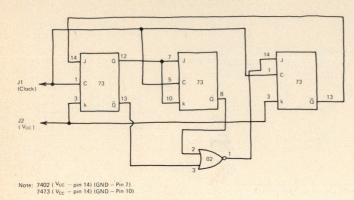

Note: 7402 (V$_{CC}$ — pin 14) (GND – Pin 7)
7473 (V$_{CC}$ — pin 14) (GND – Pin 10)

FIGURE 4–15

4–6 Using templates, draw the synchronous counter circuit shown as a sketch in Fig. 4–15. Completely label the drawing with the terminal and reference designators given.

5

Detail Drawings for Electronics

Upon completion of this chapter on detail drawings, the student should be able to:

1. Correctly illustrate multiview planes.

2. Correctly illustrate pictorial views.

3. Properly draw detail drawings of:
   a. Axial lead components.

   b. Electronic devices.
   c. Hardware and fasteners.
   d. Chamfers and slots.
   e. Printed circuit boards.

5-0

Introduction

Detail drawings are used in the electronics industry to provide the required mechanical information for both design and manufacturing. A detail drawing normally describes a single part and provides information such as shape, material, finish, tolerance, size, and often includes handling instructions. In printed circuit board (PCB) design, a detail drawing is typically required for every component, device, and mechanical fastener used in the assembly. For example, such a drawing would show a resistor's body width and length, its lead diameter, and the distance between the bent leads. In addition to this information for the individual parts and hardware, detail drawings of the printed circuit board itself are provided to show board material; its length, width, and thickness; the location and size of all drilled and punched holes; the plating finish; chamfers; slots; and all other mechanical information required for the complete fabrication of the board.

The majority of detail drawings for printed circuit boards are shown in two-dimensional form by making the appropriate projections from multiview drawings. These drawings and the specific planes of projection to obtain two-dimensional views are presented in this chapter. In addition, other types of pictorial views are described which show greater detail relative to the geometry or style of the object.

5-1

Multiview Drawings

Multiview drawings show several different views of an object and are positioned in a prescribed manner so that the relationship of each view is readily apparent. There are three viewing planes used in multiview drawings which are referred to as *planes of projection*. These are shown in Fig. 5–1a. With the use of these projections, any or all of the six views of an object may be shown. The *frontal* plane shows either the front or rear view, the *horizontal* plane shows the top or bottom view, and the *profile* plane is used to show the left or right side of the object. Unless the geometry of the object is unusually complex, showing the front, top, and right-side views is generally sufficient to convey complete detail.

Three views of a relay are shown in Fig. 5–1b. Note that each view is shown as if one were standing directly in front of each plane of projection. Because the terminals of the relay protrude from the recessed bottom of the case, this view was selected to be shown with

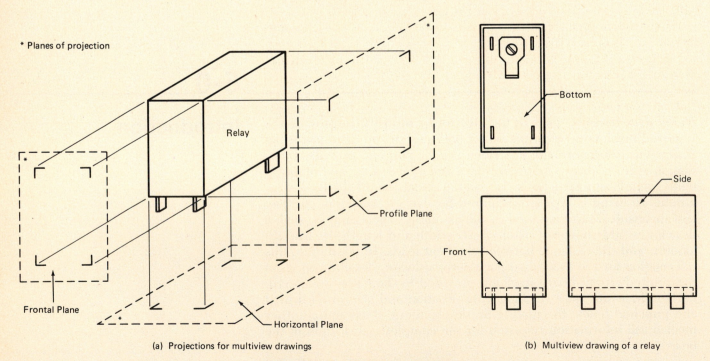

(a) Projections for multiview drawings

(b) Multiview drawing of a relay

FIGURE 5–1   Multiview drawings.

the front and side views instead of the top view. This arrangement provides more detail of both the recessed bottom and the terminals. The use of hidden lines on the front and side views show the recessed base of the relay.

It can be seen in Fig. 5–1 that when an object is shown positioned so that its sides are parallel to the planes of projection, all detail lines in these planes are true length. Of course, the size of any drawing is determined by the complexity and actual size of the object. Large parts are generally scaled down while small parts are scaled upward to show more easily the detail. In all cases, however, objects must be shown in correct proportions to avoid distorted views.

5-2

Pictorial Drawings

Although all necessary dimensional information regarding the design detail of a part can be obtained with the use of multiview drawings, the geometry of the part may not always be evident, even when three separate views are shown. To overcome this problem, a *pictorial* view may be necessary so that the overall shape of the object is more recognizable. Because three views are shown by one pictorial diagram, the use of hidden lines is rarely required.

The three principal types of pictorial drawings are the *axonometric*, *oblique*, and *perspective*. In electronic design applications, pictorial views are normally of the axonometric or oblique types. Perspective views are associated more with architecture and other allied fields and thus will not be considered here.

Axonometric drawings are of three types: *dimetric, isometric,* and *trimetric*. Of these three, the isometric is most widely used. Dimetric and trimetric views require two and three scales, respectively, when drawing their axes. This requirement of different axes in each of three views complicates the drawing of objects having circular detail. For these reasons, the discussion of pictorial views used in electronic design will be limited to isometric and oblique drawings.

An *isometric* view of an object is produced by first drawing its three axes uniformly spaced at 120 degrees. These are called the isometric axes and are all considered to be *receding*, that is, other than full scale as they lead away from the plane of projection. This means that no axis is parallel to the plane of projection. See Fig. 5–2a. The principal dimensions are then laid out in proportion to their true lengths along these axes. The lengths of all lines on any of these axes are drawn to the same scale even though each axis is receding. This results in a view having minimum distortion as well as allowing the detail of three sides of the object to be clearly viewed.

Using the ends of the three lines drawn along the axes and with *isometric* grid lines (lines drawn at 30, 90, and 150 degrees to the horizontal), a box can be constructed which will form parallel planes representing the three major sides of the object. All detail lines can then be drawn by measuring along the isometric grid lines. This in-

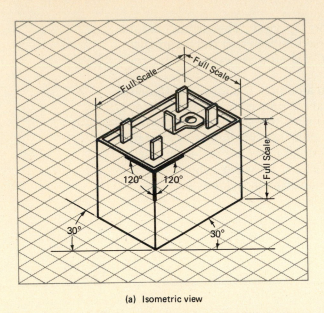

(a) Isometric view   (b) Oblique view (cabinet)

FIGURE 5–2 Pictorial views.

cludes lines which detail any portion of a side that may not be parallel to any of the three planes. (Any line that is not parallel to any of the three oblique planes is called a *nonisometric* line.) An isometric view of a relay enclosure is shown in Fig. 5–2a in one of its four possible positions.

Oblique views are more commonly used since they are the easiest type of pictorial drawing to construct. Two axes are initially drawn full scale at an angle of 90 degrees. The third axis is drawn as a receding line at a 45-degree angle from the horizontal and generally at one-half scale. See Fig. 5–2b. The term *cabinet* is used to describe this arrangement. The receding axis may be drawn at other than 45 degrees, with 30 and 60 degrees sometimes preferred. The larger the angle, the more relative space is available to show detail at the top plane of the object. Other scales may be used for the length of the receding axis, such as in the arrangement termed *cavalier*, where all three axes are drawn to equal lengths. This form, however, results in viewing distortion which is quite pronounced and thus is rarely used in electronic drafting. An example of an oblique drawing in the cabinet form is the relay shown in Fig. 5–2b.

It is sometimes the case where irregular curves or circular features need to be detailed in a pictorial view. Because of the angle of viewing, major distortion of a circle will result unless the circle is shown as an *ellipse*. Two methods of constructing an ellipse are shown in Fig. 5–3.

The *four-center ellipse*, shown in Fig. 5–3a, is commonly used in isometric views. The construction of this ellipse begins by drawing an isometric square along the receding axes with all four sides equal to the true diameter of the circle. This is shown in step I. A T-square and a 30–60-degree triangle are then used to construct lines which bisect the sides opposite points 1 and 2, as shown in step II. In step

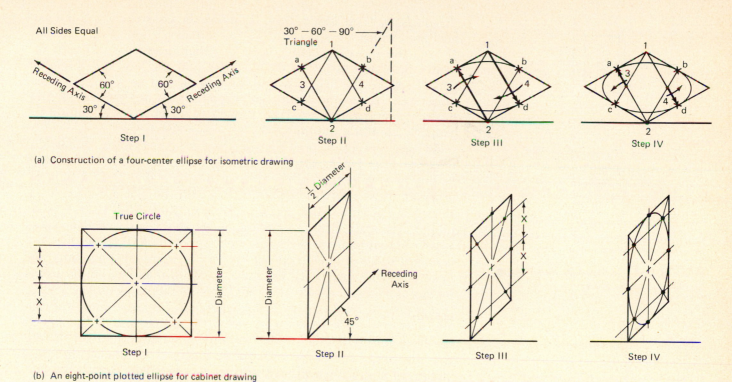

(a) Construction of a four-center ellipse for isometric drawing

(b) An eight-point plotted ellipse for cabinet drawing

FIGURE 5–3 Drawing ellipses for isometric and oblique views.

III, a compass is set to draw arcs having a radius equal to either length 1d or 2a. With points 1 and 2 serving as compass pivot points, arcs are drawn through points a and b from point 2 and through points c and d from point 1. In step IV, the compass is set to draw arcs having a radius equal in length to either 3a or 4b. Using points 3 and 4, arcs are drawn through points a and c from point 3, blending with the previously drawn arcs. Finally, an arc is drawn through b and d from point 4, again blending with the previously drawn arcs. The results of this approximate method is an ellipse having negligible distortion.

Circular detail in oblique views that must be drawn in planes that are at the same oblique angle to the plane of projection as the receding axis can be drawn by the *eight-point plotting method* shown in Fig. 5–3b. (This is the method used for the cabinet view.) In step I, a true circle with its center lines is initially drawn. A square is then constructed about the circle as shown, and diagonals between opposite corners of the square are added. The horizontal lines are drawn through the intercepts of the diagonals and the circumference of the circle. The true vertical distances between the center of the circle and the intercepts are shown with the designation x. In step II, the dimensions of the square are transferred to the plane of the receding axis which is drawn at a 45-degree angle to the horizontal. In this plane, the vertical sides are drawn equal in length to the true diameter of the circle. The length of the sides along the receding axes are equal to one-half the diameter. The center lines and the

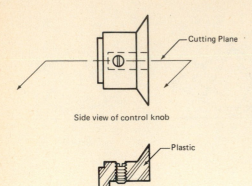

Side view of control knob

Plastic

Sectional view of control knob

FIGURE 5–4   Method of illustrating cross sectional view.

diagonals are then added. Step III shows the construction of the parallel lines which are spaced at the true distance x (obtained from step I) from the receding center line. The four intercepts between the center lines and the sides of the figure and the four interior intercepts between the diagonals and the two parallel lines which set distance x constitute the eight plotted points that are used in step IV to draw a smooth ellipse. The French curve is an excellent aid for properly blending the linework.

Another type of pictorial drawing used for conveying a clearer idea of design is the *sectional* view, such as that of a control knob shown in Fig. 5–4. This type of drawing allows more detail of the interior of a part or assembly to be shown than with the use of hidden lines. For sectional views, the position of the cutting plane is first determined. The part is then drawn showing those surfaces that would appear in that plane as sectional lines. Appendix XIII shows the conventional styles of cross-sectional lines for a variety of commonly used materials. These sectional lines are drawn thin and carefully spaced by eye as they are drawn at an angle across the surface. Any surface that is not cut by the plane is not sectioned. Sectional lines are usually drawn at a 45-degree angle from the horizontal. If this brings the lines closely parallel to any prominent visible outline of the object, the angle of the lines should be changed. Any angle between 30 and 60 degrees usually will correct this problem.

The major characteristics of properly drawn section lines are (1) drawn uniformly thin so as to show marked contrast when compared to visible outlines of the object, (2) placed far enough apart so their appearance is evenly spaced, (3) drawn so that there are no gaps between their ends and the object outline, (4) drawn so they do not extend beyond the object outline, and (5) shown in the appropriate style of crosshatching for the specific material of the section.

5-3

Detail Drawings of Axial Lead Components

The physical size and shape of most electronic components can easily and clearly be described with detail drawings. Figure 5–5 shows a two-view detail drawing of a typical axial lead component which is tubular in shape. On these components, the leads extend outward from the ends and on the same axis as that of the body. The front view is shown in Fig. 5–5a, which provides typical dimensional val-

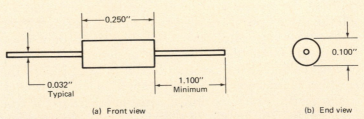

(a) Front view   (b) End view

FIGURE 5–5   Two-view drawing of a tubular component.

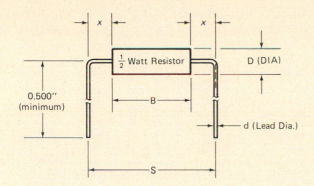

Dimension	Inches		Millimeters	
	MIN	MAX	MIN	MAX
B	0.344	0.416	8.74	10.6
D (DIA)	0.115	0.160	2.92	4.06
* X	0.090	0.120	2.29	3.05
d (Dia.)	0.026	0.036	0.66	0.91
S	0.524	0.656	13.3	16.7

* Depends on application. Minimum values for dense layouts.

FIGURE 5–6 Tabulating dimensional information on detail drawings.

ues for body length (0.250 in.), lead length (1.100 in. minimum), and lead diameter (0.032 in.) using standard dimensioning methods as discussed in Section 2–12. The width of the component could have been inclued in Fig. 5–5a with the use of extension lines, but important information may be missing. Therefore, the end view of the component is shown in Fig. 5–5b. This view establishes the fact that the component body is cylindrical (0.100 in. diameter) and not rectangular or square.

A single front view of a component may be used to describe a component completely. The axial lead component shown in Fig. 5–5a is dimensioned in Fig. 5–6 with the addition of extension lines and the letter *D*. The addition of the abbreviation DIA (diameter) further defines that the component is cylindrical and not another geometric shape.

Compare the drawing of Fig. 5–5a to that of Fig. 5–6. Notice that the leads in Fig. 5–6 are shown bent down at a right angle at a distance *X* from each end of the component body and extend to a minimum distance of 0.500 in. from the axis of the body. This configuration is called a *preformed lead* component which facilitates lead insertion through drilled holes in printed circuit boards. The manufacturers of preformed lead components provide a minimum lead length after bending. When components are assembled onto printed circuit boards, the excess lead length is cut off and discarded after the assembler has allowed sufficient length to protrude through the board. See Fig. 5–7.

A table for listing dimensional information is also provided in Fig. 5–6. Body length (*B*), body width (*D*, for diameter), bend distance (*X*), lead diameter (*d*), and center-to-center lead spacing after bending (*S*) are dimensions whose code letters are shown on the front view of the detail drawing. With the use of the table, additional information on size, tolerance, and multiple unit distances may be provided without having to crowd all of this information directly on the detail drawing. Further, other tables can be keyed to include a series of components having the same basic body styles by using additional literal coding and identifying designations to provide much more information. Thus one detail drawing of a general component

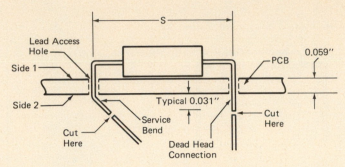

FIGURE 5–7 Assembly drawing of tubular component assembled to a PCB.

style accompanied by several tables can save time and space but still provide all the necessary information.

By listing both the maximum and minimum values of dimensions on a table (see Fig. 5–6), the user of the information is saved a great deal of time by not having to calculate these from nominal values and their tolerances that may be shown on the drawing. On some detail drawings, the maximum and minimum values are provided instead of the tolerance. These are referred to as *upper* and *lower limits* and are shown as follows:

$$\overset{0.0520''}{\underset{0.0470''}{\longleftrightarrow}}$$

The two values are placed with one above and one below an unbroken arrowheaded leader. The number above the line is always the maximum or upper limit dimensional value, and the number below the line represents the minimum or lower limit value. In the example given, the dimension is read as being greater than 0.0470 in. but less than 0.0520 in.

The information provided in Fig. 5–6 is typical for use with printed circuit board (PCB) design. For example, the center-to-center lead spacing (dimension S) is required by the PCB design draftsperson to determine the on-center spacing of lead access holes to be drilled into the board. See Fig. 5–7. It is intended that the leads from this type of component be inserted into the PCB and that the body rest flush on the board surface (side 1). It is essential that the lead access hole spacing be accurate for proper assembly. The minimum length of preformed leads of 0.500 in., as shown in Fig. 5–6, is sufficient to install them properly into a PCB and form either a dead-head connection or a service bend. These are shown in the assembly drawing of Fig. 5–7.

5-4

Detail Drawings of Devices

Detail drawings for devices are usually more complex than those for simple axial leads components. In addition, many different types may be packaged in a single case style, requiring the use of tables to con-

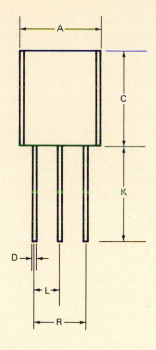

(a) Front view

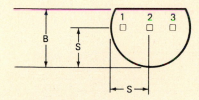

(b) Bottom (lead) view

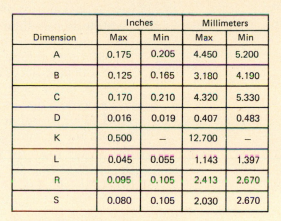

Dimension	Inches		Millimeters	
	Max	Min	Max	Min
A	0.175	0.205	4.450	5.200
B	0.125	0.165	3.180	4.190
C	0.170	0.210	4.320	5.330
D	0.016	0.019	0.407	0.483
K	0.500	–	12.700	–
L	0.045	0.055	1.143	1.397
R	0.095	0.105	2.413	2.670
S	0.080	0.105	2.030	2.670

(c) Tabulated dimensional information

Device Type	Pin Number	Lead Function
Transistor	1	Emitter
	2	Base
	3	Collector
FET	1	Drain
	2	Source
	3	Gate
SCR	1	Cathode
	2	Gate
	3	Anode
UJT	1	Base 1
	2	Emitter
	3	Base 2

(d) Lead designations for devices having TO-92 case
(bottom view)

FIGURE 5-8 Detail drawing for the TO-92 package configuration.

vey the necessary mechanical and electrical information. The detail drawing of one common case style, the TO-92, is shown in Fig. 5–8. The front and bottom (leads) views are shown in Fig. 5–8a and b, respectively. Note that all critical dimensions are coded for both views and the accompanying table of dimensional information is shown in Fig. 5–8c.

As shown in Fig. 5–8b, the three component leads are labeled 1, 2, and 3 from left to right as viewed from the bottom of the case. This common case style is used as a package for *transistors, FETs (field-effect transistors), and SCRs* (silicon-controlled rectifiers). With the leads always numbered 1, 2, and 3 in the TO-92 case, the lead designation table shown in Fig. 5–8d is required. This information is especially critical for PCB design to ensure proper device lead orientation in layout and assembly. It is important to note which view is being shown in order to distinguish and identify properly each of the three individual lead connections. To aid in this identification, a portion of the TO-92 case is flat, which serves as a key to be used as a guide for viewing the leads. As an example, assume that Fig. 5–8b represents the bottom view of a transistor. The flat side of the case is looking up and the leads are facing the observer. The lead identification is then as follows: lead 1 is the emitter, lead 2 is the base, and lead 3 is the collector, as read from left to right and using the information for a transistor in the table of Fig. 5–8d.

As a device becomes more complex, it is necessary to add more views in order to provide a complete description of the mechanical and electrical details. As shown in Fig. 5–9, three views of the TO-116 DIP (dual-in-line package) are required to completely detail and explain this IC (integrated circuit) for purposes of layout and assembly. In the top view of Fig. 5–9a, an index notch is shown at one of the short ends of the case. This notch is used as a key to aid in identifying the pins of the IC. The procedure for this identification is to rotate Fig. 5–9a so that the index notch is facing *up*. This will position pin 1 at the upper left corner of the case as viewed from the top. The pin identification count always begins at pin 1 and moves around the pins in a *counterclockwise* direction, to pins 2, 3, 4, and

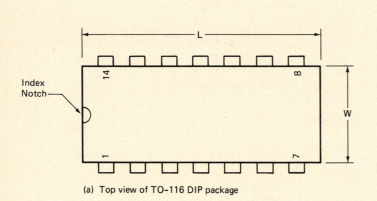

(a) Top view of TO-116 DIP package

Dimension	Inches	
	Max	Min
L	0.755	0.730
W	0.250	0.240
H	0.125	0.115
a	0.065	0.040
b	0.080	0.060
c	0.110	0.090
d	0.020	0.015
e	0.610	0.590
S	0.310	0.290
t	0.014	0.009

(c) Tabulated dimensional information

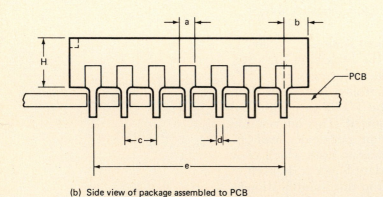

(b) Side view of package assembled to PCB

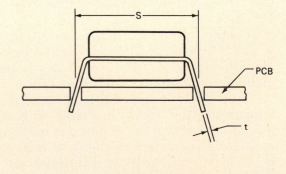

(d) End view of package assembled to PCB

FIGURE 5–9   Drawing of TO-116 integrated circuit package.

so on, down to pin 7, which is the last one in the left row. The count then continues in the same counterclockwise direction across to pin 8, which is the bottom pin in the right row of the case and then up to the last pin (14) at the upper right corner. For DIPs, the pin counting method just described is the same regardless of the total number of pins available in the package.

The side view of the TO-116 package is shown in Fig. 5–9b and a table of dimensional information is provided in Fig. 5–9c. Thus the top view, side view, and table provide all of the essential mechanical dimensions. What information is not illustrated by these views, however, is that there is an *outward deflection* or *taper* to both rows of pins extending away from the sides of the case. This can be seen only by viewing the *end* of the package, which is shown in Fig. 5–9d. Notice that this view shows the device assembled to a PCB. The reason for the taper on the pins is so they will apply pressure to the inner surface of the predrilled holes in the PCB, thus preventing the device from falling out of the board while other components are being installed or during soldering. This action of the pins can be seen in Fig. 5–9d. Dimensional information for a selection of common devices is given in Appendix XIV.

5-5

Detail Drawings of Hardware and Fasteners

In the design and fabrication of printed circuit boards, detail drawings are of paramount importance. The information provided by these drawings is essential for the positioning, placement, size, orientation, spacing, and methods of assembly of all components, devices, and hardware items. The turret terminal shown in Fig. 5–10a will serve as an example of how detail drawings are used for the design of printed circuit boards.

The use of turret terminals is one method of terminating interconnections on PCBs. They are positioned along the edge of the board and external connections are made to them. See Fig. 5–10b. While the PCB is being designed, the draftsperson would be viewing it from the *component* or *top* side, which is shown as side 1 in Fig. 5–10b. (Printed circuit board design is discussed in detail in Chapters 7 through 11.) From this view, the overall size of the shoulder of the turret terminal is one of the factors that determine the required spacing between terminals, the spacing of the mounting holes, and the alignment of the terminal centers relative to these holes. The top view of the PCB is also used for the detail of component and device positioning, shape, size, and spacing.

Detail drawings showing the *side* view are required for the fabrication phase of the PCB. See Fig. 5–10c. For this reason, drawings of this type are often called *assembly* drawings. They provide additional information relative to the final assembly of parts to the board. For example, the note in Fig. 5–10c shows that the shanks of the terminals are to be swaged from the circuit side of the board (shown as side 2) and then soldered to ensure a sound electrical connection.

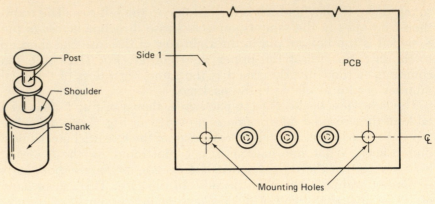

(a) Pictorial view of turret terminal

(b) Top view of turret terminal assembly

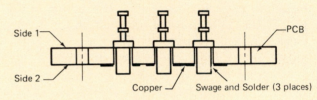

(c) Side view of terminal assembly

FIGURE 5–10 Turret terminal detail.

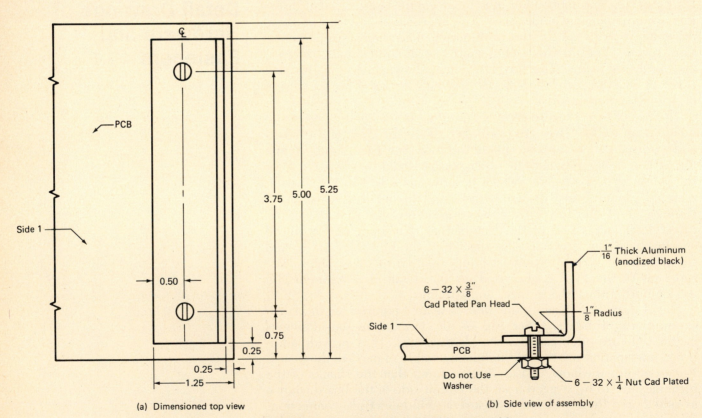

(a) Dimensioned top view

(b) Side view of assembly

FIGURE 5–11 Fasteners used in assembly.

When detail drawings are used primarily for assembly and fabrication, they are commonly accompanied by many notes specifying the methods and materials to be used.

A typical assembly drawing of machine screws used as fasteners to secure a metal bracket to a PCB is shown in Fig. 5–11. Note that the top view, shown in Fig. 5–11a, satisfies the design aspect of the drawing by providing all required dimensions but in no way provides sufficient information for the assembly and fabrication of the parts. For this reason, the side view is shown in Fig. 5–11b to provide considerably more detail to facilitate parts assembly.

For the fabrication of parts that require different styles of machine screw hardware, detail drawings such as those shown in Fig. 5–12 are necessary. The detailed information for the fabrication of a part having a *countersunk* hole is shown in Fig. 5–12a. Countersinking is a machine process much like drilling, except that the hole formed

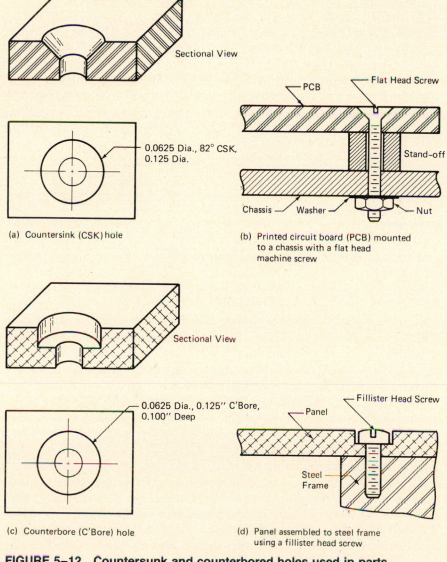

Sectional View

0.0625 Dia., 82° CSK, 0.125 Dia.

(a) Countersink (CSK) hole

PCB — Flat Head Screw

Stand–off

Chassis — Washer — Nut

(b) Printed circuit board (PCB) mounted to a chassis with a flat head machine screw

Sectional View

0.0625 Dia., 0.125″ C'Bore, 0.100″ Deep

(c) Counterbore (C'Bore) hole

Fillister Head Screw

Panel

Steel Frame

(d) Panel assembled to steel frame using a fillister head screw

FIGURE 5–12 Countersunk and counterbored holes used in parts assembly.

by the cutting tool is conical in shape. The notes on the figure indicate the diameter of the hole, the countersink angle, and the diameter of the hole at the surface of the part. Note that the abbreviation used to denote a countersink is CSK.

Figure 5–12b shows a partial assembly drawing of a PCB mounted to a subchassis element with the use of a flat-head machine screw and nut with a cylindrical *stand-off* or *spacer*. Observe that the purpose of countersinking is to prevent any portion of the screw head from protruding above the surface of the board.

The required information for fabricating a part having *counterbored* holes is shown in the drawing of Fig. 5–12c. Counterboring is a machine process by which the diameter of a predrilled hole is enlarged to a specified depth. As can be seen, the bottom of the counterbored hole is flat. The notes associated with Fig. 5–12c indicate the smaller hole diameter, the counterbore diameter, and the depth of the counterbore. This information is accompanied by the abbreviation C'Bore, which denotes a counterbored hole.

A partial assembly drawing of a panel secured with a machine screw threaded into a hole in a metal frame is shown in Fig. 5–12d. Note that the counterbored hole used in conjunction with a fillister head screw allows parts assembly without any portion of the screw head to extend beyond the surface of the panel. Additional information of machine screws, nuts, and washers is provided in Appendices XV through XVIII, respectively.

5-6

Detail Drawings of Chamfers and Slots

A complete master set of detail drawings for a printed circuit board design often includes special mechanical features. The two most common of these features for which detail is required are *chamfers* and *slots*.

Chamfering, often referred to as *beveling*, is typically required on all rigid PCBs which have been designed with closely spaced parallel conductor paths running perpendicular to the edge of the board. This edge, with its plated *fingers*, is designed to plug into an insertion-type connector, thus providing another means of making electrical connections to the board rather than using hardware such as turret terminals which increase assembly time.

The reason for chamfering the finger edge of a PCB is that it reduces the overall thickness of the board at the leading edge of the finger section to facilitate initial insertion into the connector. The bevel also minimizes the possibility of damage to the plated fingers as it is inserted. A common plating for PCB edge fingers is gold over nickel.

Refer to Fig. 5–13. The leading edge of the board's finger section is shown as it is projected along line a. The trailing edge of the chamfer is projected along line b. In this view, the detail of the chamfer is difficult to show, owing to the relative sizes of the board section and the bevel width. To provide this detail, an enlargement of the side view of the leading edge of the finger section is shown labeled method A and method B in Fig. 5–13.

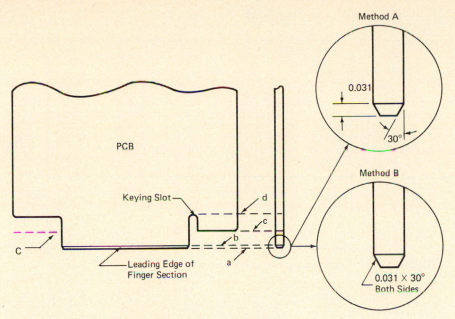

FIGURE 5–13 Chamfer of PCB finger section.

Method A shows one way to define the exact dimensions of the chamfer. Extension lines are used to project the 30-degree angle of the chamfer. Extension lines are also used to show that the chamfer is to extend upward from the leading edge of the fingers a distance of 0.031 in.

Method B is a more concise way of defining the edge detail. In this dimensioning form, the chamfer is always defined as *dimension × angle* with a leader pointing to the face of the chamfer. If both sides of the finger section are to be beveled (as shown), the words *both sides* are added to the detail.

Another feature of the drawing of Fig. 5–13 is the *polarizing* (keying) slot along one end of the finger section. This is one of several techniques to prevent the finished PCB from being incorrectly inserted into its accompanying edge connector. Note also in Fig. 5-13 that the projection line coded d is detailed on the side view as a dashed line since this is a hidden view from our perspective.

If the projection line labeled c were extended to the far left side of the board, it would show that there is an offset distance between these two sides. This configuration provides a means of introducing interference between the board edge and the connector support to further ensure that incorrect installation will not occur.

Another method of polarizing PCBs for proper installation is with a notch placed in the leading edges of the finger section. A small wedge-shaped plastic or metal key is inserted into the edge connector and the PCB notch is aligned with this key. In this type of keying, the line labeled c in Fig. 5–13 would be aligned on both sides of the finger section rather than requiring the offset as shown.

Internal slots are sometimes specified on PCBs for the mounting of specialized hardware and components. A slot is rectangular in shape with both ends rounded. See Fig. 5–14. Where slots are required,

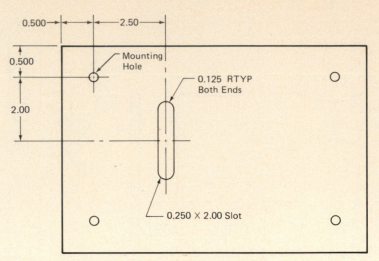

FIGURE 5–14   Positioning and dimensioning slots.

their detail will be included in a master set of PCB drawings. The length and width of the slot are shown as a note. Observe in Fig. 5–14 that the center lines of the slot are dimensioned from a tooling or mounting hole positioned near the edge of the board. (Using the tooling hole as a PCB dimension reference is explained in the next section.) The slot size is given with the use of a leader in the form *small dimension × large dimension slot*. Thus the slot shown in Fig. 5–14 is 0.250 in. wide and 2.00 in. long. To complete the detail of the slot, another leader is used on the other rounded end with the dimension in the form *dimension RTYP*. In our example, it is shown that the radius at the ends of the slot are one-half of the slot width, or 0.125 in., and is typical (the same) at both ends.

5-7

Detail Drawings of Printed Circuit Boards

The detail drawing of a printed circuit board is an engineering specification sheet which shows all of the essential information required by the manufacturer for its construction. In addition to all dimensional information and tolerances to express the overall size, shape, thickness, and location of holes, the drawings may include, in note form, some or all of the following: (1) foil thickness, (2) plating thickness, (3) substrate material to be used, (4) hole sizes, (5) hole size tolerance, (6) hole position tolerance, and (7) any other specific information unique to the board that is to be fabricated.

In this section we concern ourselves with the basic methods of detailing board outline dimensions and tooling hole locations for a typical PCB. Additional note-type information necessary for the fabrication of a PCB will be discussed in subsequent chapters.

The PCB designer must become familiar with methods of dimensioning a basic board. More often than not, the designer does not have total freedom in the size and shape of the board but rather is required to work within severe customer specifications. For example, the overall size of the board may be limited to a maximum of 3×5 in. Another restriction may be in the location required of specific tooling holes necessary for the manufacture of the board or the mounting of parts. If the PCB is to be used with a specified type and style of edge finger connector, the design for the number and arrangement of edge fingers are therefore predetermined before the designer begins. These are only a few of the many possible restrictions placed on a PCB designer.

When the engineer or technician arrives at the PCB design department, he or she typically comes with not only the complete circuit schematic diagram, but also with a detail drawing specifying the exact dimensions of the finished board. An example of this type of detail drawing, in its simplest form, is shown in Fig. 5–15. The information tells the designer what the overall size of the board will be in addition to the number and positions of the tooling holes. (These are holes drilled into the board which serve as an aid in positioning the board during its fabrication.) The method of dimensioning shown in Fig. 5–15 is termed *edge of board datum*. The lower left corner of the board serves as the 0,0 reference point from which all x dimensions are taken from the y datum line and all y dimensions from the x datum line. Using one corner of the board as a zero reference ensures that there will be no negative numbers for dimensions. However, this method of dimensioning can present problems for the

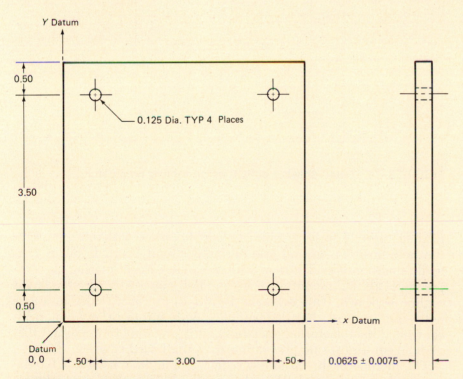

FIGURE 5–15 Detail drawing using edge of board datum as a reference system for dimensioning.

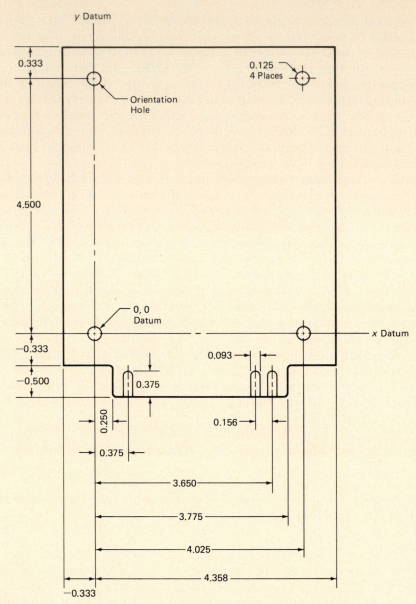

FIGURE 5–16   Detail drawing of PCB using tooling hole datum.

manufacturer in locating holes unless the x and y datum lines and the hole positions are on grid. Printed circuit board layout is normally done on a grid system with extreme care in placing all component leads and mounting or tooling holes on grid before the board outline is drawn. This generally results in the board edge locations not being on grid. For compatibility with automatic or numerical-controlled manufacturing operations, it is much more important that all hole centers are on grid rather than the board edges.

The method of dimensioning with a tooling hole used as a 0,0 datum is shown in Fig. 5–16. Selecting the hole in the lower left corner minimizes the number of negative dimension numbers. Many times, the minus (−) sign is omitted on the dimensions to the left and below the 0,0 datum. It is meaningful, however, to include the minus sign to reinforce the 0,0 reference datum point. It should be noted that sign error cannot occur on numerical-controlled equipment since the operator can register a hole location to the left or below the 0,0 datum and the negative value will automatically be entered. On occasions where a hole is not on grid, the operator must calculate and dial in the hole location dimensions.

In Fig. 5–16, the dimensions referenced from the *x* datum appear in the form that allows cumulative errors (see Section 2–12). When using a datum reference system of dimensioning, it is implied that the desired dimension is obtained by adding or subtracting the necessary accumulative numbers, but the actual location of any position is always measured from 0,0. With that said, the dimensioning technique of individual dimensions, as shown for the finger edge of the

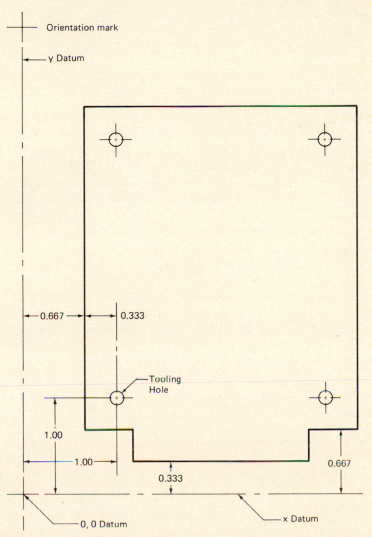

FIGURE 5–17   Drawing of 0,0 datum located on grid outside of board border.

board from the *y* datum in Fig. 5–16, is preferred to eliminate the possibility of confusion or error.

With the use of the datum reference system, the manufacturer is able to locate accurately the board edges and all other features of the circuit pattern from one tooling hole. For example, the center of the fingers shown in Fig. 5–16 is dimensioned from the tooling hole used as the 0,0 datum point.

If, for any reason, negative numbered dimensions are to be avoided, the 0,0 datum can be located *outside* the board outline. As shown in Fig. 5–17, the tooling holes have been placed on grid similar to that shown in Fig. 5–16, but the 0,0 datum is placed to the lower left and outside the board outline.

One final note on datum needs to be emphasized. The manufacturers must have a means of aligning or "squaring" the board for its fabrication. For this purpose, an *orientation hole* or *mark* is provided which is in line with either the *x* or *y* axis of the 0,0 datum. See Figs. 5–16 and 5–17. This orientation hole is added to the detail drawing either inside or outside the board outline.

_____ EXERCISES _____

5–1 Draw an isometric view of the TO-116 DIP package shown in Fig. 5–9. Convert the dimensional information in the table of Fig. 5–9c to four times actual size, using maximum values. Place the length of the IC along the receding axis.

5–2 Draw a cabinet view of the TO-116 DIP package converting maximum dimensional information in Fig. 5–9c to four times actual size. Place the length of the IC along the receding axis.

5–3 (a) Construct a four-center ellipse used in isometric drawings. The true diameter of the circle is to be 3 in. Use Fig. 5–3a as a guide. (b) Construct an eight-point plotted ellipse as used in cabinet views for a true circle also having a 3-in. diameter. Refer to Fig. 5–3b.

5–4 Draw a two-times scale sectional view along the center line of the machine screws in the assembly drawing of Fig. 5–11a as viewed from the left-hand side. Use Appendices XV and XVI for 6–32 machine screw and nut dimensions. Refer to Appendix XIII to detail the cross section of the aluminum bracket which is 1 in. wide by 1 in. high (actual size). The PCB is $\frac{1}{8}$ in. thick (also actual size).

5–5 Figure 5–16 shows the dimensional information of a typical PCB with its 0,0 datum shown as the lower left-hand tooling hole. Redraw this board outline at one-times scale but with the 0,0 datum located at $x = -2$ in. and $y = -2$ in. from the lower left-hand tooling hole. Completely dimension the board outline from the new datum.

6

Printed Circuit Boards

Upon completion of this chapter on printed circuit boards, the student should be able to:

1. Distinguish between the various classifications of printed circuit boards.

2. Be familiar with the printed circuit board materials.

3. Know the system by which copper foil thickness is determined.

4. Be familiar with the various types of printed circuit board laminates and the characteristics of each.

5. Grasp a general awareness of the arrangement of single-sided, double-sided, and multilayer printed circuit boards and the processes involved in their fabrication.

6-0

Introduction

In the field of electronic circuit design and fabrication, the use of printed circuit boards (PCBs) is relatively new. Although crude methods of manufacturing PCBs date back to the mid-1920s, the industry did not gain impetus until after World War II. Today, the printed circuit board industry employs countless numbers of people in all phases of the design and manufacture of electronic systems. For the assembly and wiring of these systems, PCBs are the most efficient and cost-effective method of production. Many new and improved methods of design and processing techniques for PCBs have evolved over the years to keep pace with the demands for higher component density. This resulted in the advancing of the basic single-sided board style to the double-sided board and, more recently, to the multilayer board. Even though the state of the art in the PCB industry has progressed dramatically, the function of the board has remained basically similar to that employed in earlier designs.

Printed circuits may be divided into two broad categories. These are the *flexible* type and the *rigid-style boards*. By far, the majority of printed circuits in use are of the rigid board variety. For this reason, the overall emphasis in this book will be the design and layout of this type of PCB.

This chapter provides general information on PCBs. In addition to the various classifications of PCBs, information is provided to aid in a basic understanding of how the designs generated in subsequent chapters are processed into finished boards. Even though the electronic draftsperson is not expected to develop a thorough understanding of the details of PCB fabrication, a familiarity with the terms used in the many processing steps is helpful for better articulation with the circuit designer.

6-1

Classification of Printed Circuit Boards

The *flexible* type of printed circuit, in many applications, serves as a replacement for conventional harnesses where individual insulated wires are bundled together to form connections in a system. Flexible printed circuits are fabricated by bonding a thin conductor foil pattern, typically copper, to a flexible thermoplastic base laminate or insulating film such as polyester or polymide. Bonding is accomplished with the application of heat and pressure for thermoplastics and with the use of adhesives for polyester or polymide. After the basic laminate has been processed to form the desired conductor pattern on the insulating film, a cover coat of the same material as the base laminate is applied. This protects the conductor pattern against dust, dirt, and moisture, making the flexible cable extremely durable and highly reliable.

To make electrical connections to the conductors in a flexible-printed circuit cable, a portion of the insulating material is removed, exposing conductors only at their intended points of interconnection.

Flexible printed circuit cables, because of their light weight and thin, flat shape, are extremely pliable and can be folded or contoured around corners to fit into tighter spaces than conventional harnesses.

Rigid printed circuit boards serve two main functions. First, their stiff construction provides a surface for mechanically supporting components, devices, and hardware. Second, the conductive foil surface can be processed into conductor path patterns to form the required electrical interconnections between all of the components and devices mounted to its surface. Even though some flexible type of printed circuit boards are made to accept the mounting of components, rigid PCBs dominate the electronics industry in this regard.

Rigid PCBs are of three major types, each described by the number of sides or layers of metal foil (made of high-purity copper) from which the conductor pattern is processed. These types are (1) single-sided, (2) double-sided, and (3) multilayer.

The *single-sided* boards are made with only one surface having a layer of conductor foil.

Double-sided boards have a layer of conductive foil on *both* surfaces. For these boards, electrical connections between conductive surfaces are made *mechanically* with the use of jumper wires run through drilled holes and soldered to the conductor pattern on each side or by *electrochemical* means, which is the application of a conductive metal plated onto the inside surface of a drilled hole to form the electrical connection. The latter method is called *double-sided plated-through holes* (DS-PTH). The *double-sided* board lends itself to higher component density since greater flexibility in conductor pattern routing (positioning) is available through the use of conductive foil on both sides of the board.

The *multilayer* PCB has both outside surfaces processed with conductive foil in addition to one or more *inner* layers of conductor planes. The use of multilayer boards is required to satisfy today's demands for high-density electronic system packaging.

6-2

Laminates for Rigid Printed Circuit Boards

Laminate is the term used to describe a sheet of unprocessed rigid printed circuit board stock. The laminate is made of three principal materials: resin, a reinforcing substrate, and copper foil. Finished laminate consists of an insulating rigid base material with a sheet of copper foil bonded to one or both sides. The thickness of the laminate is determined mainly by the amount of reinforcing material and resin used in its manufacture. A list of some available laminate thicknesses is provided in Table 6–1.

The copper foil layer is of particular concern to the circuit board designer since it will be processed into conductor paths to satisfy the electrical requirements of the system. This foil is provided in a variety of thicknesses and is specified by the industry in terms of *ounces per square foot* (oz/ft^2). As an example in the use of this designation, a laminate having a 1-oz/ft^2 copper foil will have a foil thickness of 0.0014 in. or 1.4 mils. Stated another way, a square foot of this 1.4-mil-thick copper foil would weigh 1 oz. Thicknesses of copper foil range from $\frac{1}{8}$ oz/ft^2 to as much as 15 oz/ft^2. Table 6–2 lists some common copper weights in oz/ft^2 together with their equivalent foil thicknesses in inches. The current-carrying capacity of the processed conductor paths is determined primarily by the cross-sectional area of the path (i.e., the width of the path and the thickness of the copper). This design consideration is discussed further in Section 8–8.

The choice of the type of laminate used for the design of a system is the result of a trade-off involving *performance considerations*, *manufacturing capabilities*, and of course, *cost*.

System performance considerations in the design of a PCB include mechanical variables such as board flexibility, degree of warp and twist that may be encountered, peel strength of the copper foil, and water absorption. Electrical properties include dissipation fac-

TABLE 6–1 Common Laminate Thicknesses (in.)

0.031
0.059*
0.062
0.093

*Most popular.

TABLE 6–2 Various Copper Foil Weights and Thicknesses

Weight (oz/ft^2)	Thickness (in.)
$\frac{1}{8}$	0.00018
$\frac{1}{4}$	0.00035
$\frac{3}{8}$	0.00053
$\frac{1}{2}$*	0.0007
1*	0.0014
2*	0.0028
3	0.0042
4	0.0056
5	0.0070
6	0.0084

*Currently the most common foil thicknesses being used.

tor, dielectric breakdown voltage, dielectric constant, and dielectric strength. These are some of the specifications that must be evaluated by the design engineer when selecting the type of laminate to be used.

The PCB manufacturer is concerned primarily with the processes required as they relate to the type of laminate specified. For example, the questions that the manufacturer needs to evaluate are: Do the holes have to be drilled or can they be punched? How much wear will the cutting tools be subjected to when fabricating boards with reinforcing substrates having large degrees of abrasive properties? Can special items such as dielectrics with a Teflon resin system be processed reliably?

These considerations and trade-offs are not normally within the purview of the design draftsperson. However, familiarity with common terms used to describe these materials in design specifications is considered essential. For this reason, commonly used laminates and their descriptions are listed in Table 6–3. The laminate classification or grade is specified by several organizations, the largest of which is the *National Electrical Manufacturers Association* (NEMA). Even though Table 6–3 lists only six typical grades of PCB laminates, many more less common types are designated by NEMA. The six grades included in Table 6–3 are described below.

XXXPC: a paper-base dielectric impregnated with a phenolic base. It commonly finds application in consumer products such as radios and television sets. It has high insulation resistance and low water-absorption properties. This grade can be punched at room temperature, which aids in reducing the costs of manufacturing.

FR-3: a paper-base dielectric impregnated with an epoxy resin. It has properties similar to those of grade XXXPC for insulating resistance and punchability. Flame retardants are added to make the material self-extinguishing.

TABLE 6–3   Various Laminate Materials Used in the Manufacture of Printed Circuit Boards

NEMA Classification for Laminates	Substrate (Reinforcing Agent)	Resin	Description
XXXPC	Paper	Phenolic	Holes punchable at room temperature
FR-3	Paper	Epoxy	High insulation resistance; holes punchable at room temperature
G-10	Glass	Epoxy	For general use; has the best compromise between mechanical and electrical properties
FR-4*	Glass	Epoxy	Same properties as G-10 except flame-retardant agents have been added
CEM-3	Glass	Epoxy	Possible alternative to FR-4/G-10 in certain applications
GX	Glass	Teflon	Controlled dielectric constant for high-frequency applications

*Currently the most commonly used laminate material.

G-10: glass cloth impregnated with epoxy resin. It has low dielectric losses and high foil bond strength. This laminate previously found wide use in the computer industry and in industrial electronics applications.

FR-4: identical to grade G-10 with the addition of flame retardants. For this reason, it is currently the type most often specified for applications in minicomputers, CB radios, and military and aerospace instrumentation equipment.

CEM-3: a composite material made of nonwoven glass mat and woven glass cloth impregnated with epoxy resin. Because of its performance characteristics and improved machinability, in addition to lower cost, this laminate is often specified as an alternative for applications in many price-sensitive consumer products such as radios, television sets, and tape recorders.

GX: a glass-based material impregnated with Teflon resin. Because of this laminate's ability to maintain its dielectric constant within close limits, it finds application in microwave systems.

6-3

Single-Sided Printed Circuit Boards

Single-sided printed circuit boards consist of laminates with copper foil only on one side. Thus a conductor pattern is processed only on that one side of the board. Figure 6–1 shows the basic arrangement

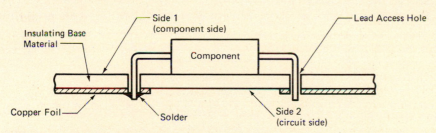

(a) Side view of PCB with component mounted

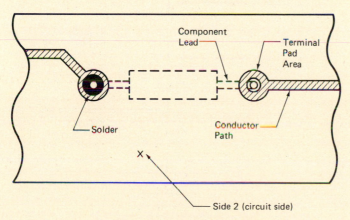

(b) Circuit side of PCB (bottom view)

FIGURE 6–1 Arrangement of single-sided PCBs.

of a single-sided PCB. Lead access holes are drilled through the insulating material and the conductor foil so that they will accept the component leads. The leads are preformed to facilitate their insertion into the drilled holes. The component body is placed in direct contact with the insulated side of the board. Note that the side on which the component body rests is labeled *Side 1* or *Component Side* in Fig. 6–1a and the side through which the leads extend is labeled *Side 2* or *Circuit Side*, sometimes referred to as the *Solder Side*. The component body normally rests flush on the board surface, which provides support for all mounted components.

To make the required electrical connections between the component lead and the conductor pattern, the leads must be soldered to the copper foil at all lead access holes. To accommodate this, the copper foil is processed into a *terminal pad area* or simply *pad* at each lead access hole position. See Fig. 6–1b. The pad is normally circular in shape (as shown) but in some cases may be square, oblong, or other geometric shape, depending on specific design requirements. The pad diameter is typically greater than the adjoining conductor path widths. This is to provide sufficient copper foil around the hole, after drilling, to solder the lead to the conductor path.

The single-sided PCB is the simplest type to design and manufacture and thus is the least expensive to use. Because it has only one foil side on which to produce a conductor pattern, however, its applications are limited to relatively simple circuit designs. More complex designs typically utilize the double-sided or multilayer PCBs, which are discussed in the following two sections.

A simplified flowchart of one of the processes used to fabricate single-sided PCBs is shown in Fig. 6–2. The process shown is called the *print-and-etch* method. Sheets of raw stock are first sheared into

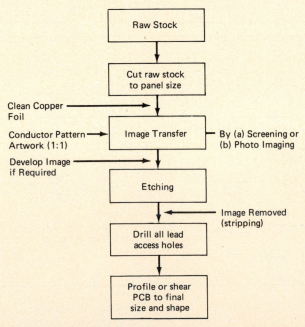

FIGURE 6–2 Simplified flow chart for typical single-sided PCB fabrication using the print-and-etch method.

the appropriate panel sizes. A 1:1 scale etchant-resistant image of the designed conductor pattern is applied to the surface of the cleaned copper foil by either screening ink or photo-imaging. The exposed copper (i.e., all the copper *except* that which is under the protective image) is then removed by being chemically milled (etched). The resist pattern is stripped from the remaining copper and the lead access holes are drilled through the pads. Finally, the board is sheared to its final shape and is ready for component mounting and soldering.

6-4

Double-Sided Printed Circuit Boards

Double-sided PCBs are more complex in their design requirements than single-sided boards but offer more flexibility in high-density packaging. These boards are manufactured with the same insulating base material as single-sided boards but are copper clad on both sides. Conductor patterns are thus processed on both the component side *and* the circuit side. Another unique characteristic of the double-sided PCB is that it can, as part of the manufacturing process, be fabricated with plated-through holes (PTHs). See Fig. 6–3. Plated holes are initially fabricated by drilling all lead access holes through the double-sided copper clad laminate. One method of producing the PTH is to process the board through a series of chemical baths which deposit a thin layer of copper over the entire board surface, and particularly the interior surfaces (barrels) of the drilled holes. This process is called *electroless copper deposition*.

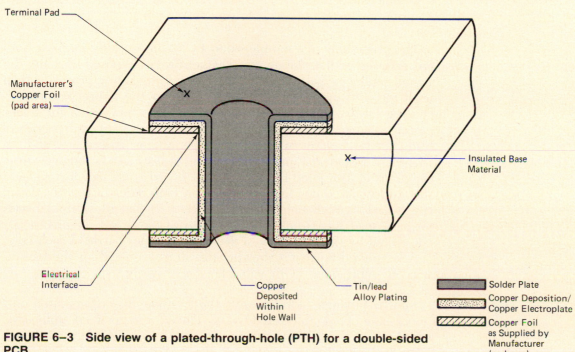

FIGURE 6–3 **Side view of a plated-through-hole (PTH) for a double-sided PCB.**

Legend:
- Solder Plate
- Copper Deposition/Copper Electroplate
- Copper Foil as Supplied by Manufacturer (pad area)

Labels: Terminal Pad; Manufacturer's Copper Foil (pad area); Electrical Interface; Copper Deposited Within Hole Wall; Tin/lead Alloy Plating; Insulated Base Material

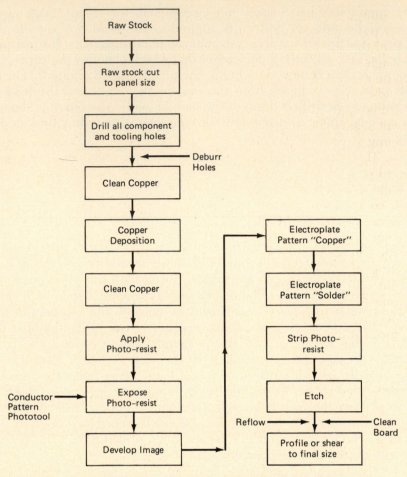

FIGURE 6–4 Simplified double-sided plated-through-hole process.

The processing of a double-sided PCB continues by masking both foil surfaces with a plating resist which leaves exposed only the designed conductor pattern area. A copper electroplating process is then used to build up the copper thickness over the entire pattern, including the barrels of all holes. The board is now ready for *solder plating*, also an electroplating process, which deposits a thin layer of tin/lead alloy (solder) over the entire conductor pattern area and the barrels of all holes. Finally, the plating-resist mask is removed from the areas of unwanted copper and the board is etched, leaving only the solder-plated conductor pattern on both sides of the board and the plated-through holes. Note in Fig. 6–3 that the terminal pads for each lead access hole on both sides of the board are electrically connected by the formation of the plated copper barrel of each hole.

A simplified flowchart showing the steps required for the fabrication of double-sided PCBs with plated-through holes is shown in Fig. 6–4.

Similar to single-sided boards, the components are mounted only to one side of a double-sided board. The advantage of a double-sided board is that conductor paths are processed on *each* side even though one side is specified as the component side. Each of the component leads pass through access holes which have a terminal pad on each side of the board. These pads are electrically connected to each other through the plated barrel. Refer to Fig. 6–5. A conductor path (w) on the component side (side 1) is electrically connected through the plated-through hole to a conductor path (y) on the circuit side of the board (side 2). When a component lead is soldered to the pad on the circuit side, solder will flow upward, filling the plated barrel and thus forming a continuous fillet of solder about the lead and the terminal pads. This results in the component lead and the two conductor paths w and y, being electrically connected.

Note in Fig. 6–5 that the conductor path labeled x on side 1 crosses over conductor y on side 2 but makes no electrical connection since they are separated by the insulating material of the board. Also, if the component body is made of a nonconductive material, such as plastic, it can rest against exposed conductor paths on side 1 without causing an unwanted electrical connection.

The ability to use two planes for electrical interconnections greatly increases the flexibility of PCB design. With the interconnections made between both sides of the board by the plated-through holes, no mechanical or special assembly techniques, such as jumper wires

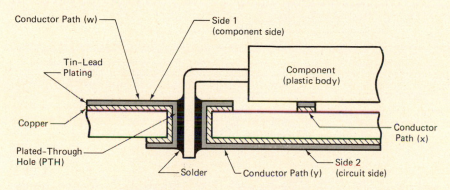

(a) Side view of double-sided plated-through-hole PCB

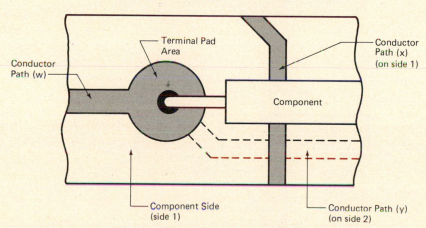

(b) DS PCB viewed from component side

FIGURE 6–5 Double-sided PCB with plated-through holes.

or eyelets, are required. It is apparent by comparing Figs. 6–2 and 6–4 that the double-sided board is not only more complex to design but requires more processes to fabricate, thus increasing production costs. This, however, is far outweighed by the fact that the double-sided board allows much more circuitry to be packaged in the same area than that capable with a single-sided design.

6-5

Multilayer Printed Circuit Boards

The *multilayer board* (MLB) is made up of a combination of equivalent single- and/or double-sided PCBs bonded together to form one integral board having two outside layers of conductive foil in addi-

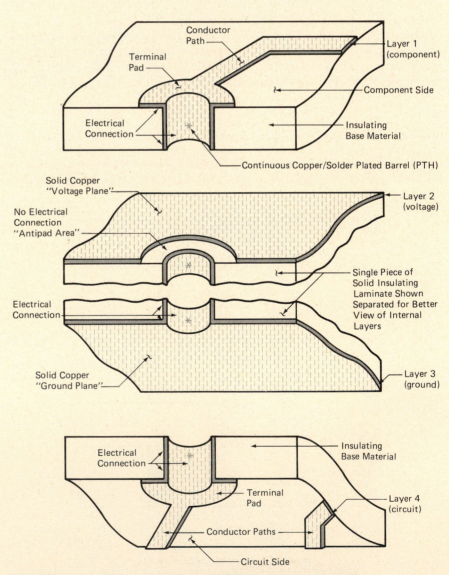

FIGURE 6–6   Exploded view of a voltage/ground type MLB.

tion to one or more internal layers. All of these conductive surfaces are used during the packaging design to form the required electrical interconnections. Multilayer boards are classified by (1) the number of conductive layers, and (2) the type of internal layer circuitry.

Multilayer boards are most commonly described by a number (3, 4, 5, etc.) which corresponds to the number of conductive layers.

An exploded view of a typical four-layer MLB is shown in Fig. 6–6. Note that there are two outer conductive layers, similar to standard double-sided PCBs and two inner layers, thus the number 4. If the inner layers of foil are processed with conductor patterns (i.e., conductive paths terminating in pads), the board is designated a *signal* MLB. Where the inner layers are used basically as unbroken sheets of copper foil, the board is termed a *voltage/ground plane* MLB. To familiarize the draftsperson with the most common multilayer design and fabrication, we will discuss the voltage/ground plane MLB.

A four-layer voltage/ground MLB is shown in Fig. 6–7. The top side, labeled layer 1 (or component side), appears identical to a single-sided board processed with terminal pads and conductor paths. The bottom side, layer 4 (or circuit side), also appears identical to a single-sided PCB with terminal pads and conductor paths processed into the copper foil. The two inner conductive layers (labeled layers 2 and 3) are made from a laminate with copper foil on both sides, similar to a double-sided board. The inner layer patterns are processed before the completed MLB is formed. After the MLB has been completed, the plated-through holes are processed through all layers in a similar manner as that used in double-sided boards. In Fig. 6–6, the laminations and the barrel of the plated-through hole are shown split for illustration of the inner copper layers 2 and 3. Layer 3 (*ground*) is a solid sheet of copper extending from the plated hole.

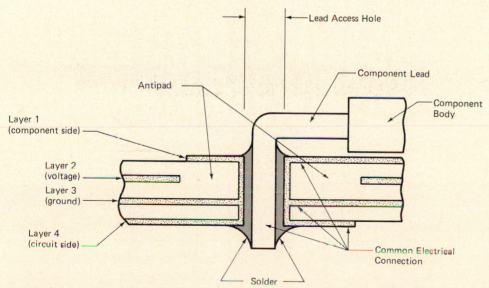

FIGURE 6–7   Cross-sectional view of the voltage/ground type multilayer board.

A component lead soldered into this hole will make electrical contact with layer 3 in addition to both outer pads on layers 1 and 4. Note, however, that layer 2 (voltage) has been processed with an insulation gap or clearance around the lead access hole so that its copper foil does not extend to the plated barrel. Thus a component lead inserted through this hole and soldered to its terminal pads on layers 1 and 4 will *not* make electrical contact with layer 2. This clearance provided around lead access holes in copper planes where no electrical contact between component leads and inner conductive layers is to be made is termed an *antipad* area.

All multilayer boards will have inner conductive layers with some antipads for isolation from plated-through holes as well as connections to others. Refer again to Fig. 6–7. Note that the barrel of the plated-through hole connects the conductive copper on layers 1, 3, and 4 while the antipad isolates the foil of layer 2 from the PTH. When the component lead is soldered, the solder will fill the barrel and electrically connect the lead to layers 1, 3, and 4 but not to layer 2 because of the antipad area.

As an application of the use of a voltage/ground MLB, refer to the digital circuit shown in Fig. 4–3. This circuit is constructed with four 14-pin DIP integrated circuits, each package of which designates pin 14 as the power connection and pin 7 as the ground connection. If this circuit was to be packaged using a voltage/ground MLB, the board would be fabricated as shown in Fig. 6–8. The voltage plane (layer 2) would have antipads about all lead access holes except for those pins numbered as 14. No antipads would appear at all holes for the

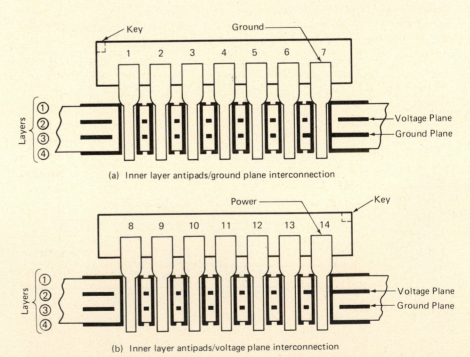

(a) Inner layer antipads/ground plane interconnection

(b) Inner layer antipads/voltage plane interconnection

FIGURE 6–8 Cross sections of a voltage/ground style multilayer board for a DIP integrated circuit.

number 14 pins so that the power plane would extend to the plated-through hole for electrical connection to these leads. Similarly, antipads would be required about all lead access holes of the ground plane (layer 3) except for those connected to the number 7 pins of each DIP package which are to be grounded. The resulting MLB provides power and ground connections on layers 2 and 3, respectively. All signal conductor paths can then be designed onto the two outer surfaces of the board.

The multilayer board is the most complex and costly to design and fabricate, but it provides distinct advantages over the double-sided PCB. When voltage and ground connections are made to inner layers, more flexibility and space are available to the designer in routing signal paths on the outer layers. Further, inner layers may also be used for signal paths in the design of high-density systems. Other advantages of the MLB are that they can provide electrical shielding with the use of an internal ground layer as well as internal heat sinking where required. Because the multilayer board plays a prominent role in today's high-density electronic packaging, it is essential that the PCB designer and the electronic draftsperson gain familiarity with this type of design. Basic design rules for a voltage/ground MLB are presented in Chapter 11.

A simplified flowchart for the fabrication of a MLB is shown in Fig. 6–9. When comparing this flowchart with that of Fig. 6–4, showing the fabrication of a double-sided board, it is apparent that the MLB requires many more processing stages which result in its higher production costs.

Following is a brief description on fabricating a four-layer board having a completed thickness of 0.062 ($\frac{1}{16}$) in. The inner layers (2 and 3) are first produced with a thin laminate (e.g., 0.020 in. with 2-ounce copper on each side). Similar to the simple print-and-etch technique, the artwork patterns are applied to each of the two inner layers by photo-imaging with precise registration to each other. These laminates are then etched to remove the unwanted copper and the etchant resist pattern is removed from the remaining copper by chemical stripping. The copper surfaces are then cleaned to prepare them for the application of a heavy oxide surface treatment, commonly called a *black oxide* treatment. This is to roughen the copper surfaces to promote adhesion and improve bond strength between layers of the MLB. The top and bottom cap sheets are then bonded to the core sheets. The cap sheets may be either single- or double-sided clad laminate, depending on how many inner layers are to be formed. For example, a four-layer MLB can be formed from only two double-sided PCBs having two inner layers, insulated from each other with the bonding sheets, and two outer layers. To continue with the fabrication of our example MLB, we will use two 0.015-in.-thick single-sided laminate clad with 1-oz copper as cap sheets. These cap sheets are bonded to the core sheets with the use of "prepreg" bonding sheets (B-stage), which are placed between them. Normally, two bonding sheets are used plus one more for each 2 ounces of copper between layers instead of one thick sheet. Table 6–4 lists commonly used B-stage and core laminate thicknesses used in MLBs. For our example, we will use three sheets of 0.002-in. B-stage ma-

TABLE 6–4 Typical Dielectric Thicknesses Available from Laminate Manufacturers

Prepreg Bonding Sheet Thickness (in.)
0.001
0.0015
0.002
0.003
0.004
0.007

Core Laminate Thickness* (in.)
0.002
0.0025
0.005
0.010
0.018
0.028
0.031
0.059
0.062
0.093
0.100
0.125

*Core laminates are available in 0.001-in. increments from 0.002 to 0.031 in.

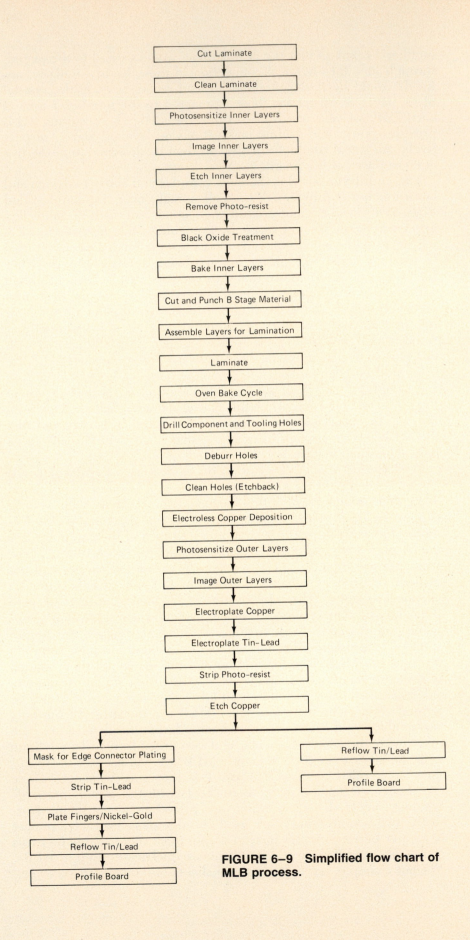

FIGURE 6–9 Simplified flow chart of MLB process.

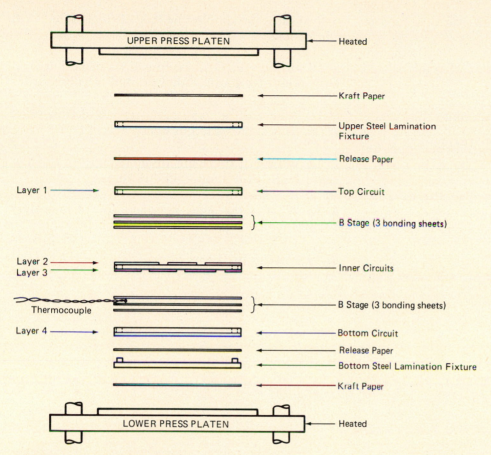

FIGURE 6–10   Lay-up detail for a four-layer board.

terial between layers 1 and 2 and three between layers 3 and 4. The top cap sheets, six B–stage bonding sheets, and the two core sheets are then laminated under pressure and elevated temperature. This will cause the B–stage material to flow and fill the voids which exist at each antipad position and bond the layers to form a homogeneous panel which appears similar to a double-sided laminate. The lay-up details for laminating a four-layer MLB are shown in Fig. 6–10.

After the MLB has been laminated, it is processed in a similar manner as that used for double-sided PCBs as described in Section 6–4.

EXERCISES

6–1 What are the three styles of printed circuit boards?

6–2 What are the two main categories of printed circuit boards?

6–3 What type of printed circuit board is used to replace a harness?

6–4 Give the two main functions of rigid printed circuit boards.

6–5 What is the most popular laminate thickness?

6–6 Determine the thickness in mils of copper foil having a weight of 0.375 oz/ft^2.

6–7 What laminate material used in the manufacture of PCBs has the best compromise between electrical and mechanical properties and is also flame retardant?

6–8 For a single-sided board, how are sides 1 and 2 distinguished?

6–9 What is the simplest method of single-sided PCB fabrication?

6–10 How are electrical connections made between both sides of a double-sided printed circuit board?

6–11 What are the main advantages of double-sided PCBs over single-sided designs?

6–12 What are the main disadvantages of double-sided PCBs over single-sided designs?

6–13 What is the most popular type of four-layer MLB?

6–14 Explain the term *antipad*.

7

Single-Sided Printed Circuit Board Design: The Preliminary Sketch

LEARNING OBJECTIVES

Upon completion of this chapter on the preliminary sketch of a single-sided printed circuit board design, the student should be able to:

1. Assemble the basic materials required to prepare a preliminary sketch of a printed circuit board.

2. Assemble the required information necessary to prepare a preliminary sketch.

3. Establish a format for the layout sketch, that is, two-side viewing and two-side color coding.

4. Be familiar with drawing common component and device body outlines.

5. Select the appropriate sizes of pad diameters.

6. Determine on-center distances between pads.

7. Properly draw pad outlines on grid intercepts.

8. Determine the optimum positioning of components.

9. Properly route conductor paths for electrical interconnections.

7-0

Introduction

After the engineer has completed the circuit schematic diagram, the printed circuit designer converts this into drawings and artworks that will result in a functional packaging arrangement which will be fabricated into a PCB. The *component layout drawing* which includes conductor path routing is first completed. From this drawing, a *taped artwork master* is produced. This becomes the primary working tool for the fabrication of the PCB. (The preparation of the taped artwork master is discussed in detail in Chapter 12.)

Component layout drawings are accurately scaled two-dimensional representations showing the exact size, shape, and position of all components, devices, and hardware that will appear on the finished board as well as the overall size and shape of the board. Also included on the layout drawing is the precise routing of all conductor paths which form the electrical connections specified in the schematic. In addition, *terminal pads* are shown at each lead access hole. These pads will be used to electrically connect (by soldering) all component and device leads to the circuit pattern. It can be seen by the description above that a component layout drawing is a composite view showing *both* sides of the PCB. It is therefore essential that these drawings be carefully constructed so that each side is clearly distinguishable from the other.

The component layout drawing is typically developed in two stages, the preliminary sketch and the finished drawing. Depending on the complexity of the circuit, the experienced designer may not need to prepare the sketch but may draw the finished drawing directly. In general, however, most electronic systems are of such complexity that they do require a preliminary working sketch to serve as a guide in preparing the final drawing. It is rare, and unnecessary, for the beginning designer to attempt a finished drawing without the aid of a sketch. The sketches not only help in establishing the overall layout plan but also result in a more technically correct drawing with fewer oversights and errors.

This chapter contains the basic information required to produce a preliminary sketch. The development of a finished component layout drawing is discussed in Chapter 8. The information provided is in three major categories: (1) how to draw the body outlines for components and devices, (2) how to use the schematic diagram to position the components and devices for a single-sided board design, and (3) how to route the electrical interconnections properly so that the circuit will function in accordance with the circuit schematic.

7-1

Materials Required for Developing a Component Layout "Working Sketch"

The term *sketch* often implies a freehand drawing, made to no particular scale. We will not be using the term in this common form for the generation of the "working sketch" for a PCB component layout drawing. This type of sketch makes use of a straightedge and circle template and is drawn to a precise scale. It needs to be emphasized that the working sketch is a guide to be followed closely when producing the finished component layout drawing. It is therefore essential that accuracy and thoroughness be maintained in all phases of generating this preliminary working sketch.

The basic implements required to develop the preliminary sketch are (1) pencils, (2) erasers, (3) straightedge, (4) decimal scales, (5) decimal and fractional circle templates, and (6) ruled grid paper.

The pencils used may be either technical pencils with thin leads, the mechanical type, or standard wooden drafting pencils with a lead hardness of at least 2H. For standard drafting pencils, appropriate sharpeners are necessary to maintain their points (refer to Section 1–3). For single-sided board design, two lead colors are required, the standard black and a contrasting color such as blue or red. The eraser used should be compatible to the leads and the type of paper used (see Section 1–4).

For drawing straight lines and making measurements, a rule with major graduations of 0.1 ($\frac{1}{10}$) in. is required. If available, a rule having minor graduations of 0.05 ($\frac{1}{20}$) in. would be extremely helpful. Fractional scales do not lend themselves to component layout drawings and should be avoided. The length of the rule will depend on the overall size of the design. Generally, a length of 12 or 18 in. is suitable.

Decimal and fractional plastic circle templates are used to draw terminal pads and round-bodied components. The templates selected should have hole sizes ranging from 0.1 to 1.00 in. and from $\frac{1}{32}$ in. to about 2 in.

The drawing media on which the sketches are to be produced should be a good-quality grid paper in a size large enough to easily accommodate the overall size of the scaled drawing. The color of the grid lines is preferably blue, although black is acceptable. The grid is a series of equally spaced vertical and horizontal lines and serves as an invaluable aid in the positioning of components and the placement of pads and conductor paths. The grid spacings should be 0.10 in. for both the vertical and horizontal lines. Grids with spacings of more than 0.1 in. should not be used in printed circuit layout. Grid paper having accented lines every 0.50 in. are very helpful in developing the preliminary sketch.

The basic materials just described for producing preliminary sketches for single-sided PCBs are shown in Fig. 7–1.

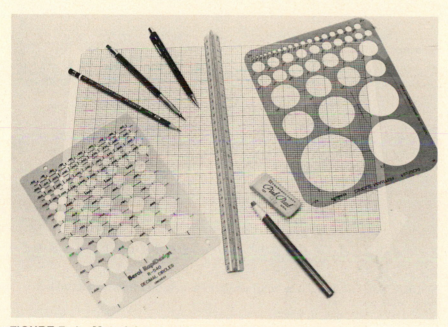

FIGURE 7–1 Material and tools required for layout sketches.

7-2

Basic Information for Beginning the Layout Sketch

In order to begin producing the layout sketch, the PCB designer must have information available relative to the circuit and the parts. Primarily, a detailed circuit schematic with all parts identified either on the circuit drawing or on an accompanying parts list is initially required. An example of a detailed schematic is shown in Fig. 7–2, which is a basic common-emitter transistor amplifier. The parts list provided with the schematic identifies each of the components.

In addition to the circuit schematic, each of the parts or data sheets, which provide the physical size and shape of all components, devices, and hardware items, need to be available. Also, electrical data sheets for all devices showing the pin arrangements are necessary.

Typically, the design engineer will provide the draftsperson with additional information regarding electrical and mechanical restrictions, such as the size and shape of the board, positioning of critical components, forbidden areas of the board for certain parts placement, required slots or cutouts, and placement of external interconnections. Because this chapter introduces the working sketch, no design restrictions will be considered. Once the basic layout skills have been developed, design restrictions and trade-offs, which al-

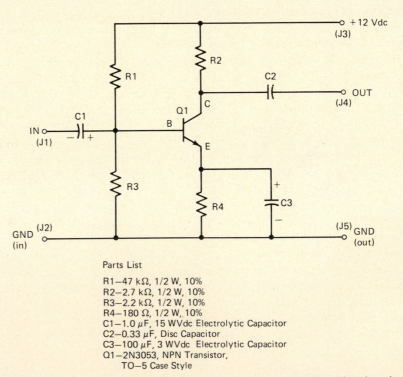

Parts List

R1—47 kΩ, 1/2 W, 10%
R2—2.7 kΩ, 1/2 W, 10%
R3—2.2 kΩ, 1/2 W, 10%
R4—180 Ω, 1/2 W, 10%
C1—1.0 μF, 15 WVdc Electrolytic Capacitor
C2—0.33 μF, Disc Capacitor
C3—100 μF, 3 WVdc Electrolytic Capacitor
Q1—2N3053, NPN Transistor,
    TO—5 Case Style

FIGURE 7–2 Standard common emitter transistor amplifier circuit schematic with parts list.

ways place limits on design flexibility, will be treated in subsequent chapters.

7-3

Format of the Layout Sketch

Recall from Chapter 6 that printed circuit boards are typically flat, rigid surfaces on which components are mounted and are electrically connected. For single-sided boards, components, devices, and hardware are normally mounted on the *insulated* side, termed the *component* side or *side 1*. The terminal pads and conductor paths are processed into the copper foil on the opposite side, referred to as the *circuit* or *foil* side or *side 2*. Because the layout sketch represents a composite view of both sides, a direction of viewing the board must be adopted before layout can begin. It is common practice for PCB designers to view the layout from the component side of the board. This is because the components are considered to be the dominant elements in a circuit. In addition, the intended layout must not only result in a functional system but also should have the components well arranged so that the board is appealing to the eye. To achieve this, the components should be uniformly spaced using all of the available board area. By careful parts placement, the resulting board will have a balanced distribution of components with no crowding and/or open areas which are void of parts. It can be seen from this discussion that the logical viewpoint for laying out a PCB is the component side. Thus initial attention will be focused on the placement of the components with the routing of the conductor paths to be considered after a balanced parts layout has been achieved.

With the direction of viewing established, it must be noted that, even though the components are positioned first, conductor routing must be considered at each step; that is, the conductor paths throughout the component arrangement must be routable. Thus the components and devices must be placed in such a position as to allow them to be conveniently connected electrically by a conductor pattern on the foil side of the board. To avoid confusion in distinguishing each of the board sides on the layout, a color-coded system needs to be established. For example, all detail appearing on the component side of the board, such as component and device bodies, hardware, and outside board edges, will be drawn using *black* lead. All detail appearing on the circuit side of the board, such as terminal pads and conductor paths, will be drawn using *blue* lead. Because this book does not have multicolor illustrations, all conductor paths in this chapter are shown as single, solid black lines connecting the circles that represent the terminal pads. It should, therefore, not be difficult to differentiate the linework on one side from that on the other.

Finally, the scale to which the drawing will be made must be considered. Industrial layouts are typically drawn to either a 2:1 or 4:1 enlarged scale. A 2:1 scale will be used in this chapter to show techniques used to make enlarged scale drawings. Scale drawings are necessary to reduce small unavoidable layout errors.

7-4

Drawing Component and Device Body Outlines

With all of the necessary information available and the format established, we can now begin developing the working sketch. Before the components can be positioned onto the sketch, we must first learn how they are shown as two-dimensional body outlines which replace the graphic symbols in the circuit schematic. These outlines will be drawn to a 2:1 scale and will be shown so as to represent the components and devices as they will appear on the layout drawing, viewed from the component side.

The circuit schematic shown in Fig. 7–2 will be used to illustrate the drawing of the component body outlines. In the parts list of the amplifier, there are four $\frac{1}{2}$-W resistors (R1, R2, R3, and R4), one tubular electrolytic capacitor (C1), one ceramic disk capacitor (C2), one vertically-mounted electrolytic capacitor (C3), one npn transistor (Q1), and five external connections [IN, GND(in), + 12 Vdc, OUT, and GND(out)].

The exact dimensions of each component must first be determined. The axial-lead-style $\frac{1}{2}$-W resistors are found to have a diameter of 0.160 in. and a length of 0.416 in. with lead diameters of 0.032 in. You will recall from Chapter 5 that this style of component is mounted to the board surface by first bending the two leads at right angles at a specified distance from the ends of the body. The leads are then passed through drilled access holes with sufficient length to make connection to the terminal pads on the circuit side of the board. It is common practice when drawing the body outline of axial lead components not to show the lead access holes or the lengths of leads between the body ends and the hole. Rather, since this is a composite view, only properly positioned terminal pads that will accept each lead are drawn beyond the ends of the body outline. The diameter of the pads and the distance between them must first be determined.

Terminal pad diameters for *unsupported* holes (not plated-through) on single-sided boards are dependent on the diameter of the lead to be inserted, the diameter of the drill hole, and the amount of copper foil required to remain about the pad after the hole is drilled. To provide the necessary clearance for our resistor leads, which have a 0.032-in. diameter, the drill hole should be from 0.010 to 0.020 in. larger in diameter. This would require a hole size of diameter between 0.042 and 0.052 in. It is general practice for unsupported holes that the terminal pad diameter be at least 0.040 in. larger than the drill hole. For our example, a pad diameter of between 0.082 and 0.092 in. would be acceptable on a 1:1 scale. Commonly available pad diameters in this range are 0.075, 0.080, 0.093, and 0.100 in. in diameter. It should be noted that after a board has been processed, the bonding of the conductor paths and terminal pads is related to the amount of foil remaining after the hole has been drilled. For this reason, it is recommended that the size of the terminal pad be made as large as possible. Making them too large, however, reduces the available space for conductor path routing. The final size selection thus becomes a trade-off between conductor spacing and maximum diameter of terminal pads.

Additional factors that must be considered at this time are (1) reducing the number of different hole sizes, which results in fewer drill bit changes during fabrication, and (2) selecting only a small number of terminal pad diameters to satisfy all lead diameters in the design, thus reducing the required inventory of pad sizes. Every effort should be made to select the fewest drill and pad sizes that will satisfy the total range of hole and pad size requirements.

In selecting pad diameters for single-sided boards, the following guidelines are recommended. For leads with diameters of less than 0.020 in., a 1:1 scale pad diameter of 0.075 in. may be used. Lead diameters of between 0.020 and 0.040 may effectively use a pad diameter of 0.100 in.

The on-center distances between pads is determined by both the component body length and the lead diameter. Since an individual pad must be provided for each lead, one is shown at each end of the body outline. The preferred distance that the leads should extend outward from the component body before the 90-degree bend is made is 0.060 in. To determine the overall bend distance X, a bend radius should be added. The minimum bend radius for each lead should be approximately 0.030 in. for lead diameters up to 0.030 in. and 0.060 in. for lead diameters of 0.031 to 0.050 in. Lead diameters greater than 0.050 in. should have a bend radius of two times the lead diameter. Refer now to Fig. 5–6. It can be seen that the formula for calculating the on-center distance (S) for terminal pads is

$$S = 2X + B \qquad (7–1)$$

where X is the bend distance (extension distance + bend radius) and B is the body length. From the discussion above, and with reference to Fig. 5.6, the *minimum* value of terminal pad spacing for axial-lead-style components using maximum lead diameter is $2(0.060 + 0.060)$ plus the body length (B).

Referring again to the $\frac{1}{2}$-W resistor shown in Fig. 5–6 and using the maximum values given for X and B in the accompanying table, we calculate the value of S as follows:

$$\begin{aligned} S &= 2X + B \\ &= 2(0.060 + 0.060) + 0.416 \\ &= 0.240 + 0.416 = 0.656 \text{ in.} \end{aligned}$$

This figure would normally be increased to 0.70 in. so that the on-center pad dimensions will fall on grid.

The terminal pads shown in Fig. 7–3a have been drawn on grid with an on-center dimension of 1.4 in. and diameters of 0.2 in. at 2:1 scale. Although the pads are shown in black, they should be drawn with blue pencil so as to more readily distinguish their true position as being on the foil side of the board.

(a) Pad spacing for 1/2 watt resistor

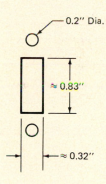

(b) Body and pad sizes for 1/2 watt resistor

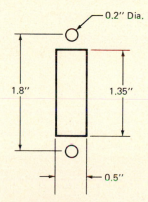

(c) Two-dimensional view of electrolytic capacitor

FIGURE 7–3 Drawing the two-dimensional component view of axial lead resistors and capacitors.

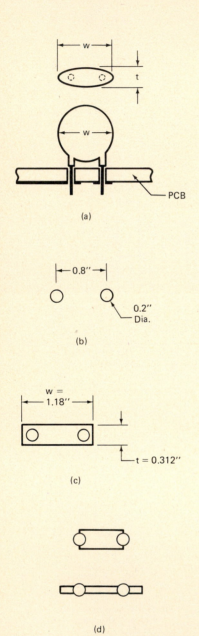

FIGURE 7–4 Drawing the two-dimensional component view of disc capacitors.

Note in Fig. 7–3 that no center lines are used in drawing the terminal pads. Simply drawing the outline of the pad with a circle template is sufficient. The underlying grid should be used as a guide to locate and center the pad. It is extremely important to center the pads on grid intercepts and construct the component body outlines off-grid as required by their dimensions. For precise artwork generation and board fabrication, it is more important for the pads to be positioned accurately on grid than it is for the component body outlines. (There are occasional exceptions to on-grid pad locations such as for disk capacitors, discussed in Section 8–2.)

When mounted to the component side of a PCB, axial lead components normally rest flush against the insulated board surface. As such, their two-dimensional body outlines, as viewed from the component side (side 1), would appear as rectangles. Using the adopted color-code system, all of the linework for side 1 would be drawn with black lead. Since our layout will be drawn to a 2:1 scale, each full-size dimension must be multiplied by a factor of 2 before drawing the body outlines. For a $\frac{1}{2}$-W resistor, the maximum dimensions for the body length and diameter given in the table of Fig. 5–6 would be doubled, resulting in a rectangle drawn as close to 0.32 in. × 0.83 in. as the grid will permit. This is shown in Fig. 7–3b. Note that the grid lines provide a useful guide for measuring and drawing the body outlines by simply counting squares. When using a 0.1 grid to show the outline of this resistor, the body is to be laid out to a rectangle of just over 3 square (0.32 in.) by a little more than 8 square (0.83 in.). Note also that the body outline is neatly drawn and reasonably centered between the two terminal pads using a straightedge even though the layout is termed a *sketch*.

Capacitor C1 in Fig. 7–2 is the electrolytic type with axial leads and with a body diameter of 0.25 in. Its body outline, shown in Fig. 7–3c, is drawn similar to that of a resistor, the only difference being the dimensions of the lead diameter, body diameter and length. Because its maximum lead diameter is slightly less than that of a $\frac{1}{2}$-W resistor (0.031 in. versus 0.032 in.), the diameter of its pads may be the same as those used with the resistor (i.e., 0.2 in. at 2Xscale). Also, since the lead diameter is 0.031 in., the bend radius will be 0.060 in., the same as that for the resistor. The total body length of the electrolytic capacitor is 0.625 in. Therefore, the on-center pad spacing is $S = 2(0.060 + 0.060) + 0.625 = 0.865$ in. To be on grid, this becomes 0.90 in. between pad centers at a 1:1 scale. As shown in Fig. 7–3c, the body outline is centered between 1.8-in. spaced pads and is shown as a rectangle of approximately 0.5 × 1.35 in. at a 2:1 scale. Note the + sign inside the body outline near one end. Because electrolytic capacitors are polarized, they need to be correctly assembled onto the board. The + sign serves as a key for proper installation.

Another component commonly used in electronic circuits is the radial lead ceramic type *disc* capacitor, which is shown in Fig. 7–4a properly mounted to a PCB. Capacitor C2 in Fig. 7–2 is a disc type. It is seen from its side view that the body is approximately circular and that it has two unpolarized leads which extend directly through the board without bends. Above the side view shown in Fig. 7–4a is the two-dimensional shape of the disc capacitor as viewed from the top or as it would appear on the component layout drawing and viewed from side 1. Accurately drawing this complex shape, which tapers almost to points at its outside edges with maximum body width at its center is time consuming and unnecessary for PCB layout. Rather, the less difficult shape of a rectangle has been adopted even though it does not exactly represent the body shape.

To construct the component view of disc capacitor C2, the on-center lead spacing is first measured and found to be 0.375 in. At 2X scale, this dimension becomes 0.750 in. This is then increased to 0.80 in. for the on-grid locations of pad centers. The lead diameters of the component are 0.025 in. With this information, the pads, at 2X scale, will be drawn with 0.20-in. diameters on grid. This is shown in Fig. 7–4b. Remember that these pads would normally be drawn with blue pencil. The diameter (w) and thickness (t) of the capacitor are next measured and found to be 0.590 in. and 0.156 in., respectively. At 2X scale, these dimensions become 1.18 in. and 0.312 in. The two-dimensional view of the component is thus represented by a rectangle which is slightly over 3 square by approximately 12 squares and centered between the terminal pads. Note in Fig. 7–4c that the terminal pads are on grid and are *inside* the body outline. This is not always the case. Other types of orientations of body outlines to terminal pads are shown for different sizes of disc capacitors in Fig. 7–4d.

Capacitor C3 in Fig. 7–2 is an electrolytic type with a tubular shape having two radial leads protruding from the *same* end of the body. This style is designed to be mounted vertically onto the PCB with its leads extending straight into the access holes. This orientation is similar to that of the disc capacitor except that the base of the electrolytic type is seated flush with the surface of the board. This is shown in Fig. 7–5a. As with other PCB-mounted components, the pad positions and size will first be determined. The 1:1 on-center lead spacing and lead diameters are found to be 0.2 in. and 0.025 in., respectively. Thus the pads will be drawn on grid at 2X scale with a 0.4-in. on-center dimension and diameters of 0.2 in. This is shown in Fig. 7–5b. The case diameter is measured to be 0.335 in. and at 2X scale, it becomes 0.67 in. Using a 0.7-in. hole in a circle template, the body outline is centered and drawn around the two terminal pads. See Fig. 7–5c. As seen from the component view, this capacitor appears as a circle since it is vertically mounted. Because it is polarized, a + sign is placed beside the appropriate pad outside of the body outline to ensure correct orientation onto the board.

The five external connections to the points labeled IN, GND(in), +12 Vdc, OUT, and GND(out) in Fig. 7–2 will each be made with turret terminals for our example design problem. You will recall that a pictorial view of a turret terminal and its assembly to the board is shown in Fig. 5–10. These terminals are a common method of providing points for soldering wires to make electrical connections between the PCB circuit and other external points. Each of the five turret terminals will require a pad in order to make electrical connections to a path on the circuit side of the board. The two-dimensional representation of a turret terminal is quite simple, as shown

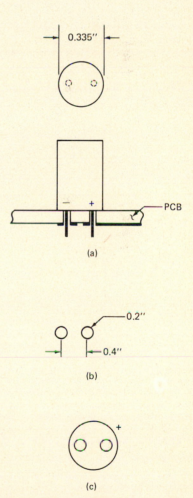

FIGURE 7–5 Drawing the two-dimensional component view of an electrolytic capacitor/radial leads.

FIGURE 7–6   **Drawing the two-dimensional representation of a turret terminal pad.**

by Fig. 7–6a. The inner circle represents the diameter of the turret section, the next larger circle the shoulder, and the outer circle represents the terminal pad. Actually, all that is required is to show the terminal pad diameter for each of the terminals. This is shown in Fig. 7–6b. Appropriate labeling on the layout sketch will make it clear which of the pads will be used for making external connections requiring turret terminals. In addition, their larger diameters and locations near the board edges make them easy to identify.

To serve as an example, we will use a turret terminal with a shank diameter of 0.1 in. If an access hole of 0.01 in. greater than the shank diameter is selected for ease of installation, a drill diameter of 0.11 in. would be required. Refer to Appendix XIX for a table of common drill bit sizes. It can be seen from the table that a No. 35 drill bit would be used for this hole. To satisfy the requirement of selecting a pad diameter that will be 0.040 in. larger than the drilled hole, a pad with a 1:1 scale diameter of 0.150 in., having a 2X scale diameter of 0.300 in., would be required. This is shown in Fig. 7–6b. This size pad would be shown for all external connections requiring the installation of turret terminals having a shank diameter of 0.1 in.

The final component of Fig. 7–2 to be drawn is the transistor Q1. The body outline and pad arrangement for transistors are more complex in their drawing than the components discussed thus far. Although all of the critical dimensions can be measured with a scale, this is a type of component where dimensional information is best obtained from the manufacturer's specification sheets. Refer to Fig. 7–7a or Appendix XIV for the manufacturer's outline drawing of the TO-5 transistor case specified for Q1. Note that the maximum lead diameter is 0.019 in. Therefore, a 1:1 scale pad diameter of 0.075 is required. When drawing the pads to a 2X scale, the diameter used will be 0.150 in.

The transistor leads are numbered 1 (*emitter*), 2 (*base*), and 3 (*collector*). These leads are positioned in a triangular arrangement on a 0.2-in.-diameter dotted-lined circle with the base lead centered but elevated a distance of 0.1 in. above the in-line leads of the emitter and the collector. The emitter lead is closest to the case tab and the collector lead is farthest away from the tab with a separation of 0.2 in. The three terminal pads that will accept the transistor leads for straight entry into the board are drawn with 0.15-in. diameters at a 2X scale and laid out at the dimensions shown in Fig. 7–7b. Note that the arrangement of the pads is inverted from that of the bottom view shown for this case style. This is necessary to represent the true orientation of the device as viewed from the top of the case

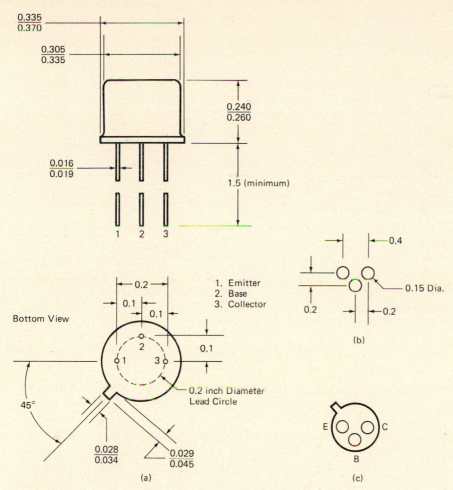

FIGURE 7–7 **Drawing the two-dimensional component view of a T0-5 style transistor using manufacturer's specifications.**

as it is positioned on the component side of the board. Care must always be exercised in arranging pad positions for devices from the component view since the manufacturer's specifications commonly provide lead designations from the lead side or bottom view of the device.

Referring again to the specifications for the TO-5 case, the largest body diameter is specified as 0.37 in. at a 1:1 scale or approximately 0.75 in. at a 2X scale. This circular body is drawn concentric to the dotted-line lead circle. For ease of drawing, the device tab is shown as a square approximately 0.1 by 0.1 in. and is positioned closest to the pad numbered 1 (emitter). This square is drawn outward from the body outline at a 45-degree angle to the center of the horizontal grid line which passes through the centers of the emitter and collector pad circles.

The complete outline drawing of the TO-5 case with leads coded and with the identification tab is shown in Fig. 7–7c. Note from this view that the emitter, base, and collector are read in a counterclockwise direction from the tab.

It is not possible to demonstrate here the drawing of body outlines for all styles of components and devices and it is hardly necessary. As emphasized in this section, the basic objective is to draw the top view of the part as seen from the component side of the board. As much detail as necessary is shown to illustrate the important features of the part. Drawing excessive detail not only may be a waste of time but also may cause confusion.

In summary, terminal pads are first positioned and drawn on grid after their scaled dimension (in our case, 2X) and spacing have been determined. The component body outline is then drawn in its relative position to its pads. The following sequence should be followed: (1) calculate the pad diameter from the lead diameter and the hole size; (2) determine the 2X pad spacing for the component or device; (3) position and draw the 2X diameter terminal pad on grid for each lead; and (4) center and construct the body outline in its proper position. Finally, because the pads will be processed from the copper foil on the circuit side of the board, they should be drawn in a contrasting color, normally *blue*. The body outlines, which appear on the component side of the board, should be drawn with *black* lead.

7-5

Positioning Components on the Layout Sketch

Now that you have become familiar with the drawing of common components and devices typically mounted on a PCB, we will begin the discussion of the most creative and interesting aspect of printed circuit design. This is the positioning of the parts on the board and the routing of the electrical connections (conductor paths) between terminal pads in accordance with the requirements of the circuit schematic. The typical concern faced by beginning designers is that after several parts have been tentatively positioned on the board, connecting paths between some of their pads will be unroutable. That is, there may be no possible routes available to run a conductor path between pads that must be connected without crossing another drawn path that is not associated with it. This problem, however, will be eliminated as proper layout skills are developed. This also emphasizes the need for drawing the preliminary sketch prior to preparing the final component layout. The sketch is used to resolve any problems of components being positioned so as to create unroutable paths.

It needs to be understood initially that printed circuit design is a creative task and, as such, does not lend itself to any single "correct solution." There are many acceptable layout practices, some with obvious reasons for specific component positioning and some where the reasons are somewhat obscure but are based on past practices that have proven successful. For this reason, the printed circuit designs presented in this and subsequent chapters represent one of several acceptable layouts. Your unique designs may be just as acceptable and correct as long as the basic guidelines initially established are not violated.

For our first design, we will construct a preliminary sketch of the amplifier circuit shown in Fig. 7–2 using the skills for drawing component body outlines developed in the preceding section. A statement of the design problem follows.

DESIGN PROBLEM

Using a 2:1 scale and a 0.1-in. grid system, design a component layout sketch of the amplifier circuit of Fig. 7–2. There are no restrictions on final board size or shape except that the finished layout maximizes the total board area. There are also no restrictions on the position or orientation of any external connections. All terminal pads are to be drawn on grid. The component view should show all necessary labeling using black lead to show the component side and blue lead to represent all pads and conductor paths on the foil side of the printed circuit board.

It is seen from the design problem that the designer has few restrictions, especially in component positioning and external connections. This is not typical, but it is felt that your first design should provide layout practice without rigid requirements. In subsequent chapters, restrictions will be imposed on example designs to typify more clearly what is generally encountered.

We begin the design with the circuit schematic shown in Fig. 7–8, which is essentially the same as that shown in Fig. 7–2 with the addition of specific points identified with letters and numbers to serve as guides for locating portions of the circuit as they are discussed.

For purposes of initial positioning of components and hardware to provide ease of conductor path routing, there is no easier way to begin than to locate the parts as they appear on the circuit schematic. In your mind's eye, replace each of the component and device sym-

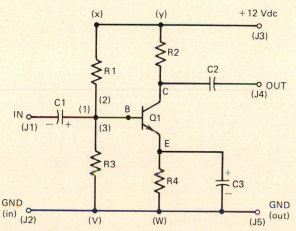

FIGURE 7–8 Schematic labeled for use as guide in parts placement and conductor path routing.

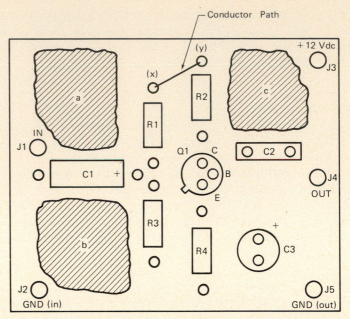

FIGURE 7–9 Initial component layout using schematic as a guide.

bols on the schematic with its appropriate body outline and terminal pads of proper size and position. When drawn on 0.1-in. grid paper to a 2:1 scale, the result may appear as shown in Fig. 7–9. All component body outlines and terminal pads are laid out to a 2:1 scale with the bodies drawn with black lead and the pads with blue lead. All of the parts are labeled with the same designations as on the schematic. Note that the parts are positioned approximately as they appear in the schematic. For example, R1 is directly above R3 and to the left of R2, which is above Q1. Some components, like capacitor C1, require special orientation. This capacitor must be positioned with its negative side, which is not labeled, closest to the external terminal marked IN and its positive terminal, labeled +, facing toward Q1. The input, output, ground, and power terminals are positioned toward the ends of the board, which follows their position on the schematic. Remember that all pad centers are to be placed on grid.

As an example of providing a conductor path, note that points *x* and *y* at the upper ends of R1 and R2, which appear in Fig. 7–8, are to be electrically connected. On the layout sketch, this is shown by drawing a straight blue line between the terminal pads labeled *x* and *y* in Fig. 7–9. Observe that there are no crossover leads in the circuit schematic. Since out parts placement follows that of the schematic, it is possible to continue making all of the electrical connections as shown on the schematic. Even though the design is routable as shown, it does not meet the specified requirement of maximizing the total board area. The bold outline drawn around the positioned parts in Fig. 7–9 represents the board outline. Notice the large unused areas of the board. These areas are shown as a, b, and c. This gives the appearance that the board is much larger than required for this design. To overcome this problem, we will slightly reposition some of the parts until a more reasonable balance is achieved. Compare Fig. 7–9 with Fig. 7–10. To begin, C1 will be rotated 90 degrees with the GND (in) and IN terminal pads positioned directly above and below this capacitor while still remaining centrally positioned. Also, note that for better component balance, moving Q1 to the right will allow R2 and R4 to be brought inward so that they will be adjacent to R1 and R3. This will reduce the overall width of the board. C3 can then be placed to the right of R4 bringing it directly under Q1. Finally, C2 is rotated 90 degrees and positioned to the right of R2 and above Q1. Wherever possible, centers of component bodies should be aligned as was done with R1, R2, R3, and R4. There are trade-offs, however, in order to reduce the total required space and to improve the overall appearance of the board. For example, C1 is positioned midway between the top and bottom edges of the board. C2, Q1, and C3 are spaced vertically to avoid needless crowding, but this does not allow C2 to center on R2 nor C3 to center on R4. These types of trade-offs in rearranging components help to optimize the overall layout in regard to uniform distribution. Note also that no body outlines are positioned on a diagonal to the horizontal and vertical grid lines. This is to prevent the generation of wasted board space. In addition, any component placed at a diagonal would look out of place compared to the other vertically and horizontally placed parts. The final component layout sketch incorporating all of the techniques discussed is shown in Fig. 7–10.

Even though parts have been rotated and repositioned, two important points need to be emphasized. First, no part has been moved any great distance from its initial position in the schematic. The resulting path routing distances are thus as short, or in some cases shorter, than in our initial parts placement. Keeping interconnecting paths as short as possible is normally a prime requirement of printed

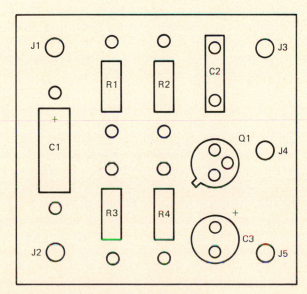

FIGURE 7–10 Final component layout sketch.

circuit design. Second, each component move or rotation has resulted in reducing the amount of required board space, generating a more compact layout which is also required in a good design.

Once the components have been repositioned to result in an improved layout, the external connections on both ends of the board can then be moved inward to reduce the required length of the board.

We will now summarize how optimum component placement is accomplished. We first establish an initial point of reference by positioning the parts as they appear in the circuit schematic. The schematic is a concise representation of all electrical connections between parts. It contains minimum crossovers and, as such, part placement based on this scheme also results in minimum crossovers. The parts may then be rearranged to (1) reduce path routing problems, (2) result in a more compact design to reduce wasted board space, and (3) improve the appearance of the board by arriving at a uniform distribution of parts. In general, conductor path routing is simplified when the component layout parallels that of the circuit schematic.

7-6

Conductor Path Routing

The final process in the completion of the component layout sketch is to interconnect correctly the terminal pads associated with each of the components and devices in accordance with the circuit schematic. To continue in our layout of the amplifier circuit shown in Fig. 7–8, conductor paths between the pads shown in Fig. 7–10 will be drawn.

The width of the conductor paths is primarily related to (1) the maximum current they are expected to handle, (2) circuit density, and (3) fabrication limitations. Because it is the purpose of this chapter to introduce the basic techniques for routing conductor paths, we will focus only on this topic. The factors mentioned will be discussed in subsequent chapters so that the appropriate conductor path widths and spacings may be determined. For our example layout, we will select a conductor width of 0.025 in. at 1:1 scale, which would require a tape width of 0.050 in. at the 2X scale.

Conductor paths are drawn on the sketch with a straightedge and are shown as single lines, typically 0.5 or 0.7 mm thick. Blue lead should be used to indicate that these paths are on the foil side of the board together with the pads. Since the conductor paths form the electrical connections, they must originate at a pad or at the intersection of other paths and terminate at a pad or intersect with another path. The conductor paths should be drawn on grid lines whenever possible since the pads are centered on grid. They should extend toward the pads as if to bisect them. For clarity, the conductor line drawn should cut across the pad circumference and extend approximately one-half way to its center point. See pad a in Fig. 7–11. Changes in path direction are drawn in 45-degree angle steps as shown in the paths between pads a and b and between pads b and d in Fig. 7–11. This accurately simulates a method of taping on the artwork that will be discussed in Chapter 12. When routing a con-

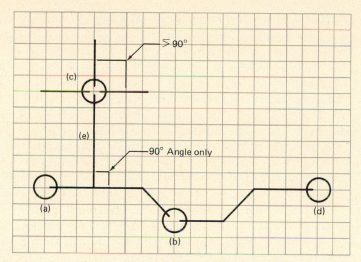

FIGURE 7–11 Techniques for routing conductor paths.

ductor path that must intercept another path for electrical connection, it is drawn only at right angles, as shown in the path labeled e in Fig 7–11. Again, the actual taping also follows this practice. Finally, no more than four paths should originate or terminate at one terminal pad. Space permitting, paths entering a pad should be separated by an angle equal to or greater than 90 degrees. See pad c in Fig. 7–11.

The task of conductor path routing begins with drawing as many of the short interconnect points as possible, leaving the longer paths until later. The circuit schematic of Fig. 7–8 shows an electrical connection between points x and y. A single solid blue line is drawn horizontally between the pads labeled x and y to form this connection. This is shown in Fig. 7–12a. As each connection is made on

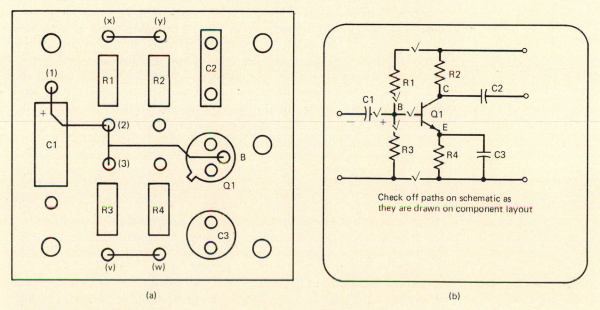

FIGURE 7–12 Routing conductor paths using the schematic as a guide.

the layout sketch, it should be checked off or lined out on the circuit schematic. This is shown in Fig. 7–12b and is done to ensure that all connections have been correctly made in accordance with the schematic. The procedure is repeated for the connecting points labeled v and w.

A grouping of short connections is labeled as 1, 2, 3, and B on the schematic. This indicates that a single electrical connection is to be made from the + side of C1 (point 1) to the bottom pad of R1 (point 2) and the top pad of R3 (point 3) as well as to the base of Q1 (point B). Points 2 and 3 are first connected with a single vertical line. The connection between point 1 can then be made to either point 2 or 3 since they are now shown as electrically connected. However, it is recommended that this connection be made to the pad of point 2 since it is closer to the + side of C1 and thus will result in a shorter path. See Fig. 7–12a. Note that this path is partially routed through the body outline of C1. This is normal practice since the conductor path (in blue) is on the foil side of the board and the component body is on the component side, which prevents mechanical or electrical interference problems. Even though conductor paths are generally made to follow grid lines, long diagonal paths are acceptable as long as they contribute toward high circuit density and leave no large unused board areas. To complete this connection, a line is drawn to the base pad of Q1, which is perpendicular to the path between points 2 and 3. It is important to note that this path was positioned approximately midway between the bottom pads of R1 and R2 and the top pads of R3 and R4. Following is an explanation of this design criteria. The conductor paths and terminal pads represent copper conductors which are separated by an insulating path or gap. As such, conductor path positioning is always planned so as to maintain as uniform an insulating gap as possible between adjacent paths and terminal pads as allowed by proper grid positioning. The insulating space between conductor paths and pads and between adjacent paths is of major concern in PCB layout. Needless crowding between conductors is to be avoided. A conductor path may have unequal insulation spaces on either side as it is routed between other conductors, but this should not result in a severe difference in their proportions. (The minimum spacing between conductors will be discussed in subsequent chapters.) Conductor paths must not be routed along the very edge of the board. A space must be allowed between the outside edge of the path and the board edge to prevent damage to the path when the board is processed into its finished size.

The finished layout sketch is shown in Fig. 7–13 properly labeled and coded.

Following is a summary of the basic steps used in conductor path routing:

1. Route conductor paths along vertical and horizontal grid lines.
2. Change path direction by using short 45-degree line segments.
3. Draw conductor path lines so that they extend slightly inside the pad.
4. Avoid crowding.
5. Begin path routing by drawing the shortest and most direct routes.

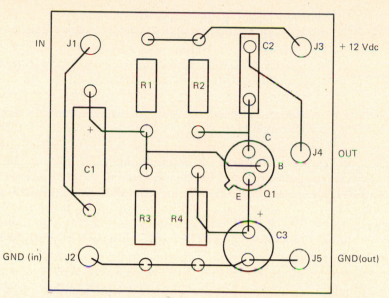

FIGURE 7-13 Final component and conductor pattern layout sketch.

6. Keep track of each path drawn by checking it off on the circuit schematic.
7. Draw path to path intersections as perpendicular lines.
8. Keep angles of paths entering common pads equal to or larger than 90 degrees.

For this first introduction to PCB layout, common design problems were not considered. This was so that we could focus on acquainting the beginning designer on representing parts, the placement of parts on the board, and basic conductor path routing. In the next chapter, the procedures that have been discussed here will be expanded to include typical restrictions and requirements normally encountered in single-sided layout sketches which are more challenging to the designer.

EXERCISES

7-1 Construct a 2X scale two-dimensional view of an axial $\frac{1}{4}$-watt resistor. Refer to Fig. 7-3a and b and use the following information: the body diameter and length of the resistor are 0.1 in. and 0.25 in., respectively. The lead diameter is 0.230 in. Using Appendix XIX and the criteria for determining overall pad size, calculate the appropriate pad diameter and drill hole size to be used. Be sure to locate the pad centers on grid.

7-2 Draw a 2X scale component and conductor pattern layout for the resistor ladder network shown in Fig. 7-14. Use a 0.100-in. grid and consider the minimum 1X scale path widths and spacings to be

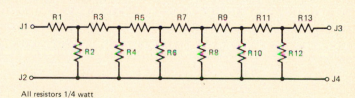

All resistors 1/4 watt

FIGURE 7-14

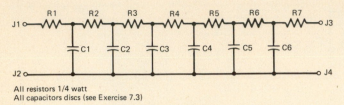

All resistors 1/4 watt
All capacitors discs (see Exercise 7.3)

FIGURE 7–15

0.025 in. There are no mechanical restrictions as to the size and shape of the printed circuit board. Refer to Fig. 7–6 for the dimensional information for determining the pad diameters to be used for external connections J1 through J4. There are also no restrictions in regard to the positioning of these connections except that they should be located near the periphery of the board with their centers on grid.

7–3 Refer to Fig. 7–4 and using the following information, construct the 2X scale two-dimensional view of a radial lead disc type of capacitor. Be sure to locate pad centers on grid. The diameter of the capacitor is 0.6 in., its thickness 0.1 in., and its lead diameters are 0.019 in. The on-center lead spacing is 0.4 in. Using Appendix XIX and the criteria for determining overall pad size, calculate the appropriate pad diameter and drill hole size to be used.

7–4 Draw a 2X scale component and conductor pattern layout for the filter network shown in Fig. 7–15. Use a 0.100-in. grid and consider the 1X scale minimum path widths and spacings to be 0.025 in. The layout is to be constructed within the board outline shown in Fig. 7–16. Refer to Fig. 7–6 for the pad diameters used for external connections J1 through J4.

7–5 Draw a 2X scale component and conductor pattern layout for the 555 timer circuit shown in Fig. 11–6. Use a 0.100-in. grid and consider the 1X scale minimum path widths and spacings to be 0.025 in. The layout is to be drawn within the board outline shown in Fig. 7–17 providing a 0.1-in. forbidden area. The pad diameters for J1 through J4 are to be as shown in Fig. 7–6. To draw the two-dimensional component outline for the diode and the 555 IC, refer to Figs. 8–5a and 9–4, respectively. Use the offset configuration for the IC. Be sure to locate all pad centers on grid.

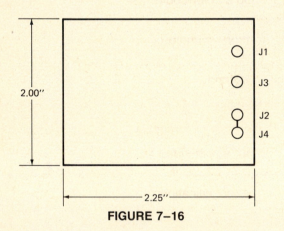

FIGURE 7–16

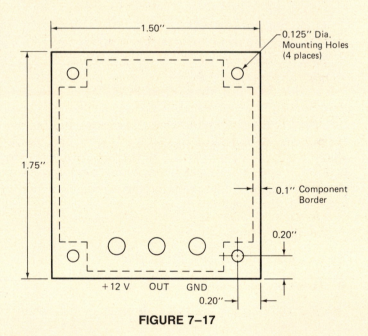

FIGURE 7–17

8

Single-Sided Printed Circuit Board Design: The Finished Component and Conductor Layout Drawing

LEARNING OBJECTIVES

Upon completion of this chapter on the finished drawing of a single-sided printed circuit board design, the student should be able to:

1. Use printed circuit board design templates for drawing component body outlines.

2. Properly orient component and device case outlines.

3. Draw outlines to their appropriate scales.

4. Understand all physical and mechanical restrictions of the board design.

5. Select an appropriate grid system.

6. Select an appropriate reduction scale.

7. Determine optimum conductor width.

8. Develop a color-coding system for distinguishing different path widths.

9. Determine optimum conductor spacing.

10. Be familiar with the available pad diameters and their selection criteria.

11. Provide for lead bend allowance for locating pads.

12. Properly show jumper wires on the drawing.

13. Lay out single-sided printed circuit designs.

14. Develop a checklist to determine if all electrical and mechanical specifications on the finished drawing have been met.

8-0

Introduction

The basic layout procedures for printed circuit board design were introduced in Chapter 7. In order to make this introduction as meaningful as possible to the beginning designer, a relatively simple circuit with only a few components was used to demonstrate basic techniques. In addition, many restrictions, limitations, and trade-offs common to PCB design were not specified, in order to emphasize these techniques. The beginner was thus able to concentrate on converting a circuit schematic into a component layout sketch from which a finished drawing, required for board fabrication, could be produced.

This chapter is intended to expand on these introductory techniques by considering a more complex circuit schematic. The next circuit to be designed has a larger parts count and thus will result in a more challenging design problem. In addition, some typical electrical and mechanical restrictions are imposed to further demonstrate practical PCB design problems. In order to expand on techniques to solve these problems, additional topics presented include printed circuit design templates, grid system selection, board outline drawings, corner brackets, reduction scale, conductor width and spacing requirements, pad size selection, lead bend allowance, and color codes for pad size and conductor path width. Although several of these topics were introduced in Chapter 7, they will be expanded on here to provide the reader with a better understanding of how to produce high-quality designs under a variety of restrictions and specifications.

To assure that no design specification is overlooked, the material presented in this chapter will allow the designer to develop a checklist of considerations which will be applicable to all levels of circuit complexity. Any design problem can then be matched against this list to verify that all mechanical and electrical requirements are satisfied.

8-1

Negative-Acting One-Shot Multivibrator Circuit

The circuit used in this chapter to demonstrate a more complex single-sided PCB component layout will be a *negative-acting one-shot multivibrator*. The circuit schematic and the parts list are shown in Fig. 8–1. Recall that a similar circuit was presented in Fig. 4–2 to show the drawing of a complex analog circuit schematic. A comparison of the two schematics will show that some minor changes have been made to the circuit shown in Fig. 8–1. These changes will be described in the discussion that follows.

In this chapter we describe the step-by-step procedure for designing the component layout drawing for the entire circuit shown within the dashed lines in Fig. 8–1. External connections will be made at the points numbered 1, 2, 3, 4, and 5. Point 1 is the input terminal and labeled IN. Point 2 is the output terminal, labeled OUT. The dc power is supplied to points 3, 4, and 5. The +15-V terminal of the external split power supply is connected to point 3 and the −15-V terminal to point 4. The external ground connection is made at point 5, which serves as the ground for the entire circuit. Thus all connections shown with the symbol $\bigtriangledown$ are connected together electrically and also to point 5.

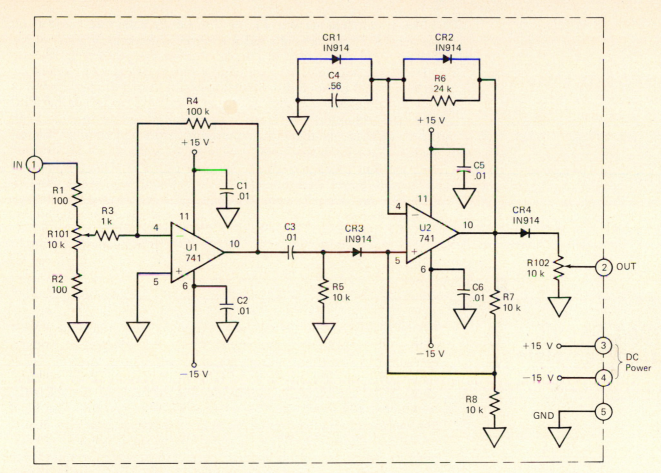

All resistors are 1/2 watt ± 10% unless otherwise specified.
All capacitors are in microfarads unless otherwise specified.

Parts List

U1, U2 — 741 Op-Amp, TO–116 style case
R1, R2 — 100 Ω, 1/2 W, 10%
R3 — 1 kΩ, 1/2 W, 10%
R4, — 100 kΩ, 1/2 W, 10%
R5, R7, R8 — 10 kΩ, 1/2 W, 10%
R6 — 24 kΩ, 1/2 W, 10%
R101, R102 — 10 kΩ, trim pots
C1, C2, C3, C5, C6 — 0.01 μF disc capacitors
C4 — 0.56 μF disc capacitors
CR1, CR2, CR3, CR4, IN914 diodes, DO–35 style case

FIGURE 8–1 Negative-acting one shot multivibrator.

The operation of the circuit of Fig. 8–1 is described in Section 4–3. Whereas the integrated circuit devices labeled U1 and U2 in Fig. 4–2 are 301 operational amplifiers, those in Fig. 8–1 are the 741 type. For these devices, the +15 Vdc is connected to pins 11 and the −15 Vdc is connected to pins 6. The inverting input pins are numbered 4 and the noninverting input pins, 5. The outputs are taken from pins 10. Other than the pin connections for the 741 op-amp being different from those for the 301 device, the circuit operation is basically the same as described for the circuit shown in Fig. 4–2.

The circuit schematic of Fig. 8–1 shows that the total parts count consists of 22 components and 5 external connections. Each of the 741 op-amps is packaged in a 14-pin dual-in-line package (DIP) with a TO-116-style case. There are eight $\frac{1}{2}$-watt resistors with axial leads; six radial lead ceramic disc capacitors; four 1N914 diodes, each in a DO-35-style case which also have axial leads; and two adjustable 10K-ohm potentiometers which are to be mounted to the printed circuit board. Of these components, only the resistors and ceramic disc capacitors have been discussed previously in Section 7–4 for drawing their two-dimensional outlines. The drawing of the body outlines for the diodes, potentiometers, and the 741 op-amp in a 14-pin TO-116-style case has not yet been described and will be considered in the next section.

8-2

Printed Circuit Design Templates

Let us briefly review the method of drawing two-dimensional body outlines to represent components. First, the component body is measured to determine the length and width (or diameter). These dimensions are then multiplied by the scale factor (2X for our previous design problem) to determine the overall size to which the component will be drawn. Consideration is then given to lead bend allowance and spacing for the location of terminal pads and finally, to the pad diameters. Where a circuit has a small number of components which are simple in their shape (rectangular or round) the use of a straightedge and circle template for drawing their body outlines is adequate. This method was presented in some detail in Chapter 7. The multivibrator circuit of Fig. 8–1, however, requires more complex component body outlines to be drawn. If the straightedge and circle template were the only instruments available, it would be a time-consuming task to draw these more complex outlines. For this purpose, *printed circuit design templates*, such as the set shown in Fig. 8–2, are used to quickly and accurately draw complex component body outlines. These templates are made of a durable plastic material which contains a variety of cutouts for component body outlines, pad sizes, and positions. The drawing of these outlines is done with a one-or two-step tracing technique. Printed circuit design templates are typically available in 1:1, 2:1, and 4:1 scales. The 0 to 6 in. scale, and the printed values for hole size, lead circle, and on-center spacings are 1:1 scale values on the templates shown in Fig. 8–2. However, the cutouts are made to specified scale size. A 2:1 scale size is shown in Fig. 8–2a and b.

The template shown in Fig. 8–2a contains the outlines of a selection of common device packages, resistors and capacitors. The TO-92, TO-18, and TO-5 package outlines together with their pad patterns are in the upper section of the template. Also shown are three common round IC pad patterns for the 8-, 10-, and 12-pin case styles. The outline circle labeled TO-5 is intended to be used for all of the round IC pin patterns shown. For example, a TO-99 case with eight leads has the same diameter case as the TO-5 style having three leads.

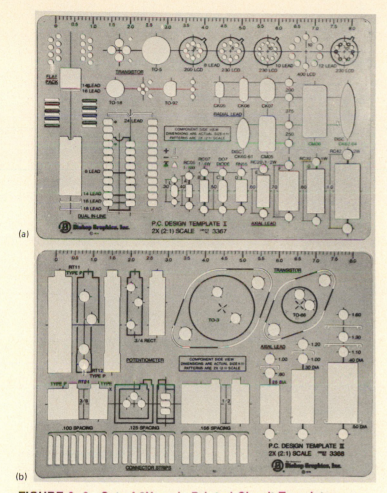

(a)

(b)

FIGURE 8–2   Set of 2X scale Printed Circuit Templates.

The square cutout for drawing the keying tab is located to the left of the TO-5 outline. This would be located outside the lead circle beside pin 1, 8, 10, or 12, depending on the device. The manufacturers case styles, listed in Appendix XIV, should always be consulted to avoid drawing errors.

The upper right section of the template shown in Fig. 8–2b contains the pin arrangements and case outlines for the TO-66 and TO-3 package styles. These devices find wide application in circuits that require large power dissipation, such as in dc regulators and power transistors. Care must be exercised when using the template to draw a transistor for a component view. When viewed through the case, the lead positions are thus reversed, as discussed in Chapter 7. This reversal technique is simply to turn the template over to trace the pad positions and then to code the pin arrangements properly. The template is then moved to center the body outline about the pads.

For the layout of the multivibrator circuit, our specific interest is the TO-116 (14-pin DIP) case style which is required for the 741 op-amp. The component view of this package, with the symbol for an op-amp superimposed inside the outline, is shown in Fig. 8–3a. This figure aids in identifying the pins as to their electrical function and also in orienting the device on the layout for routing conductor paths.

FIGURE 8–3   14-pin dual-in-line package (DIP).

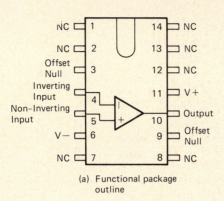

(a) Functional package outline

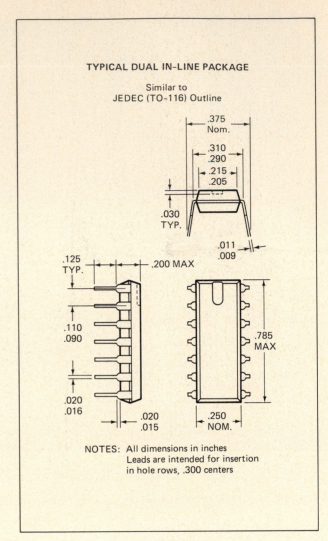

(b) TO-116 style case (mechanical outline)

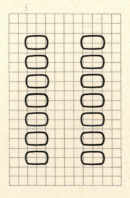

(c) Pad outline 14-pin DIP package

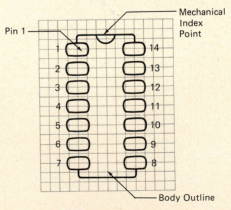

(d) Complete package representation (component view)

However, to obtain the dimensional information to draw the pad arrangement and body outline, Fig. 8–3b is required, which provides this type of mechanical information. All 1:1 scale dimensions are given in inches and specify maximum and minimum values. It can be seen that the plastic case is approximately 0.785 in. long and 0.250 in. wide, with a maximum height of 0.200 in. Two rows of pins are spaced at a distance ranging from 0.290 to 0.310 in. apart, with a typical value of 0.300 in. The pins in each row are separated by a typical on-center distance of 0.100 in., having a range of 0.090 to 0.110 in. The width of each pin is specified as 0.016 in. minimum to a maximum of 0.020 in. Each row has seven pins which run along the edges of the case. When the case is viewed from the top (component view), the *mechanical index point* or *notch* at one end of the case aids in locating pin 1. Those index points may be half-circle depressions in the middle of one end of the case or as a *dot* (full-circle depression) in the top of the case at a corner beside an end pin. The index system and method of pin number arrangement are required because the pin numbers are not shown on the case. To identify pin numbers, the case is viewed from the top with the dot in the upper left corner or the half circle at the top end when viewed vertically. This index orientation places pin 1 at the top of the left row of pins. The pin numbers are then read counterclockwise starting with pin 1 through pin 14. That is, pins 2, 3, 4, and so on, are counted down to pin 7, which is the last pin in the left row. Continuing in a counterclockwise direction, go across to pin 8, the bottom pin in the right row, and continue up to pins 9, 10, and so on, until pin 14, which is the top pin in the right row, is reached.

It can be seen from Fig. 8–3 and from the device description that attempting to draw the pad arrangement and body outline of the TO-116 case style using only a straightedge and circle template can be tedious and time consuming. The use of the device template facilitates quick and accurate work.

To use the template for drawing the TO-116 package, the 14 pad cutouts (left side of Fig. 8–2a) are first located over the correct grid intercepts and traced. See Fig. 8–3c. The template is then shifted to center the body cutout over the pads and the ends of the body, including the index point, and drawn. This is shown in Fig. 8–3d. Notice that the sides of the body outline are not drawn between or through the pads. Only the ends of the outline are drawn to contact the end pads (numbers 1 and 14 and 7 and 8). Lines extending across or between pads would be confused with conductor paths. This two-step operation with the use of the device template automatically sets the appropriate pad size for 0.1 in. (1:1 scale) pin centers and accurately separates each row of pins by the required 0.3 in. In addition, it defines the body outline size and shows the position of the index key.

The design templates shown in Fig. 8–2 provide additional outline cutouts for common layout applications. For example, three different widths and on-center spacings for common gold finger designs used for drawing connector strip pattern layouts for edge connectors are provided. In addition, several axial-style tubular capacitor and ceramic disk capacitor sizes and shapes including pad locations are

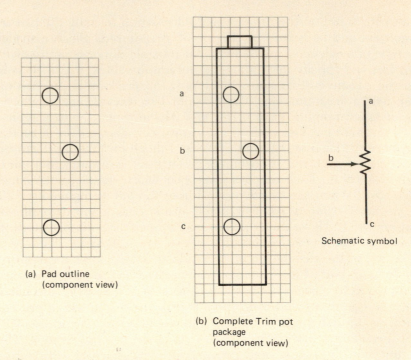

(a) Pad outline
(component view)

(b) Complete Trim pot
package
(component view)

Schematic symbol

FIGURE 8–4 Two-dimensional view of trim potentiometer.

included. Note also that the special elliptical shape for disc capacitors is shown which more closely resembles the top view of this component. However, the rectangular representation shown in Chapter 7 for disc capacitors is acceptable. Observe from the imprinted pad spacing dimensions on the template for disc capacitors that the 1:1 dimensions do not allow both pad centers to be placed on 0.100-in. grid intercepts. These dimensions are for designs that require perfect vertical entry into the board. For this requirement, one of the pads would be placed on grid so as to keep the number of off-grid pads to a minimum. However, where exact vertical lead entry is not a design specification, the template may be shifted slightly to allow both pads to be placed on grid.

Two potentiometers (R101 and R102) are required in our multivibrator circuit. The dimensions of these potentiometers are 0.28 in. wide, 0.32 in. high, and 1.25 in. long. It is designated on the template (Fig. 8–2b) as RT11. Note on the template that a wide variety of potentiometer-style cutouts are provided. The *top* view of these is represented by the template. After the pads have been drawn the cutout of the body outline is centered about the pads and traced. See Fig. 8–4b.

The template of Fig. 8–2a also shows cutouts for the body outlines of the $\frac{1}{8}$-, $\frac{1}{4}$-, $\frac{1}{2}$-, 1-, and 2-W resistors in addition to the DO-7-style diode case. These are found in the lower right corner of the template. Note for the layout of these components that both pads and the body outline of each can be traced without having to move the template. Note also that the on-center dimensions for the pads are printed at a 1:1 scale (1.0, 0.8, 0.6, 0.5, and 0.4), while the cutout sizes of pads and bodies are 2X scale. Observe that the pad spacing for the ½-W resistor is given as 0.6 in. instead of the 0.7-in. spacing that was used in Chapter 7 for this component. The explanation for this dimensional difference will be found in Section 8-11, which deals with lead bend allowance when tighter pad spacing is desirable for improved component positioning and conductor path routing.

There is no cutout on the template for the 1N914 diodes used in the circuit. The pad sizes and spacing and the body outline size will be determined using the procedure outlined in Chapter 7, which makes use of the mechanical drawing details shown in Fig. 8–5a. At 2X scale, the pad size for the 0.020-in.-diameter leads will be 0.200 in. With reference to Eq. 7–1, the on-center spacing of the pads will be $2(0.060 + 0.030) + 0.180$ in. or 0.36 in., which must be increased to 0.40 in. for pad centers to be on grid for a 1:1 scale and to 0.80 in. at 2X scale. The body outline at 2X scale will measure approximately 0.150 in. by 0.40 in. and will be centered between the pads. Thus the component view of the diodes will resemble a rectangle with the addition of a keying bar at the cathode end to ensure correct assembly to the board. This is shown in Fig. 8–5b.

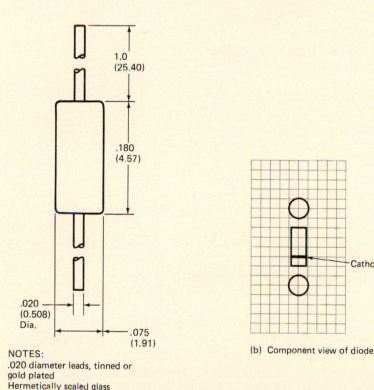

1.0
(25.40)

.180
(4.57)

.020
(0.508)
Dia.

.075
(1.91)

Cathode Coding Bar

(b) Component view of diode

NOTES:
.020 diameter leads, tinned or gold plated
Hermetically sealed glass
Package weight is 0.14 grams

All dimensions in inches (bold) and millimeters (parentheses)

(a) Mechanical detail for the TO-35 style case

FIGURE 8–5 IN914 diode outline.

In concluding this section on the use of PCB layout templates, it is important to mention a word of caution in their use. Even though the cutout to be used is labeled with the component or device to be drawn, it may not have the pad size or lead spacing required by the specifications. For this reason, each cutout to be used should be checked to ensure that all dimensions specified in the design restrictions are satisfied by the tracing techniques employed.

We will make use of templates in describing the layout throughout the remainder of this chapter. As mentioned, a straightedge and circle template will be used to lay out the 1N914 diodes because of the limitations of the template.

8-3

Information Required Prior to Beginning the Design

A great deal of specific information is required by the printed circuit designer before he or she is able to begin the component layout sketch and the finished drawing. This information would normally be provided by the circuit design engineer and begins with a detailed circuit schematic with an accompanying parts list so that all parts are completely identified. See Fig. 8–1. All external connections must be specified as to their number, location, and type. For example, if a finger-type edge connector is specified, the on-center spacing and number of fingers must be designated. In addition, all information relative to any electrical restrictions of specific fingers must also be available.

The physical and mechanical restrictions of the design are also specified by the design engineer. Heat-sensitive components are normally identified so that they may be positioned as far away as possible from components that generate heat. Critical lead lengths (common to high frequency circuits) are specified so that the designer knows that a component placement problem and/or a conductor-width problem may exist in the area of that component. Further, *forbidden areas* of the board are usually clearly outlined. This information tells the designer how close to the board edge components or conductor paths can be placed without running into fabrication problems, yet still allowing sufficient space for mounting all of the parts. Positions of internal board tooling and mounting holes as well as their dimensions are defined in addition to those for interior slots or cutouts.

Finally, if the design specifies a particular size and shape of the finished board, the required detail drawings are also supplied by the design engineer.

It can be seen from the discussion above that a great deal of detailed mechanical and electrical information must be made available to the PCB designer before the design of a component layout drawing can begin. The number and type of restrictions specified determines, to a large extent, the degree of difficulty and the time required to complete the design. In the next section we outline the design restrictions for the multivibrator circuit to be designed.

8-4

The Multivibrator Design Problem

A complete circuit schematic and parts list for the multivibrator circuit is shown in Fig. 8–1. The mechanical parts information for all components and devices is detailed in Section 8–2. We will now specify the design problem, which will include all of the electrical and mechanical restrictions.

DESIGN PROBLEM
Design a single-sided component layout drawing for the circuit shown in Fig. 8–1. The PCB stock is to be type FR-4 with a thickness of 0.059 in. clad on one side with 2 oz/ft^2 copper. The final layout is to be a component view to include all necessary labels and codes. The layout is to be restricted to the area shown in the detail drawing of Fig. 8–6. All parts are to be placed in such a way as to minimize any wasted board space, yet maintain uniform distribution and balance.

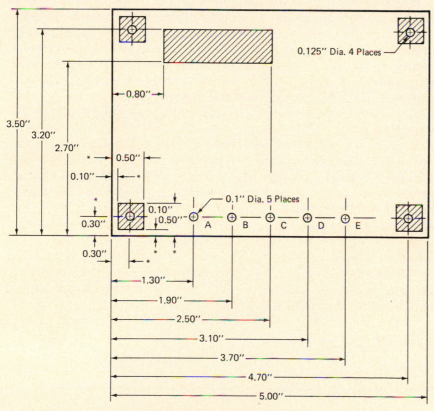

* Dimension at each corner of board

A — INPUT (IN)
B — GROUND (GND)
C — (−15 Vdc)
D — OUTPUT (OUT)
E — (+15 Vdc)

Electrical orientation of external connections

FIGURE 8–6 Dimensional and electrial specifications for multivibrator printed circuit board.

No component body or conductor path is to be placed within a minimum of 0.20 in. from any board edge. The minimum spacing between all electrical conductors (paths and pads) is to be 0.025 in. The position and orientation of all external connections and mounting holes are to be placed in accordance with Fig. 8–6. Pads for external connections are to be made to accept 0.045-in.^{2} wire-wrap pins.

The ground connections are to be divided into three separate groups with an individual path provided for each group to the external ground connection point. This is to be done as follows: (1) the ground points for all of the decoupling capacitors (C1, C2, C5, and C6) will be connected together with a separate return path to the external ground point; (2) all ground points associated with input circuitry (bottom end of R2 and pin 5 of U1) will be connected and returned to the ground point; and (3) all ground points associated with the output circuitry (bottom ends of R8 and R102) will be connected and returned to the ground point.

The dc power supply terminals are to be capable of handling a maximum current of 100 mA.

It can be seen from the statement of the design problem that a number of mechanical and electrical restrictions have been specified. These are more typical of a PCB design than is the simple design presented in Chapter 7. Note that some of the design specifications defined in the layout requirements in Chapter 7 were omitted here. For example, the specific grid system and the reduction scale were not indicated. It is common practice to leave these types of decisions to the designer since he or she must select grid systems and reduction scales which will satisfy the requirements. Again, the conductor path width was not specified. However, the amount of current that the power supply is expected to handle was stated. Remember that the current-carrying capacity of a conductor path is determined mainly by the cross-sectional area of the path (i.e., width and thickness). Since the thickness of the copper foil was specified, the width and spacing of the paths need to be evaluated as well as pad diameters.

Because of the complexity of the multivibrator circuit and the restrictions placed on the design, many of the considerations introduced in Chapter 7 now must be dealt with more thoroughly. In addition, before the layout can begin, some of the specifications must be evaluated as to their impact on the overall design.

8-5

Selection of a Grid System

In Chapter 7 our simple design circuit was laid out accurately and effectively using a suitable grid system as a reference plane upon which to work. The precise placement of component body outlines, the exact positioning of terminal pads, and the accurate routing and spacing of conductor paths were laid out with relative ease.

Standard 0.100-in. (10 × 10 lines per inch) and 0.050-in. (20 × 20 lines per inch) grid systems commonly used for PCB layout drawings are shown in Fig. 8–7. Grid systems are typically printed with line widths of 0.004 in., with wider widths used for accenting. Grid systems with two and even three different line widths on one sheet are available. Accented lines may appear on the 2nd, 4th, 5th, 10th, 20th, or 50th line with a 0.006-in. width. Heavier accents are of a 0.008-in. line width. These accented lines can aid in the layout of a PCB design.

For greater accuracy, the grid system selected should be printed on a stable base material such as glass or polyester film (e.g., Mylar). Although glass grids are the most stable, the electronics industry uses the much less expensive polyester film. These films with printed grids are available in thicknesses of 0.004 and 0.007 in. The 0.007-in. thickness is the most stable in regard to dimensional changes due to variations in temperature and humidity. Typically, this film is stable to ±0.002 in. over a length of 36 in. when maintained in the recommended environment of 70°F and 50% relative humidity. For an increase of 10°F and an accompanying increase of 20% in relative humidity, this film will not increase in length more than 0.015 in. over a distance of 36 in. (See Section 1–1 for a more detailed discussion of dimensional stability.)

Grids printed on polyester film are available in a variety of sizes up to approximately 40 × 60 in. The film is supplied in rolls or in flat form. The rolled type needs to be left flat for several days before using. The grid lines are available in regular black or a photographically nonreproducible blue, brown, or black. If a layout is to be drawn directly onto the surface of the grid sheet, one of the filterable colors is required. However, since a precise grid system on polyester film is relatively expensive, a more economical method is to use it in conjunction with plain film. The grid sheet is first taped to a work surface such as a drafting table or better yet, to a light table. The plain sheet of film, with the mat surface facing upward, is then taped over the grid sheet. Thus the total design drawing can be completed on the less expensive plain film and the grid sheet may be used many times over with this overlay technique. The preliminary sketch need not use a polyester sheet but can be produced on an even less expensive sheet of vellum overlaid onto the grid sheet.

The selection of an appropriate grid system for a specific design problem depends on several factors. Because the grid system is used for the quick and precise location of lead access hole positions for components and devices at the intercept of the grid lines, the system selected should conform to the typical 1:1 scale lead spacings of the components to be laid out. For example, IC DIP packages are manufactured with center-to-center lead spacings of 0.100 in. and row-to-row spacings of multiples of 0.100 in. Single-sided PCB designs incorporating this type of device package should therefore be laid out on grids with spacings of 0.100 in. It would not make sense to use grid spacings of 0.250 in., for example, since this would result in many lead positions not falling on grid. Typical lead spacings of most components at a 1:1 scale are rounded off to the nearest $\frac{1}{10}$ in. Using a finer grid would complicate the rapid location of grid intercepts. The use of a coarser grid results in there not being sufficient grid intercepts with which to work. Finally, since path routing is

0.1 inch grid (10 x 10 lines/inch, heavy accent on 4th line)

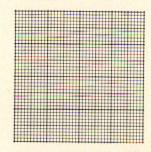

0.05 inch grid (20 x 20 lines/inch, heavy accent on 10th line)

FIGURE 8–7 Typical grid systems used for PCB layouts.

generally not as tight on single-sided layouts as they are for double-sided and multilayer board designs, a 0.100-in. grid is suitable for maintaining reasonable path spacings. For these reasons, we will use a 0.100-in. grid system for the design of the multivibrator circuit.

8-6

Reduction Scale

It was shown in Chapter 7 that the design of a layout drawing is normally constructed to a scale larger than the finished printed circuit board. The taped artwork master, which is produced from the layout drawing, is made at this same enlarged scale. The major reason for working at an enlarged scale in PCB layout drawings is to reduce any error in the manual placement of terminal pads. The manual placement of a pad on a specific grid intercept could be off position by as much as 0.015 in. because of hand and eye coordination. If the layout drawing (and the taped artwork master) were produced at a 1:1 scale, the true positional error of that pad would be the same 0.015 in. When a 2:1 scale is used, however, this error can be reduced by a factor of 2. This is done by photographically reducing the 2:1 taped artwork master to the required 1:1 size for the fabrication of the actual board. This reduction results in the previous 0.015-in. error to be only 0.0075 in.

A further advantage to working with enlarged scales is that the working area is larger and the positioning and drawing of body outlines becomes more manageable. In addition, we will show later that conductor path tapings are easier to route at an enlarged scale.

A realistic and practical scale must be selected that will satisfy the accuracy of specifications and yet can be produced on the available equipment. Too large a scale can create problems in the availability of suitable-size grid sheets or in the photographic equipment that will be used to handle the enlarged layout. To illustrate this point, assume that a final board size is to be 12×16 in. If the scale selected for this design is 4:1, the layout would have to be an overall size of 48×64 in., which is larger than stock-size grid systems. In addition, it would require a very large copy camera to accommodate this large layout.

A 4:1 scale is generally used only for smaller boards with high component density. In this case, this enlarged scale is necessary for the degree of accuracy required in this type of design. The 4:1 scale is discussed in more detail in Chapter 11. For our example layouts in the next several chapters, the boards are not small, nor are the circuits particularly dense. For these reasons, the 2:1 scale will be adequate for our designs.

It is important that the scale to which a PCB layout is made be shown on the drawing and on the taped artwork master. This is to aid the technician in the photographic reduction process. Placement of the reduction scale onto the layout drawing is shown in Fig. 8–8. Two half-circles, called *reduction marks*, are placed outside the board edges with their vertical straight edge facing each other and spaced

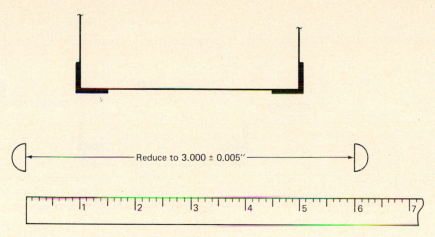

FIGURE 8–8   Reduction scale layout.

the *actual* distance apart. Arrowheads are positioned between the half-circles and point to the vertical edges to indicate this distance. The numerical values placed between the arrows indicates the resulting *reduced* distance. In Fig. 8–8, the actual distance between the half-circles is shown to be 6.000 in. and the indicated reduced distance required is 3.000 in. together with a tolerance of ±0.005 in. The reduction scale is an integral part of the artwork. The spacing between the reduction marks is not necessarily the size of the board outline but is selected to result in whole numbers for ease of camera setup. As large a separation as practical should be used for the sake of production accuracy.

8-7

Board Outlines and Mounting Holes

To represent the edges of the finished printed circuit board in the layout drawing, *corner brackets* are used. For a square or rectangular board, the four corners are shown with the use of these brackets by drawing short lines along right angles to one another at each corner along the outside edges of the board. This is shown in Fig. 8–9a. The corner brackets are used as guides to define the size and shape of the board but are not an integral part of the board. The finished board is profiled (sheared or routed) to the inside edges of the brackets. For this reason, the size of the brackets is not critical. Typically, the bracket lengths, labeled x in Fig. 8–9a, are drawn between 0.5 and 1.0 in. long. The outside line that forms the width (w) should be placed on an adjacent grid line for ease of drawing. Note in Fig. 8–9a that the ends of the corner brackets are closed with short line segments drawn perpendicular to the boards edge.

The outside edges of a finger-type extension (tongue), along one side of a board, are represented as shown in Fig. 8–9b. Note that corner brackets are used to define both the finished width and depth of the tongue. The additional depth below the corner brackets and

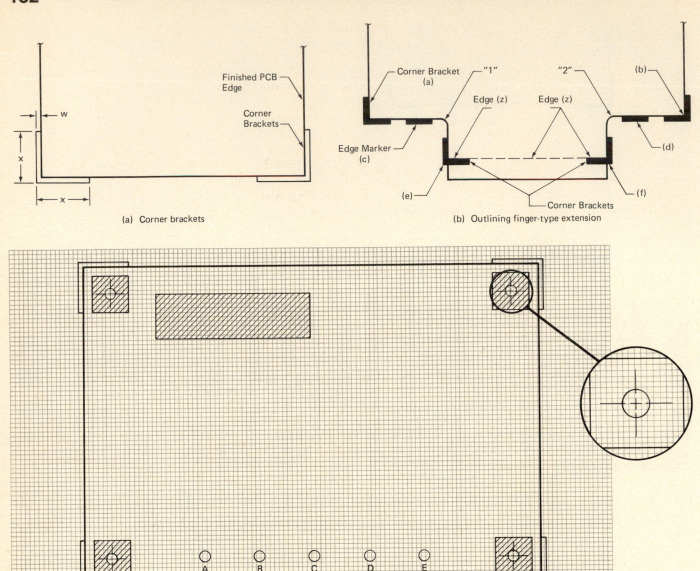

(a) Corner brackets

(b) Outlining finger-type extension

(c) 2:1 scale board outline of multivibrator PCB

FIGURE 8–9 Initial board outline configurations.

dashed line is for manufacturing and is removed after the finger section has been processed. The corner brackets in Fig. 8–9b have been shaded for emphasis only. This shading is not normal practice on layouts. The brackets labeled a and b show the outside edges of the major board. Two *edge markers*, labeled c and d, are added along the side of the board from which the tongue extends to highlight the finished shape of this edge. The corner brackets, labeled e and f, define the finished depth and width of the tongue. The finished depth is shown by the dashed line and the inside edges of the corner brackets, labeled z. Note that the inside corners of the board, shown as points 1 and 2, are to be rounded. Note, also, that neither the edge markers (c and d) nor the ends of the corner brackets (e and f) extend into either radius corner. If they did extend into the corners, this would indicate a square inside corner rather than the rounded ones shown.

A 2:1 scale representation of the board outline using the dimensional drawing information of Fig. 8–6 for the multivibrator circuit design is shown in Fig. 8–9c. The board outline is defined with corner brackets on a 0.100-in. grid system with the correct reduction symbols and dimensions. All of the forbidden areas have been added and drawn to the 2:1 scale.

Note in Fig. 8–9c that four 0.025-in-diameter mounting holes are shown with the use of *targets*. A target is shown in the blow-out. This is the conventional symbol used for locating the center of holes which will not be used as part of the circuit, that is, holes that are not to be plated or where no component lead will be inserted.

8-8

Determining Conductor Width

Before the selection of the appropriate conductor width for a specific layout can be made, several design considerations need to be evaluated. The current-carrying capacity of a conductor path is dependent on its cross-sectional area (i.e., its width times its thickness). With ever-growing circuit complexity, the design trend is toward extremely high density packaging. This results in the demand for tighter conductor patterns and component positioning.

Where space allows, wider conductor paths than those required to handle the circuit current should be selected. As conductor widths and the spacings between them are reduced to their absolute minimum, board processing becomes more critical. Reduced yields of boards having no rejection defects after fabrication will result, causing higher manufacturing costs. Minimum allowable conductor width and/or spacing appearing anywhere on the board creates these fabrication problems. The weakest-link theory applies here. That is, high reliable-board yield is jeopardized even if maximum conductor widths and spacings are used on 99% of the board and absolute minimum values on only 1% of the board. It is therefore essential that minimum conductor widths and spacings not be used as a trade-off for higher density and should be avoided.

Conductors may be grouped into two categories, each described by their circuit function. Conductors that make interconnections between the components and devices mounted onto the board are referred to as *signal paths* and typically handle currents in the milliampere ($mA = 10^{-3}$ A), microampere ($\mu A = 10^{-6}$ A), and picoampere ($pA = 10^{-12}$ A) ranges. Because of the small currents that signal paths are required to deal with, their minimum widths are determined only by the board manufacturer's equipment capabilities, which is typically down to 0.010 in. (10 mils). The second category of conductors, called *power* and *ground* paths or *buses*, are normally required to handle larger currents in the range of 100 mA to tens of amperes, depending on the circuit application.

As will be shown, a 10-mil (1 mil = 0.001 in.) conductor width is capable of handling approximately 1.3 A of current with a copper thickness of 2 oz/ft^2 (0.0028 in.) and 0.75 A for a copper thickness of 1 oz/ft^2 (0.0014 in.), each with a maximum temperature rise above ambient of 10°C. By comparing these currents with the relative order of magnitude given for signal paths, it can be seen that the typical 10-mil minimum conductor width, when used for signal paths, can handle much larger currents than normally expected. This emphasizes the statement that signal path conductor width is determined more by the manufacturer's capabilities than by its current-carrying requirements. Conductor width, then, must be selected primarily by the current it is expected to handle, without jeopardizing reliability of manufacturing. Even though PCB manufacturers are capable of fabricating 8-mil conductor widths and even smaller, reduced yields of acceptable boards result in volume production. For this reason, the 10-mil signal path width is an acceptable compromise for volume production, resulting in reduced costs per board.

Before the PCB designer can determine the required width for power and ground bus conductors, specifications as to the maximum current that each is expected to handle must be provided. In addition, the designer needs to know the specified temperature rise, above ambient, of the conductor when it is carrying this maximum current. The circuit design engineer provides this information in the detailed specifications that accompany the schematic. With this information, the PCB designer may determine the minimum conductor width from the graph shown in Fig. 8–10. To best illustrate the use of this graph, we will use the following example. Assume that power and ground bus paths are specified to handle a maximum of 7 A. In addition, these conductors should not increase in temperature more than 10°C beyond the ambient temperature when carrying this maximum current. Assume also that the PCB specified has a copper thickness of 2 oz/ft^2. The top of the graph of Fig. 8–10 shows various characteristic curves of temperature rise above ambient temperature versus current for the conductor. The temperature rise is expressed in degrees Celsius for various currents expressed in amperes. Using the current scale on the left side of the graph, locate the 7-A point. This is shown as point 1 on the graph. Project a horizontal line until it intercepts with the 10°C temperature-rise curve (point 2). Next, project a vertical line from point 2 downward, passing completely through the bottom graph. This lower graph shows the characteristic curves of various conductor thicknesses expressed in oz/ft^2 of copper (on the finished board) as a function of conductor widths in mils versus cross-sectional area in square mils. Locate the intercept point of the vertical line drawn from point 2 to the lower graph's characteristic curve for the specified conductor thickness. In our example, the vertical line intercepts the 2-oz/ft^2 curve at point 3. Finally, a horizontal line is projected from point 3 to the vertical axis of this graph. The intercept of this line and the vertical axis determines the re-

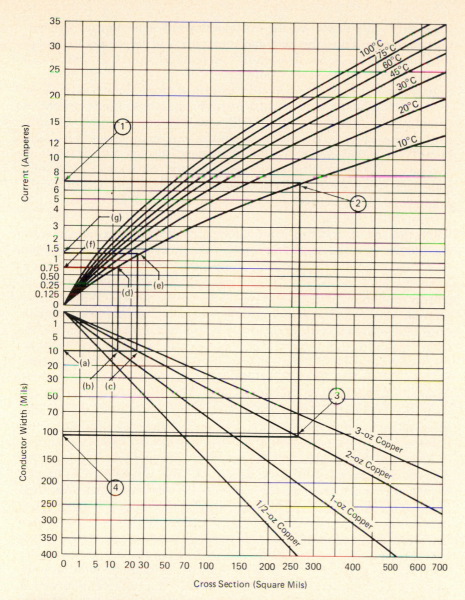

FIGURE 8–10 Current capacity of PCB copper as a function of conductor thickness, width, and temperature.

quired minimum conductor width. From Fig. 8–10, this is read as 105 mils (0.105 in.).

The graph of Fig. 8–10 can also be used in reverse, that is, for determining the current-carrying capacity for a given conductor width. To illustrate this, recall that it was mentioned that a 10-mil conductor width of 2 and 1 oz/ft^2 copper thickness will handle 1.3 and 0.75 A, respectively, with a maximum of a 10°C temperature increase. We will use these data to show how the graph is used in reverse order from that just described to arrive at these figures. A horizontal line is drawn on the bottom graph from the 10-mil-conductor-width point (a) and projects through the 1- and 2-oz/ft^2 curves at points (b) and (c). From these two intercepts, vertical lines are projected upward to the 10°C temperature curve at points (d) and (e). Horizontal lines are finally drawn to the vertical current scale to find the current readings. Thus the resultant currents will be found to be approximately 0.75 A for the 1-oz/ft^2 projections, point (f), and 1.3 A for the 2-oz/ft^2 projections, point (g).

It is general practice to avoid the use of more than two or three different conductor path widths in a given layout. For example, it would not be practical to use 0.040- and 0.050-in. conductor widths in the same layout. As long as minimum conductor spacing is maintained, the 0.050-in. width should be used throughout. The use of several conductor widths in the same layout can lead to confusion, especially when the conductor taping is done by hand. It is difficult, if not impossible, to distinguish tape widths that differ by only a few thousandths of an inch.

On the layout drawing, conductor paths are drawn as single standard-width pencil lines. Recall from Chapter 7 that the color blue was recommended for drawing these paths as seen from the component side of the board. This single-color selection implies that all of the conductor paths are to be of the same width. Where more than one path width is to be used on the same drawing, a more elaborate color-coding system is required in order to distinguish between paths of different widths. One common method is to draw conductor paths having the same width with the same color and providing a legend on the layout drawing to indicate which color represents a specific width. This type of legend is shown in Fig. 8–11. Note that it is not necessary to include the written word on the line of an actual drawing but simply to draw the line using the indicated color.

From the specifications of our multivibrator circuit, the maximum current that the power supply or ground path is expected to handle is 100 mA. Since the signal paths will be carrying much less than 100 mA, no conductor in the system will be required to carry more current than can easily be handled by a 1:1 scale path width of 0.010 in. However, in conformance with standard practice, a 0.050-in. width (1X scale) for all power and ground paths from the external connections will be represented. In addition, the signal paths will be indicated as a 0.20-in. width (1X scale). Because the multivibrator circuit is not a highly complex design, severe conductor spacing problems will not be encountered to justify the use of path widths narrower than 0.020 in.

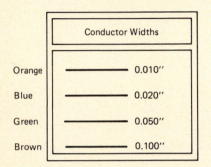

FIGURE 8–11   Conductor path width color code legend.

8-9

Determining Conductor Path Spacing

The insulation spacing that exists between adjacent conductor paths, terminal pads, or between conductor paths and pads is called the *air gap*. Examples of air gaps are shown in Fig. 8–12. The minimum air gap allowed on a finished PCB depends on (1) manufacturing limitations, (2) maximum specified peak voltages between conductors, (3) whether the conductors are to be uncoated or coated with a protective coating, (4) the altitude at which the system will be operated, and (5) whether the air gap is on the internal or external surface of

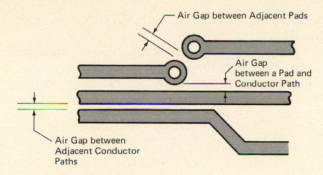

FIGURE 8–12 **PCB insulation spacings.**

a multilayer board. Most of this information is listed in Table 8–1. It should be pointed out that the recommended spacings shown in the table are conservative values. Many electronic industries that are not engaged in military contracts tend to reduce these distances, as can be done safely, in order to increase the circuit density on the boards. In today's technology, boards with conductor spacings of 0.010 in. are routinely fabricated. Spacings of 0.008 and even 0.006 in. with a controlled yield are not uncommon. The PCB designer should use the recommended minimum spacings listed in Table 8–1, however,

TABLE 8–1   Spacing Specifications and Standards

Voltage Between Conductors DC or AC Peak (V)	Uncoated 0–10,000 ft. Alt.*	Uncoated Above 10,000 ft. Alt.*	Coated† and Internal Layers‡	Voltage Between Conductors DC or AC Peak (V)
0			0.005 in.	0
15			(0.13 mm)	15
16	0.015 in.	0.025 in.	0.010 in.	16
30	(0.38 mm)	(0.64 mm)	(0.25 mm)	30
31			0.015 in.	31
50			(0.38 mm)	50
51		0.060 in.	0.020 in.	51
100	0.025 in.	(1.52 mm)	(0.51 mm)	100
101	(0.64 mm)			101
150		0.125 in.		150
151		(3.18 mm)		151
170			0.030 in.	170
171	0.050 in.	0.250 in.	(0.76 mm)	171
250	(1.27 mm)	(6.35 mm)		250
251				251
300		0.500 in.		300
301	0.100 in.	(12.70 mm)	0.060	301
500	(2.54 mm)		(1.52 mm)	500
500+	0.0002 in./V (0.0051 mm/V)	0.0010 in./V (0.0030 mm/V)	0.00012 in./V (0.00305 mm/V)	500+

*IPC-ML-910A.

†Coating per MIL-I-46058.

‡MIL-STD-275D and IPC-ML-910A.

Source: Data from Military Standards MIL-STD-275C and MIL-STD-1495.

unless these have been relaxed by the design engineer or by company standards.

For the multivibrator design, a minimum conductor spacing of 0.025 in. will be selected since the level of component density in this circuit does not demand a smaller air gap. The general rule should be to provide as wide a space above the minimum as the circuit density will allow.

8-10

Determining Pad Diameters for Single-Sided Designs

The criteria used for the selection of the correct terminal pad diameter for a specific component lead were discussed in some detail in Section 7–4. We will quickly review this selection process and then expand on it for purposes of developing a table to be used in determining the practical minimum pad size for a given lead diameter.

Recall from Section 7–4 that pad size selection begins by obtaining the maximum lead diameter of each component from actual measurement or from manufacturers' data sheets. See Fig. 8–13. The drill hole clearance for easy lead insertion should be 0.005 to 0.010 in. Therefore, the drill hole diameter should be typically 0.010 to 0.020 in. larger than the maximum lead diameter. The minimum pad diameter for *unsupported* (unplated) holes should be such that the width of the copper remaining after the pad has been drilled is approximately 0.020 in. Thus the miminum pad diameter must be at least 40 mils larger in diameter than that of the drill hole. See Fig. 8–13.

For single-sided boards, the general equation that may be used for the calculation of the minimum pad diameter is

$$\text{pad diameter (min.)} = \text{lead diameter (max.)} + 2(\text{drill clearance}) + 2(\text{drilled pad width}) \tag{8-1}$$

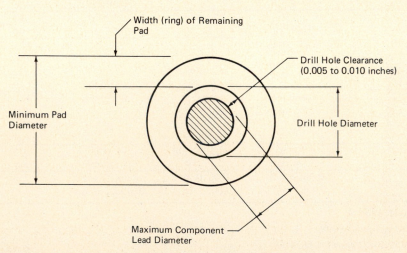

FIGURE 8–13 **PCB pad and clearance nomenclature.**

TABLE 8–2 Guide to Pad Diameter Selection Versus Various Lead Diameters and Drill Hole Clearances

Maximum Component Lead Diameter (in.)		Actual Pad Diameter (in.)		
Drill Hole Clearance, 10 mils (0.010 in.)	Drill Hole Clearance, 5 mils (0.005 in.)	1:1 Scale	2:1 Scale	4:1 Scale
N/A*	N/A*	0.050	0.100	0.200
N/A*	0.012	0.062	0.125	0.250
0.010	0.020	0.070	0.140	0.280
0.015	0.025	0.075	0.150	0.300
0.020	0.030	0.080	0.160	0.320
0.025	0.035	0.085	0.170	0.340
0.030	0.040	0.090	0.180	0.360
0.033	0.043	0.093	0.187	0.375
0.040	0.050	0.100	0.200	0.400

*N/A, not applicable.

With the use of Equation 8–1 and typical pad diameter sizes that are available from printed circuit graphic arts manufacturers, Table 8–2 was constructed. A comparison of the dimensional information given in this table reveals two considerations that deal with aiding in obtaining higher board density as the result of reducing the drill hole clearance for a given lead diameter. Let us take a 0.040-in. lead diameter as an example. If we adhere to the 20-mil copper width and use a 10-mil hole clearance, a 0.100-in. pad diameter would be required. However, if we reduce the clearance hole to 5 mils, a 0.090-in. pad diameter would suffice. Again, you will note that a 0.100-in.-diameter pad is required for a 0.040-in.-diameter lead with a 10-mil clearance, but this same pad will support a larger lead (0.050 in.) using a 5-mil clearance. Thus it is seen from Table 8–2 that by reducing the size of the drill hole clearance, a pad size may be reduced *or* a larger-diameter lead may be used with the same-size pad. Table 8–2 also provides the pad diameters for 2:1 and 4:1 scaled layouts.

We will now construct a working table listing lead diameters, clearance hole diameters, closest available drill bit sizes, and pad diameters for the multivibrator circuit of Fig. 8–1. Refer first to the *theoretical* table shown in Table 8–3a. Note that each of the components and the terminals are listed together with their lead diameters. In this table, an ideal drill size is given which would provide a 10-mil drill hole clearance. This drill hole size is found by simply adding 0.020 in. to the lead diameter to result in a 10-mil clearance. To find the closest available bit, we refer to Appendix XIX. Finally, with the use of Equation 8–1 or Table 8–2, the minimum pad diameter for each lead size is obtained. The use of the equation does not lend itself to the determination of the pad sizes for the 741 op-amp, however. Printed circuit graphic arts manufacturers provide a wide variety of sizes and shapes of IC pad patterns. For the 741 op-amp the 14-pin arrangement having 80 × 100 mil rectangular pads is recommended for a 0.040-in. drill hole.

With Table 8–3a completed, it is now possible to fabricate the PCB.

TABLE 8–3 Tables Used to Determine the Most Practical Drill Hole, Drill Bit, and Pad Sizes for PCB Designs

(a) Theoretical Values

Component or Device	(Maximum) Lead Diameter (in.)	Ideal Drill Hole Size (in.) (10-mil Clearance)	Closest Available Drill Bit	Selected Pad Diameter (in.)	
				1:1 Scale	2:1 Scale
½-W resistor	0.032	0.052	No. 55	0.093	0.187
Disc capacitors	0.025	0.045	1.15 mm	0.085	0.170
10-kΩ trim pot	0.025	0.045	1.15 mm	0.085	0.170
IN914 diode (DO-35)	0.020	0.040	No. 60	0.080	0.160
741 0P-amp (TO-116)	0.020	0.040	No. 60	0.080 × 0.100†	0.160 × 0.200
0.045-in. wire-wrap stakes	0.064 (diagonal)	N/A*	No. 53‡	0.100	0.200

*N/A, not applicable.

†Rectangular IC pattern with 0.080 in. × 0.100 in. pads.

‡Drill hole for wire-wrap stakes are chosen at 0.004-in. diameter smaller than diagonal value.

(b) Practical Values

Component or Device	(Maximum) Lead Diameter (in.)	Selected Drill Hole Size (in.) (≈5-mil Clearance)	Closest Available Drill Bit	Selected Pad Diameter (in.)	
				1:1 Scale	2:1 Scale
½-W resistors	0.032	0.042	No. 58	0.093	0.187
Disc capacitors	0.025	0.035	No. 65	0.093	0.187
10-kΩ trim pot	0.025	0.035	No. 65†	0.093	0.187
IN914 diode (DO-35)	0.020	0.030†	No. 65†	0.093	0.187
741 0P-amp (TO-116)	0.020	0.030†	No. 65†	0.070 × 0.125‡	0.140 × 0.250
0.045-in. wire-wrap stakes	0.064 (diagonal)	N/A*	No. 53§	0.100	0.200

*N/A, not applicable.

†No. 65 drill bit diameter = 0.035 in. (results in 7.5-mil clearance for 0.020-in. lead diameters).

‡Cut pad pattern: 0.070 in. wide by 0.125 in. long.

§Drill hole for wire-wrap stakes are chosen at 0.004-in. diameter smaller than diagonal value.

However, note that four different drill sizes are specified (No. 55, 1.15 mm, No. 60, and No. 53). Since more drill bit changes result in higher manufacturing costs, the PCB designer should, wherever possible, reduce the number of drill bits required by combining those whose sizes do not differ greatly. Remembering that the recommended clearance for a lead is between 5 and 10 mils, we will reduce the drill hole sizes listed in Table 8–3a having 10-mil clearances to provide a 5-mil clearance. These reduced hole sizes are listed in Table 8–3b, which is designated as the *practical* table. The closest available drill sizes are again selected for this table. Comparing the two tables, you will observe that a No. 58 drill bit (0.042 in. diameter) will be used for the resistor leads instead of the larger No. 55 bit. Also, a No. 65 drill bit will be used for all the other component leads. Appendix XIX tells us that the closest drill bit size for the diode and op-amp leads on the practical table is 0.75 mm. However, by using a No. 65 bit for these leads, the clearance will be 7.5 mils, which is again within the recommended limits. The result of this exercise is that we have reduced the number of required drill bits from *four* in the theoretical table to *two* that will satisfy the drill hole sizes for all of the components and devices. Note also that with the reduced hole size for the op-amp leads, a thinner pad style is specified in the practical table (0.70 × 0.125 in.). This arrangement will be the *cut pad* style, which is shown in Fig. 8–3c and d.

The next practical consideration will be to reduce the number of pad sizes to minimize the inventory and to simplify the coding system. Note in Table 8–3b that the 0.085-in.-diameter pads for the disc capacitors and trim pots and the 0.080-in. pads for the diodes have all been increased to 0.093 in. The result of this is that there will be a wider ring of copper remaining after the hole is drilled with negligible impact on reducing the available space for conductor routing. It can be seen that the original *four* different pad sizes have been reduced to *two* (0.093 and 0.100 in.) along with the IC pattern. The column at the extreme right of each table lists the closest available 2X pad sizes.

A final consideration is the drill hole and pad size diameters for the wire-wrap stakes to be used for the external connections. These stakes are square in shape and measure 0.045 × 0.045 in. See Fig. 8–14a. An end view of the stake pressed into a hole is shown in Fig. 8–14b. Because the square stake is to be press-fitted into a round hole, the hole diameter must be smaller than the diagonal dimension of the stake by a few thousandths of an inch. This is the reason for the entry N/A in Table 8–3a and b for drill hole size. The relationship between the hole size formed by a No. 53 drill bit and the 0.064-in. diagonal dimension of the stake is shown in Fig. 8–14c. Finally, for the pad diameter to be 0.040 in. larger than the drill hole (0.060 in.), a 0.100-in. pad is required.

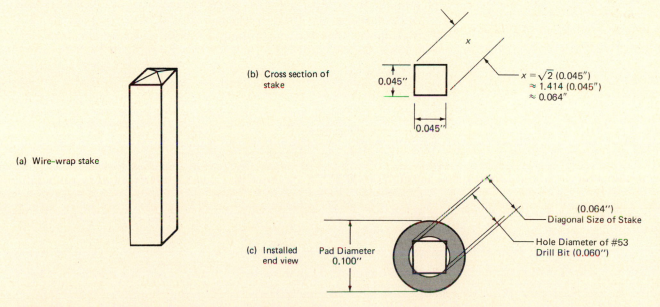

FIGURE 8–14   Hole and pad sizes for wire-wrap stakes.

8-11

Lead Bend Allowance for Single-Sided Designs

Recall that the determination of the proper point at which to bend the leads of an axial component for locating the center-to-center spacing of pads was introduced in Section 7–4. This information, together with manufacturers specifications, will now be used to construct a table listing the size of each pad and the required spacing for each component in the multivibrator circuit. Refer to Fig. 8–15, which shows an axial component with the proper lead bend for insertion into a PCB. The pad center, with respect to the end of the component body, is specified as distance X, which is the sum of the lead length labeled L and the radius of the bend R. In equation form, this is

$$X = L + R \qquad (8-2)$$

Note that this equation is a slight variation of the distance X shown in Fig. 5–6. In order to reduce valuable board space by positioning a larger number of parts closer together, the distance between pads for a component needs to be made as short as possible. For this purpose, Eq. 7–1 will be rewritten in Equation 8–3 to include the L and R dimensions in the calculation of on-center pad spacing S.

$$S = B + 2(L + R) \qquad (8-3)$$

Where tight board space is not a consideration, the guidelines discussed in Section 7–4 are appropriate for determining on-center pad spacing.

A table will now be constructed to more accurately determine minimum bend distance (X) for a range of lead diameters when tighter component spacing is a requirement. See Table 8–4. The previously determined on-center pad spacing for the diodes and resistors will be recalculated to see if a reduction in this distance is possible. Table 8–4 shows that the recommended radius (R) for the 0.020-in.

TABLE 8–4 Minimum Bend Distances for Axial Lead Components

Range of Lead Diameters, d (in.)	L_{min} (in.)	R_{min} (in.)	X_{min} (in.)
0.015–0.019	0.030	0.020	0.050
0.020–0.029	0.030	0.045	0.075
0.030–0.039	0.030*	0.060	0.090
0.040–0.050	0.060	0.060	0.120
>0.050	0.120	$2d$	$0.120 + 2d$

*0.060 in. preferred, but 0.030 in. is acceptable if required for tighter spacing.

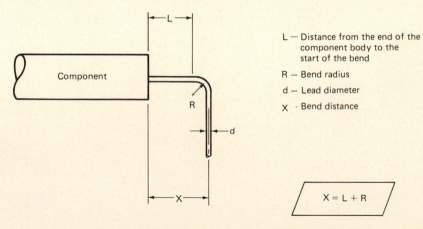

L — Distance from the end of the component body to the start of the bend

R — Bend radius

d — Lead diameter

X - Bend distance

$$X = L + R$$

FIGURE 8–15 Formula to determine component lead spacing.

TABLE 8–5 Tabulated Dimensional Information for Multivibrator Parts

Component	Body Width (in.)	Body Length, B_{max} (in.)	Lead Diameter, d (in.)	$X_{min} = L_{min} + R_{min}$ (in.)	$S = B_{max} + 2(L_{max} + R_{max})$ (Eq. 8.3) (in.)	Practical Value for S (in.)		Pad Diameter (in.)	
						1:1 Scale	2:1 Scale	1:1 Scale	2:1 Scale
Axial									
½-W resistors	0.160	0.416	0.032	0.090	0.596	0.6	1.2	0.093	0.187
IN914 diode (DO-35)	0.075	0.180	0.020	0.075	0.330	0.4	0.8	0.093	0.187
Radials									
Disc capacitors	0.150	0.400	0.025	N/A*	0.25	0.3	0.6	0.093	0.187
10-kΩ potentiometers	0.250	0.750	0.025	N/A*	Staggered (Fig. 8.4)	Same	×2	0.093	0.187
Device:									
741 op-amp (TO-116)	0.250	0.785	0.020	N/A*	0.10 × 0.30 (Fig. 8.3)	Same	×2	0.070 × 0.125	0.14 × 0.250

*N/A, not applicable.

diode lead is 0.045 in. Using Equation 8–3 to calculate the pad spacing, we find that $S = 0.18 + 2(0.030 + 0.045)$ or 0.330 in. To be on grid, this value is increased to 0.040 in. Because this was the spacing previously determined for the diodes, it is not possible to reduce this value.

We will now repeat the process above for the ½-W resistors having a lead diameter of 0.032 in. Thus $S = 0.416 + 2(0.030 + 0.060)$ or 0.596 in. This is increased to 0.60 in. at 1X scale or 1.2 in. at 2X scale. Recall that the values of S calculated previously were 0.70 and 1.4 in. Thus the on-center spacing can be reduced by 0.20 in. at 2X scale. This reduction was brought about because, instead of using the preferred distance of 0.060 in. for dimension L, we used 0.030 in., which is the minimum allowed and used only where a tightly positioned component layout is required. Note also that 1.2 in. is the pad spacing on the 2X template for a ½-W resistor.

All of the dimensional information discussed thus far for the components and devices to be used in the multivibrator circuit has been brought together and shown in Table 8–5. We will refer to this table as the layout of the multivibrator continues in the following section.

8-12

Component Layout Drawing of the Multivibrator Circuit

With all of the dimensional information for all of the parts used in the multivibrator circuit calculated and tabulated, the PCB layout drawing may begin. A 0.100-in. polyester grid sheet is first taped to a light table. A sheet of high-quality drafting vellum is placed over the grid sheet and taped in place. Begin the layout by first drawing the board outline from the dimensional drawing of Fig. 8–6 to a 2:1 scale. The following are then drawn onto the layout: corner brackets, targets for the four 0.125-in.-diameter mounting holes, and five external connection pads using a 0.200-in.-diameter pad (from Table 8–3b) for each. Referring again to Fig. 8–6, crosshatched lines are

Layout Legend				
Symbol	Pad Diameter (1:1 scale)	Final Hole Size	Tool Bit	Remarks
● *	0.093″	0.042″	#58	Blue
○	0.093″	0.035″	#65	No color
⊘ +	0.100″	0.060″	#53	Green
⬭	0.070″ × 0.125″ Oval	0.035″	#65	For 14 pin DIP Package (only IC's)

Conductor Width	Color	Remarks
0.020″ (20 mils)	Draw Blue	Signal Paths
0.050″ (50 mils)	Draw Green	Power Supply (voltage and ground)

* ● Color blue to represent #58 drill size

\+ ⊘ Color green to represent #53 drill size

FIGURE 8–16   Layout legend to be included with Figure 8–9c.

used to draw in the forbidden areas around each mounting hole and the rectangular area in the upper left corner of the board. The appropriate reduction scale is positioned below the board outline. To complete the initial phase of the layout, a *layout legend* is constructed to the right of the board outline using the dimensional information for components and pads from Table 8–3b and from the discussion on conductor widths in Section 8–8. The layout to this point would appear as the 2:1 scale drawing of Fig. 8–9c but accompanied by a layout legend such as that shown in Fig. 8–16. Note that the legend provides separate tables for the pad and conductor information. Also, by using the color-coding techniques shown in Fig. 8–11, the drill size for individual pads, the conductor widths, signal paths, power supply paths, and ground paths can all be coded on one drawing.

For the pad sizes listed, the legend provides information as to their shape or diameter and the drill hole size. Recall from Table 8–3b that the 0.093-in. pads would be drilled with a No. 65 bit for the diodes, disc capacitors, and potentiometers and with a No. 58 bit for the resistors. The legend shows that the holes drilled with a No. 58

bit will be shaded in with the color blue and those drilled with a No. 65 bit will be left uncolored.

Two different path widths are differentiated on the legend. The signal paths will be 0.020 in. wide (1:1 scale) and drawn in *blue*. The power supply and ground paths will be 0.050 in. wide and drawn in *green*. The use of a color-coding system of this type aids in producing the taped artwork master.

The order of drawing sections of the sketch is left to the designer. The peripheral information, such as corner brackets, reduction scale, and legend, could be added after the component placement and conductor routing have been completed.

With the board outline and forbidden areas drawn, the parts placement on the sketch may begin. Using the schematic diagram of Fig. 8–1 and Table 8–5, a logical approach to positioning must first be planned. For the beginning designer, it may be helpful for this more complex circuit to use a separate sheet of 0.10-in. grid paper and a printed circuit template to sketch out roughly the approximate component positions. As pointed out in Section 7–5, positioning components as they appear on the circuit schematic is a convenient starting point. Inspection of the schematic (Fig. 8–1) shows that it can be divided into two major groups, with one IC device dominating each group. For purposes of identification, we will refer to group 1 as all those parts to the *left* of the coupling capacitor C3 and group 2 as those parts to the *right* of C3.

Recall from Fig. 7–9 that parts placement for that layout resulted in uniformly spaced columns of components with varying degrees of alignment along adjacent columns. We will apply the same technique of parts positioning here. In group 1, it is seen that five parts could be positioned to the left of U1. Using the technique of vertical positioning, three columns of components can be formed. The first column will be made up of R1 and R2, the second just to include potentiometer R101 since it is physically much larger than the $\frac{1}{2}$-W resistors, and the third column to consist of R3 and R4. R3 will be placed below R4 since its position is lower in the schematic. The IC device U1 is then placed in a vertical position, forming a fourth column. It is common practice to place all decoupling capacitors (C1, C2, C5, and C6) as close as possible to the ends of the IC to which they are associated. For example, C1 will be placed at the top of U1 and C2 at the bottom of U1. In addition, these capacitors should also be placed in a horizontal position to avoid blocking any path routing through the center of the device outline.

We will place the coupling capacitor C3 into group 2. It can be seen that at least five components can be placed in vertical columns to the left of U2. These are C3, CR3, R5, CR1, and C4. It initially may appear logical to also consider placing CR2 and R6 to the left of U2. Notice, however, that the location of the combinations CR2 with R6 and CR1 with C4 appear at the top of the schematic, where there is a fairly large usable area in the same relative position of the board. For this reason, these four components will be placed on the board in the same positions as they appear on the schematic. This reduces the number of parts to be placed to the left of U2 to three. To maintain a symmetry of parts with those associated with aU2, R8 will be placed below CR3. Thus, in group 2, the first vertical column of parts to the right of U1 will be made up of C3 and R5 and the next column will be comprised of CR3 and R8. Note that to this point, starting from the extreme left of the board, six columns have been formed. The seventh column will be U2 positioned vertically with C5 placed at one end and C6 at the other. The eighth column will consist of CR4 and R7, while the ninth and last column will be formed by potentiometer R102 from which the output is taken. The resulting sketch generated from this planning process is shown in Fig. 8–17. Note that no border outline or peripheral detail is shown. Only the relative positions of the parts is drawn, to allow the designer to make a determination of an effective layout before drawing the exact component locations on the vellum.

The approximate sketch just described with the use of the template should be drawn lightly so that erasures may be made easily if a slight repositioning of parts is required to result in an improved layout.

After the designer is satisfied that the initial planning sketch is satisfactory, the component layout is drawn within the board outline of Fig. 8–9c. This is shown in Fig. 8–18. Note that the nine columns of parts have been uniformly distributed inside the outline. Their central placement allows sufficient space for conductor paths to be routed along the top, bottom, and sides of the boards in addition to routing them between the columns.

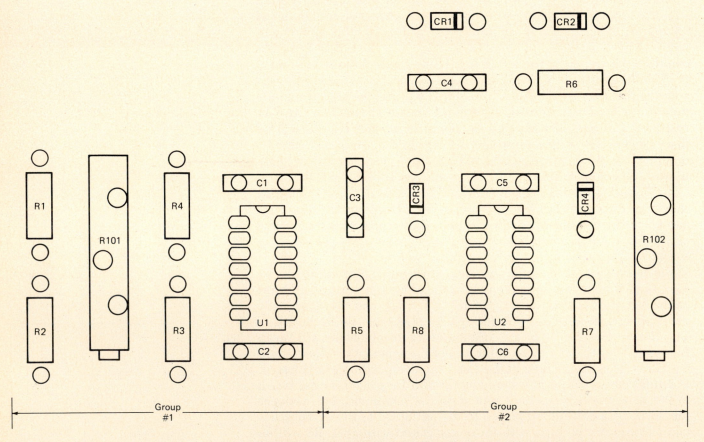

FIGURE 8–17 Initial parts placement for the multivibrator circuit.

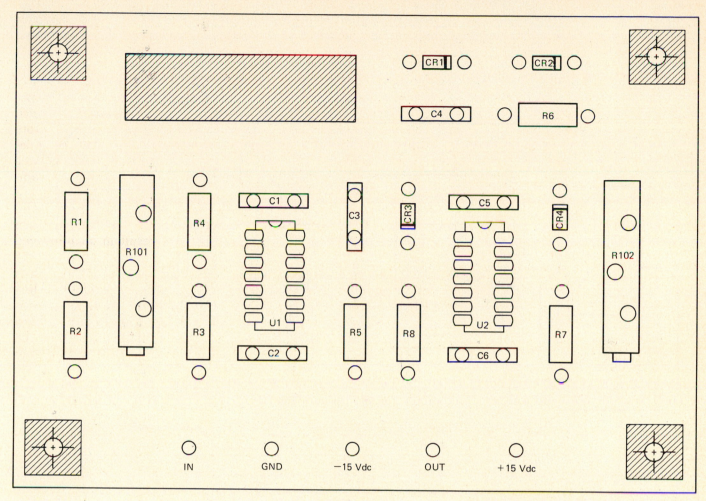

FIGURE 8–18 Parts placement with board outline of Figure 8–9c added.

With the completion of the parts placement, conductor path routine will be done in two phases. The first phase is to draw all of the signal paths and the second phase will be the drawing of the power supply and ground paths.

As was described in Section 7–6, the drawing of signal path interconnections starts with the shortest paths being drawn first. Remember to check off each connection on the schmatic as it is drawn on the layout drawing. See Fig. 7–12. Again, recall from Section 7–6 that paths are routed along vertical and horizontal grid lines, changing path directions at 45-degree angles and also keeping all air gaps as large as possible with none less than 0.025 in. Following these guidelines, all of the signal paths are completed on the drawing. You will notice that as each successive path is added, a maze develops which results in it becoming more challenging to route paths from one pad to another *without crossing* other paths or violating the minimum air gap spacing. Remember that drawing paths through body outlines is permissible.

It is common practice to designate the first external connection pad in the bottom left side of the board as the signal input connection. The circuit schematic dictates that a path from this pad to the lower end of R1 must be provided.

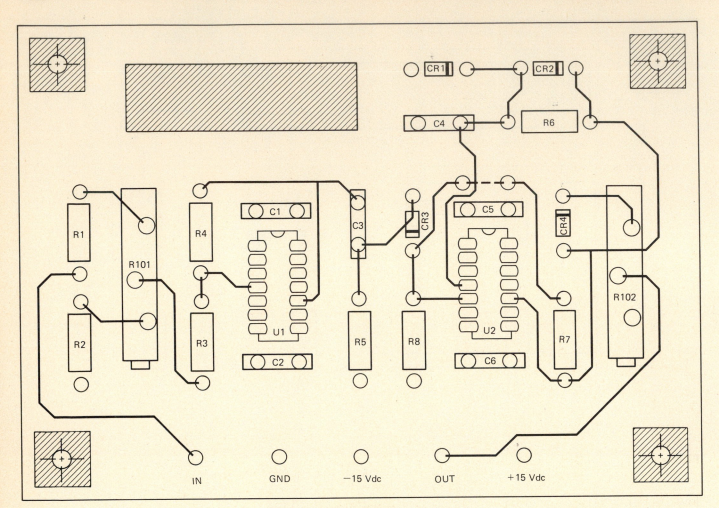

FIGURE 8–19   Signal path routing.

The layout drawn to this point, including all of the signal paths, is shown in Fig. 8–19. Note that two additional pads have been placed above C5 and connected by a dashed line. This is the symbol used when a *jumper wire* is required. See Fig. 8–20a. This jumper is conveniently placed above C5 to allow signal paths to be routed without crossing (in this case, path C4 to pin 4 of U2 and path CR3 cathode to R7). On the actual board, the jumper is formed into a bridge using a short length of solid bus wire which is installed into its own pair of pads as shown in Fig. 18–20b (cross-sectional view) and 18–20c (component view).

The second phase of conductor path routing will involve placement of those connections associated with the power supply and ground connections. (The order of the phases is not significant. They may be done in reverse order to that described.) We will begin by routing a path between the No. 11 pin pads of the ICs and their decoupling capacitors to an external connection point. It is common practice to route the wider conductor paths (+15 Vdc, −15Vdc, and ground) toward the outside edges of the board whenever possible. These paths may also be routed through the center of the ICs, that is, between the two parallel rows of pads. This routing scheme al-

(a) Component layout symbol for jumper

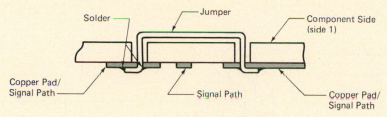

(b) Sectional view of installed jumper

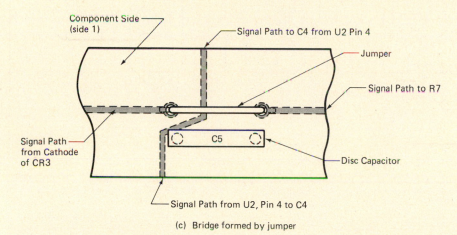

(c) Bridge formed by jumper

FIGURE 8–20 Sectional view of jumper used to aid in path routing.

lows for convenient connections to the decoupling capacitors and the ICs. The external connection pad at the bottom right of the board will conveniently serve as the +15 Vdc terminal. See Fig. 8–21. A 0.10-in. (2X) path will be run from this point up the right side of the board and along the top just above CR1 and CR2. From this position, one path will run downward through CR2 and R6 toward U2. Another path will extend around the left of CR1 and C4 and downward to skirt around the forbidden area toward U1. Care must be taken that the air gap between this path and the forbidden area is a minimum of 0.020 in. As these two vertical paths approach the ICs, it is found from Fig. 8–19 that although one can be conveniently connected to C5, the other is blocked by a signal path between R4 and C3. To overcome this problem, another jumper wire is added just above C1 on the *component* side of the board. Again, two appropriately positioned pads are placed on the conductor side to allow connection of this jumper wire. The bridges formed by the jumper wires allow access of the power bus under the signal paths to the decoupling capacitors C1 and C5 and the ICs, as shown in Fig. 8–21. Because jumper wires have 0 ohms resistance, their introduction into a circuit will in no way affect its electrical characteristics. Jumpers should not be used indiscriminately because each constitutes another part to be installed onto the board. However, they are acceptable when required for an improved conductor path layout.

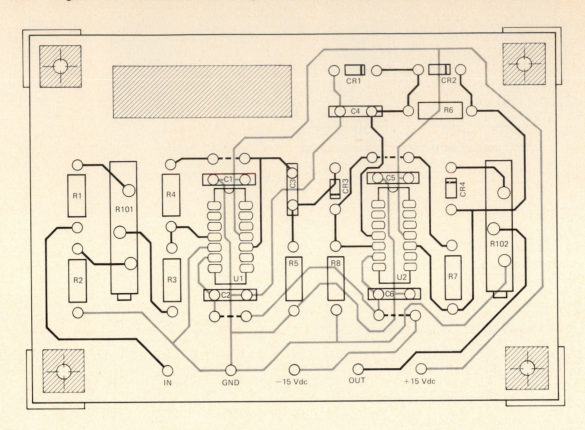

Reduce to 6.000 ± 0.005 inches

FIGURE 8–21   Final layout of the multivibrator circuit with all signal and power connections made.

The external connection pad directly below U1 is selected as ground since this position represents a convenient point to route the three separate paths from the group of ground points described in the problem statement. It will be found that after the addition of the ground paths, the −15-Vdc paths are prevented from being routed without crossing ground paths. By adding two jumper wires just below the decoupling capacitors C2 and C6, the −15-Vdc bus can be bridged over the ground paths and connected to those two capacitors as well as to the No. 6 pin pads of both ICs.

With the addition of the power and ground bus wiring, the component and conductor layout drawing is completed. Note that all decoupling ground connections from C1, C2, C5, and C6 return to the external ground circuit independently from those associated with the input circuitry of the ICs and also from those of the output circuitry. This technique of providing separate ground paths of the circuit to a single external ground connection is referred to as *star grounding*.

After the layout sketch has been completed, including component keying, labeling, and path and pad coding, it should be carefully examined to see that all electrical and mechanical specifications have been satisfied. Because of the exacting nature and detail involved in a PCB layout, it is good practice to have the design given a final check by someone other than the originator.

Following is a check-off list which will serve as an aid in determining if the complete layout drawing of the multivibrator circuit meets specifications and conveys the required electrical and mechanical information to the manufacturer of the printed circuit board.

Electrical information

1. Number and type of each component checked against the circuit schematic.
2. All signal paths correctly routed.
3. All power bus lines correctly routed.
4. All ground lines correctly routed.
5. Star grounding technique.
6. Power bus connects to decoupling capacitors before making connections to the No. 11 pin pads of the ICs.
7. Signal paths coded for 0.040-in. width at 2X scale.
8. Power bus coded for 0.100-in. width at 2X scale.
9. Ground lines coded for 0.100-in. width at 2X scale.
10. ICs and diodes keyed for proper electrical connections.
11. Components labeled and checked against the schematic for correct positions.
12. External connections labeled for proper wiring.

Mechanical Information

1. Drawing made to proper scale with complete reduction scale information.
2. Correct board size and shape.
3. Corner brackets to define the board outline.
4. Correct location and size of forbidden areas.
5. Correct location and size of mounting holes, including targets.
6. Correct location and size of external connection pads.
7. Correct codes for pad sizes and drills to be used for each.
8. Correctly scaled component shapes and sizes.
9. Correct pad spacing for each component and device.
10. No air gaps less than 0.025 in. at 1:1 scale.
11. No space of less than 0.200 in. between any path (or pad) and the forbidden area or board edge.
12. Balanced component layout.

After all checks have been made and the necessary corrections completed, the finished drawing may be transferred to the plain polyester sheet. The sheet of vellum on which the accurate preliminary sketch has been drawn is first removed from the light table. The polyester grid sheet is then removed. The completed preliminary sketch is taped to the light table with care to ensure that the vellum is laid flat and has no wrinkles. The plain polyester sheet is taped over the drawing with the mat side up. With a technical pen or pencil, templates, and a straightedge, the drawing is traced to produce the finished single-sided component and conductor pattern layout.

EXERCISES

8–1 Draw the 2X scale two-dimensional body outlines for the components found in the regulated power supply circuit shown in Fig. 4–9. Use Fig. 8–5a for necessary dimensional information to determine pad spacing for the diodes in the bridge (CR1 through CR4) and the zener diode VR1. Use printed circuit templates such as those shown in Fig. 8–2 to draw U1 (14-pin DIP), transistors Q1 (T0-3) and Q2 (T0-5), potentiometers R2 and R5 ($1\frac{1}{4}$ rectangular), resistors R3, R4, and R6 ($\frac{1}{2}$ W), and axial lead capacitor C2 (0.38 dia.). Refer to Fig. 8–22 to draw the two-dimensional body outline for C1.

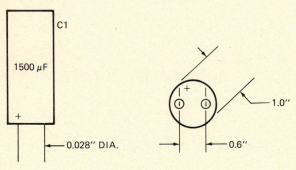

FIGURE 8–22

8–2 Using Fig. 8–10, calculate the minimum path width for the power and ground bus for the regulated power supply circuit shown in Fig. 4–9. The maximum value of current will be 1 A. Assume the weight of copper foil to be 1 oz/ft^2 with a temperature rise of 10°C.

8–3 Draw on a 0.100-in. grid the 2X scale component and conductor pattern layout for the regulated power supply circuit shown in Fig. 4–9. Only components to the right of points a and b will appear on the board. There are no mechanical restrictions as to board shape or size. External connections for points a and b and J1 and J2 are also not restricted

in regard to position except that they should be located near the periphery of the board. Four $\frac{1}{8}$-in. (1X scale) mounting holes are to be located $\frac{1}{4}$-in. (1X scale) from each board edge at all corners. The forbidden area will be 0.2 in. between edges of the component area and the board edge with a 0.6 in. × 0.6 forbidden area about each mounting hole. The minimum signal path width and spacing will be 0.010 in. with power and ground bus widths equal to the value calculated in Exercise 8–2. Include corner brackets and reduction scale.

8–4 Draw the component and conductor pattern layout for the digital circuit shown in Fig. 4–3 using the dimensional drawing of Fig. 8–23. All specifications with regard to grid, conductor width and spacing, corner brackets, and reduction scale apply as outlined in this chapter. There is no restriction to the number of jumpers required to complete the wiring of the layout.

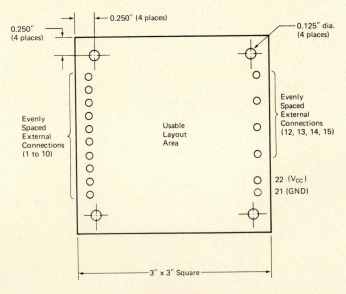

FIGURE 8–23

9

Double-Sided Printed Circuit Board Design: Analog Circuits

Upon completion of this chapter on the double-sided printed circuit design of analog circuits, the student should be able to:

1. Use via holes to simplify conductor path routing.

2. Determine proper pad diameters.

3. Select the proper grid system and reduction scale.

4. Determine lead spacing for high-density designs.

5. Properly place, orient, and space components and devices.

6. Be familiar with automatic component insertion requirements.

7. Determine path widths and spacings for high-density designs.

8. Develop a color-coding system for distinguishing conductor paths on each side of a board.

9. Perform a package feasibility study for analog circuit layouts.

10. Understand all of the physical and mechanical design restrictions.

11. Design double-sided analog circuit layouts.

12. Develop a checklist to determine if all design specifications on the finished drawing have been met.

9-0

Introduction

As pointed out previously, one of the major limitations of the single-sided PCB is that only one side is provided with copper foil from which electrical interconnections are formed. The double-sided board overcomes this limitation in that *both* sides of the board can be processed with a copper conductor pattern.

One of the most important features of a double-sided board is the *plated-through hole* (PTH). Refer to Fig. 6–3. The cross-sectional view of the PTH shows a pair of pads, one on side 1 and one on side 2, joined electrically by the plated barrel within the hole. The plated copper barrel covers the entire surface of the hole wall, making one continuous metallic connection over the surfaces of the top and bottom pads. This technique of electrically interconnecting conductors on both sides of the board simplifies the design of the conductor pattern by increasing the number of possible routes in making connections. It also reduces the overall board size for a circuit design when compared to a single-sided layout. This is due to the fact that more efficient utilization of the component side of the board for both parts placement and path routing is realized. Thus components can be placed closer together, resulting in a higher-density package.

It can be seen that double-sided PCB design meets today's high-technology demands, which emphasize increased component density and smaller board size. This is accomplished primarily by (1) the use of plated-through holes (also called *feed-through* or *via holes*), (2) reducing the size of the terminal pads, and (3) routing conductors on both sides of the board, thus reducing the copper paths required for any one side.

The use of feed-through holes allows the designer to effectively transfer the route of a conductor path from one side of the board to the other at the desired point. Recall from Chapter 8 that jumper wires were required to do this on single-sided boards to avoid conductor paths from crossing.

For single-sided board design, recall that the terminal pad diameter was specified as a minimum of 0.040 in. plus the hole diameter. This minimum value was required to ensure that the bond strength between the pad and the insulating base material is not weakened. On double-sided boards using plated-through holes, the continuous plated barrel extends over the top and bottom pads, which maintains an improved bond strength. As a result, pad diameters can be reduced when compared to those used in single-sided design, allowing more area for conductor path routing. Reducing pad diameters also aids in reducing the required space between components.

In this chapter we present detailed information on feed-through holes, pad diameters for high-density designs, component spacing, and conductor width and spacing for double-sided designs. In addition, a layout density survey will be examined to determine design feasibility and packaging density. To demonstrate the techniques and procedures discussed, we will lay out an analog circuit of a dc motor position control on a double-sided board.

9-1

Feed-Through Holes

Plated-through holes are referred to as *feed-through* or *via holes* when they are used to simplify the conductor path routing as well as to allow more effective use of the total board area. They are used on both double-sided and multilayer boards. The terms *feed-through* and *via* are used interchangeably when a plated-through hole is used to connect a conductor path on one side of the board to the other side rather than being used for component lead connections. The electrical connection between both sides of the board is made by the plated barrel as well as the solder plug which typically fills the hole completely when the board is wave soldered. Figure 9–1a shows how a feed-through hole contributes to improved conductor path routing so as to result in a higher-density board. This figure shows a cross section of a double-sided board utilizing two via holes. Note that two

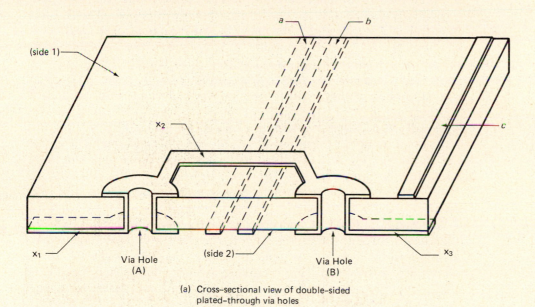

(a) Cross–sectional view of double–sided
    plated–through via holes

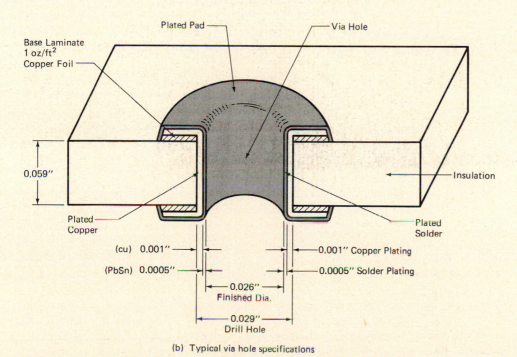

(b) Typical via hole specifications

FIGURE 9–1   Feed-through or via holes on double-sided PCBs.

parallel paths, labeled a and b and shown as dashed lines, have been designed to appear on side 2. The conductor path labeled x_1, on the left side of the board, also on side 2, is to be routed completely across to the right side of the board to path x_3. This routing cannot be done on side 2 since conductors a and b stand in the way of both paths x_1 and x_3. With the use of via holes, path x_1 is provided with a bridge over paths a and b so as to satisfy the design. The path labeled x_1 first crosses over to side 1 through via hole A to path x_2, which bridges over paths a and b. If path x_2 were allowed to continue to the right, it would cause it to make an undesired electrical connection with path c, which is also on side 1. Therefore, via hole B is placed to the left of path c and allows path x_2 to provide electrical continuity with path x_3. Because the bridge is on side 1, it is electrically insulated from paths a and b through the base material of the board. It can thus be seen that via holes are used when any electrical connections from one side of the board to the other are required for improved path routing.

Recall from Chapter 8 that the bridge shown in Fig. 9–1a could have been made with the use of a jumper wire, as shown in Fig. 8–20. However, via holes require less space and much less assembly time than do jumper wires. They are, therefore, used exclusively in double-sided designs to simplify conductor routing quickly and efficiently.

All feed-through holes are processed at the same time as are all the plated-through holes. However, the drill hole size required to form a feed-through hole is typically smaller than the smallest of those which are to be used for a component lead. The reason for this is that a via hole does not have to support a lead and, as such, its pad diameter is typically the smallest on the board.

The specifications of a typical via hole would be the following:

1. *Drill hole size:* approximately one-half the thickness of the printed circuit board stock. (One-third the board thickness is also common.)
2. *Pad diameter:* approximately 0.024 in. larger in diameter than the finished hole size.

Figure 9–1b illustrates the use of these specified values. This sectional view of a via hole is processed into a PCB having a stock thickness of 0.059 in. The hole was initially drilled with a No. 69 drill bit having a diameter of 0.029 in. (see Appendix XIX). After drilling, the board was processed with copper plated onto the hole wall to a thickness of approximately 0.001 in. (1 mil) and an additional 0.0005-in. ($\frac{1}{2}$ mil) thickness of solder. The resultant hole, after plating, is 0.026 in. or approximately 0.003 in. smaller than the actual drill hole size. A pad diameter of 0.050 in. will be 0.024 in. larger than the finished via hole. This size pad, although smaller than those used for component leads, is typical for via holes.

Because no leads are to be used with via holes, no hole size tolerance is required in the specifications. The reason for the specification that the pad diameter be 0.024 in. larger than the finished hole size will be explained in Section 9–2.

For the circuit designs described in this and succeeding chapters, via holes used with 0.059-in. PCB stock will be provided with 0.050-in.-diameter pads (1:1 scale) and No. 69 (0.0292-in.) drill holes. The use of the 0.050-in. pads for via holes greatly reduces the amount of required space when compared to the pads used for component leads, which are required to be 0.040 in. larger than the drill hole size. This reduction results in increased package density by allowing more board area around the pads to be used for conductor path routing. In addition, components may be positioned closer together without violating minimum space requirements.

9-2

Determining Pad Diameters for Double-Sided Designs

There are several factors that determine the proper pad size for use in a double-sided board design incorporating plated-through holes. Refer to Fig. 9–2. Note that the maximum lead diameter that is to be used in this example is 0.016 in. We will first determine the selection of the appropriate drill bit size and the lead clearance required for proper insertion. Recall that for single-sided boards with no plated-through holes, the recommended clearance is from 5 to 10 mils or a drill bit size of from 10 to 20 mils larger in diameter than the maximum lead diameter. The recommended lead clearance for plated-through holes is a minimum of 0.010 in. Therefore, the drilled hole will be at least 20 mils larger in diameter than the lead to be inserted. Even with high-density packaging as the primary objective, the PCB designer must allow this 0.010 in. of clearance before the plating process. As shown in Fig. 9–1b, the plating process results in a layer of copper and solder onto the hole walls, which reduces the drill hole size by approximately 0.003 in. For the example

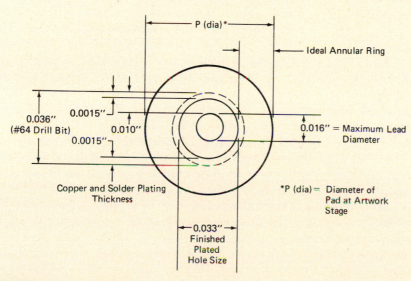

FIGURE 9–2 Pad diameter and specifications for a typical double-sided PTH.

shown in Fig. 9–2, a No. 64 drill bit, which has a 0.036-in. diameter, must be used to provide the 0.010 in. clearance before plating. With a finished plated hole size of 0.033 in., this will give the 0.016-in. lead a sufficient clearance of approximately 0.007 in.

Recall that the minimum pad diameter of 0.040 in. for unsupported holes in a single-sided board was established in order to maintain the bond between the foil and the base material. In a plated-through hole, this specification can be relaxed since the pads of each hole are supported by the plated barrel. See Fig. 9–1b. This results in a more secure foil bond, allowing a considerable reduction in the minimum pad diameter required.

The selection of an acceptable pad diameter is related to the hole position to pad position relationship after the board is processed. As in most manufacturing processes, the fabrication of a double-sided PCB involves many tolerances which could have a cumulative effect. Refer to Fig. 9–3, which shows a selection of four finished hole size-to-pad size relationships. The same 0.033-in.-diameter finished hole size is shown in all of the examples.

A pad whose diameter is 0.030 in. larger than the finished hole is shown in Fig. 9–3a. After the fabricating processes, it is probable that in a few areas of the board, a misalignment between the pad and the hole center due to cumulative tolerances could be as much as 0.010 in. This results in a worst-case minimum *annular ring* of 0.005 in. as shown in Fig. 9–3a.

In Fig. 9–3b, a pad diameter 0.024 in. larger than the finished hole is shown. Again assuming a misalignment of 0.010 in., a worst-case minimum annular ring of 0.002 in. will result randomly in some areas of the board. Figure 9–3c shows a pad diameter that is only 0.020 in. larger than the finished hole diameter, resulting in some pads whose outside edge touch the edge of the hole. This condition is called the *point of tangency* or simply *tangency*, which results in there being no annular ring of copper or solder on the board at this point.

In Fig. 9–3d, a pad is shown whose diameter is less than 0.020 in. larger than the hole diameter. Again, with an overall processing misalignment of 0.010 in., a *breakout* will occur at some holes on the board. This is when the pad does not completely surround the hole and no annular ring of copper or solder exists along a part of the edge of the hole.

As can be seen from this discussion and from Fig. 9–3, the smaller the pad diameter for a given finished hole size, the more severe the worst-case condition of annular ring becomes. Therefore, the choice of pad diameter is based on what is acceptable on the finished board for an annular ring that determines hole integrity. This decision is normally one of management policy which dictates tolerance values. For example, the management policy may be to design for an 0.002-in. annular ring. It is then the responsibility of the designer to make all pads 0.024 in. larger in diameter than the finished hole size.

It needs to be emphasized that the *worst-case* representations shown in Fig. 9–3 are just as the name implies. After a board has been completed, the majority of the plated holes will have larger annular rings than the worst cases shown. However, the weakest-link

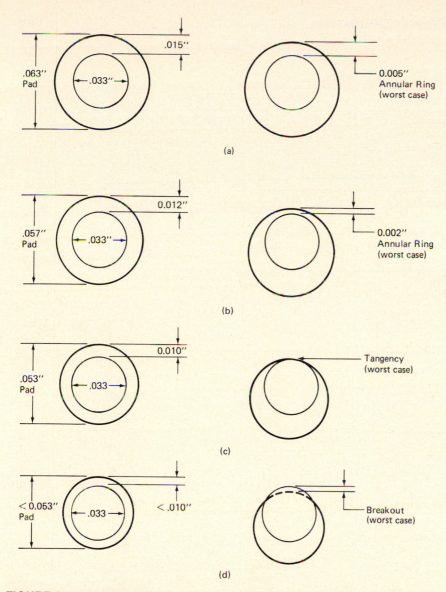

FIGURE 9–3 Various annular ring conditions resulting from pad diameter selection and worst case cumulative misalignment.

theory also applies here. This determines the amount of yield of boards that meet specifications. Only one hole not meeting the minimum annular ring requirement can result in that board being rejected.

The left column of Fig. 9–3 shows the ideal result expected in the manufacture of PCBs. The right column represents the worst case of hole-to-pad center relationship. This is the result of tolerances of materials, tools, and machines used in manufacture which, taken cumulatively, could reduce the annular ring remaining on any given hole on the board. Some of the factors that contribute to the reduction of the annular ring are drill positioning error, drill hole size and tolerance, drill programming error, temperature and relative humidity, tolerance on pad diameters, registration of artworks to drilled

TABLE 9–1 Pad Diameter Selection Information for Double-Sided PCB Designs*

Available Pad Diameter (1:1 Scale)		Annular Ring		Tangency	Breakout
		0.005	0.002		
0.050	Finished hole size (max.)	0.020	0.026	0.030	>0.030
	Max. lead diameter	N/A	N/A	N/A	N/A
0.062	Finished hole size (max.)	0.032	0.038	0.042	>0.042
	Max. lead diameter	0.015	0.021	0.025	>0.025
0.070	Finished hole size (max.)	0.040	0.046	0.050	>0.052
	Max. lead diameter	0.023	0.029	0.033	>0.033
0.075	Finished hole size (max.)	0.045	0.051	0.055	>0.055
	Max. lead diameter	0.028	0.034	0.038	>0.038
0.080	Finished hole size (max.)	0.050	0.056	0.060	>0.060
	Max. lead diameter	0.033	0.039	0.043	>0.043
0.085	Finished hole size (max.)	0.055	0.061	0.065	>0.065
	Max. lead diameter	0.038	0.044	0.048	>0.048
0.090	Finished hole size (max.)	0.060	0.066	0.070	>0.070
	Max. lead diameter	0.043	0.049	0.053	>0.053
0.093	Finished hole size (max.)	0.063	0.069	0.073	>0.073
	Max. lead diameter	0.046	0.052	0.056	>0.056
0.100	Finished hole size (max.)	0.070	0.076	0.080	>0.080
	Max. lead diameter	0.053	0.059	0.063	>0.063

*All dimensions are in inches. N/A, not applicable.

panels, and undercutting of pads in the etching process. It is seen that one of the responsibilities of the PCB designer is to select pad diameters for a specific worst case of annular ring.

To satisfy the demands of high-density packaging, the tendency is to press for a specification of tangency and/or breakout being acceptable. The objective then becomes to select the smallest pad diameter, for a given lead, which will take up minimum space on the board. On the other hand, reliability of boards dictates that they be manufactured with minimum annular rings of 0.002 to 0.005 in. A typical trade-off is a 0.002-in. annular ring with an occasional breakout to achieve both high-density requirements and reliability.

Using the information provided in this section, Table 9–1 has been constructed to serve as a guide to the PCB designer in making pad diameter selections for double-sided plated-through-hole designs. The pad diameters listed in the table are the nine available sizes, ranging from 0.050 to 0.100 in. in diameter. After the level of board quality has been decided (e.g., 0.002 in. annular ring for worst case), the designer may select the appropriate pad diameter for a specific lead diameter or finished hole size with the use of Table 9–1. To illustrate the use of the table, assume that we are to determine the appropriate pad diameter for a lead diameter of 0.032 in. if a 0.002-in. annular ring is specified. Using the vertical column labeled *0.002"* *Annular Ring*, you will find that the closest entry for a maximum lead diameter of 0.032 in. is 0.034 in., with a finished hole size of 0.051 in. Projecting these values to the left column of the table will designate a pad diameter of 0.075 in., which will satisfy this design requirement.

9-3

Selection of a Grid System and Scale

Recall from Chapter 7 that a 0.100-in. grid system was satisfactory for laying out a low-density printed circuit board. Again, in Chapter 8, although more design restrictions were imposed, which resulted in tighter placement of a large number of parts, the 0.100-in. grid was found suitable and no problems were encountered for on-grid positioning of pads. In this and succeeding chapters, we will demonstrate that a grid system with half-increments of the 0.100 in. is required in order to result in on-grid positioning of pads so that higher-density designs will be achieved.

Recall that the 0.050-in. grid (20×20 lines per inch) was shown in Fig. 8–7 and has combinations of medium and heavy accents on the 2nd, 5th, 10th, and 20th lines. With this system, dimensional information can be positioned more easily and accurately than on the 0.100-in. grid. For example, the design specifications may require mounting holes with an on-center spacing of 6.250 in. In order for these holes to be located on grid, the 0.050-in. grid system would be necessary for accuracy of placement.

The accent lines aid in on-grid positioning of device leads, such as those for the dual-in-line package. In using a 0.100-in. grid system, an accent on every 2nd line simplifies lead hole placement at a 2X scale since the pad placement at this scale is 0.200 in. between pin centers and 0.600 in. between row centers. For high-density packaging, the 0.050-in. grid with the accent on the 2nd line (0.100-in. increments) would be used. This would result in the location of on-grid pad locations at 2X scale by using alternate accent lines or 0.200-in. increments. It can be seen from this discussion that the designer selects a grid system and accent line arrangements which are compatible with the design density, dimensional values, and scale size.

For the degree of accuracy required for the double-side designs in this and the following chapter, a 2:1 scale will again be used with a 0.100-in. grid system subdivided with a 0.050-in. grid. This will result in drawings which can easily be accommodated on standard-size drafting paper.

9-4

Lead Spacing for High-Density Designs

After the proper pad diameters for component leads mounted to a high-density board have been selected using the criteria established in Section 9–2, the next task is to determine the correct lead spacing. For axial lead components, Equation 8–3 also applies for double-sided high-density designs. To limit the required board space, however, the dimension L in Fig. 8–15, which is the distance that the lead is brought out of the component body, will be the 0.030-in. value rather than 0.060 in. Equation 8–3 and Table 8–4 are thus used to determine the minimum on-grid lead spacings.

For radial lead components, such as transistors and TO-5 case style ICs, the leads are fed straight into the board without bending and the device may be flush mounted onto the board surface. However, depending on the specifications of pad size and air gaps between them, it is not always possible for pads to be positioned for perfectly straight entry into the board. In addition, high-density packaging design may require a conductor path to be routed between the pads of a radial lead component. In this case, the pad positioning will have to be altered from the usual straight-through arrangement if there is not sufficient air gap to allow a path to be positioned through them.

The lead orientations for straight-through insertion of the TO-5, TO-99, and TO-100 case styles are shown in Fig. 9–4a. The leads are positioned on the circumference of circles having specific diameters for each case style. These are called *lead circles*. If the diameter of the lead circle is increased, the space between their pads will also increase. Observe in Fig. 9–4b that by changing the lead circle diameters for each of the devices from 0.200 in. to 0.400 in., the pad-to-pad clearance distance increased by almost four times for the TO-5 case and two to three times for the other syles shown. With this increased air gap between lead pads, a larger pad diameter may be used and still realize an increased air gap.

The increase of the lead circle diamter is the result of offsetting the leads before entry into the board. With this offset, the device cannot be flush mounted to the board. If support for the devices is specified, small spacers called *spreaders* are used through which the leads are passed before the device is mounted to the board.

It can be seen from this discussion that the distance between the pads of TO-5-style devices is a trade-off. In the interest of high-density packaging, the device leads should make straight entry into the boards and be provided with the smallest possible pad diameters. If it is necessary to increase the lead circle diameter for purposes of increasing the clearance between pads, this change should be limited to the smallest increment possible and the pad sizes held to their smallest acceptable diameter.

To draw the transistor in the TO-5 case style, the printed circuit template is positioned so that the pattern will be centrally located in the area allotted with the best orientation of the base, emitter, and collector pads for optimum path routing. The lead circle center for either straight or offset entry and pad centers are then placed on grid and traced.

For the drawing of any TO-5-style IC pad arrangement (8, 10, or

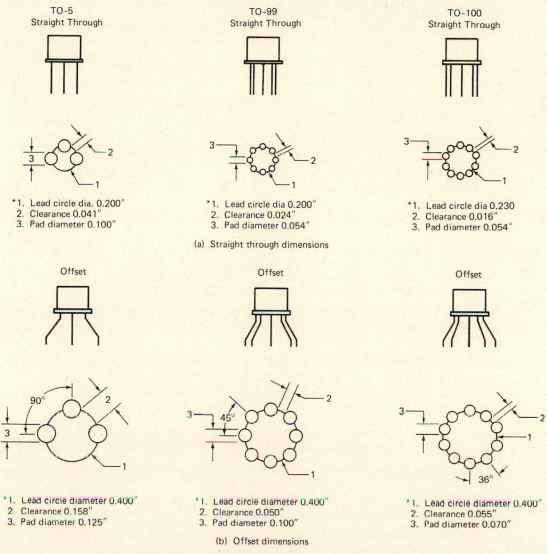

TO-5
Straight Through

*1. Lead circle dia. 0.200″
2. Clearance 0.041″
3. Pad diameter 0.100″

TO-99
Straight Through

*1. Lead circle dia 0.200″
2. Clearance 0.024″
3. Pad diameter 0.054″

(a) Straight through dimensions

TO-100
Straight Through

*1. Lead circle dia 0.230
2. Clearance 0.016″
3. Pad diameter 0.054″

Offset

*1. Lead circle diameter 0.400″
2. Clearance 0.158″
3. Pad diameter 0.125″

Offset

*1. Lead circle diameter 0.400″
2. Clearance 0.050″
3. Pad diameter 0.100″

Offset

*1. Lead circle diameter 0.400″
2. Clearance 0.055″
3. Pad diameter 0.070″

(b) Offset dimensions

*Dimensions are for 1 : 1 scale (for 2 : 1 scale multiply all values by 2).

FIGURE 9–4   Pad arrangements for several TO-style device packages.

TABLE 9–2 Pad Configurations for Several TO-Style Device Packages*

Mounting Configuration	Case Style	Lead Circle Diameter† (in.)	Clearance (in.)	Pad Diameter (in.)
Straight through		0.200	0.024	0.054
Offset	8 lead	0.300	0.037	0.078
Offset	(TO-99)	0.350	0.040	0.093
Offset		0.400	0.050	0.100
Straight through		0.200	0.015	0.046
Straight through	10 lead	0.230	0.016	0.054
Offset	(TO-100)	0.350	0.040	0.070
Offset		0.400	0.055	0.070
Straight through		0.200	0.015	0.037
Straight through	12 lead	0.230	0.010	0.050
Offset	(case 604)	0.350	0.020	0.070
Offset		0.400	0.023	0.080
Straight through	3 lead	0.100	0.015	0.056
Offset	(TO-18 or TO-92)	0.150	0.026	0.080
Straight through		0.200	0.078	0.062
Straight through		0.200	0.066	0.075
Straight through	3 lead	0.200	0.047	0.093
Straight through	(TO-5)	0.200	0.041	0.100
Straight through		0.200	0.032	0.109
Straight through		0.200	0.016	0.125

*All dimensions are for 1:1 scale. (For 2:1 scale, multiply all values by 2.)

†Pad diameters are compatible with multipad preforms shown in Section 12-4.

12 pin), the device outline is again centrally positioned in its allotted area. The lead circle center for either straight or offset entry is then placed on grid and the pads are traced. The best location for pin 1 for routing the conductor paths can then be determined.

The clearances for various lead circles and pad diameters for TO-style transistors and ICs are given in Table 9–2. The lead circles and pad diameters listed are compatible with those found on multipad preforms which are discussed in Section 12–4 and used for making taped artwork. Since the accurate layout of multipad patterns is extremely important, no attempt should be made to locate and draw the pads individually. The printed circle template becomes a very important tool for obtaining these accurate results.

9-5

Component Placement, Orientation, and Spacing

Because the basic techniques for component placement discussed in previous chapters also apply to high-density design, it will be helpful if we review them here. Recall that all tubular components, such as resistors, electrolytic capacitors, and diodes, should be oriented so that their bodies are mounted parallel to a board edge and all on one axis wherever possible. Since this orientation cannot always be

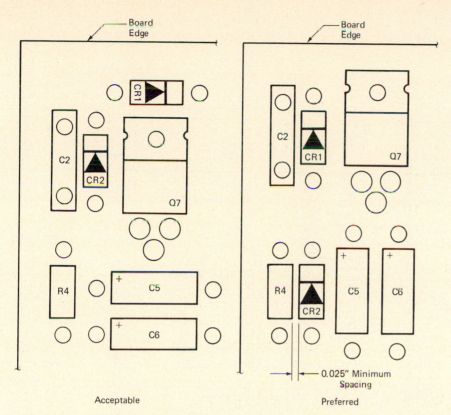

Acceptable   Preferred

FIGURE 9–5 Component placement on PCBs.

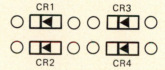

(a) Acceptable—all diode cathodes to left

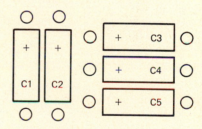

(b) Acceptable—all capacitor groupings with + sides together

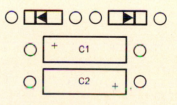

(c) Not desirable—diode cathodes and + sides of capacitors in various orientations

FIGURE 9–6 Orientation of polarized components on PCBs.

maintained, the second axis should be at a 90-degree angle from the first. In addition, components clustered in subgroups should be aligned with their bodies centered, again, wherever possible. Both acceptable and preferred component positioning are shown in Fig. 9–5, with the minimum clearance between component bodies on high-density boards. Note that no axial component body edge should be closer than 0.025 in. (1:1 scale) to another. This is a typical value, however, which can be compromised by the engineer.

Polarized components should be oriented in the same direction. For example, the positive ends of electrolytic capacitors or the cathode ends of diodes should be positioned so that these polarities are in the same direction. See Fig. 9–6.

The dual-in-line IC packages are rectangular in shape and have 14, 16, 18, 22, 24, 28, 40, 48, or as many as 64 pins. This package should be oriented so that its body is parallel to a board edge and in vertical or horizontal alignment with adjacent IC cases. In addition, their alignment should be such that pin 1 of all the ICs is oriented in the same direction. When placed beside each other, there should be a minimum spacing of 0.150 in. between adjacent hole centers when using 0.125-in.-wide cut pads. A minimum of 0.025 in. should separate IC packages when they are positioned end to end for dense designs. An example of these density requirements for a typical DIP arrangement is shown in Fig. 9–7. Note that the 0.150-in. spacing between hole centers leaves only a 0.025-in. air gap between adjacent IC pads. This space does not allow vertical routing of even one

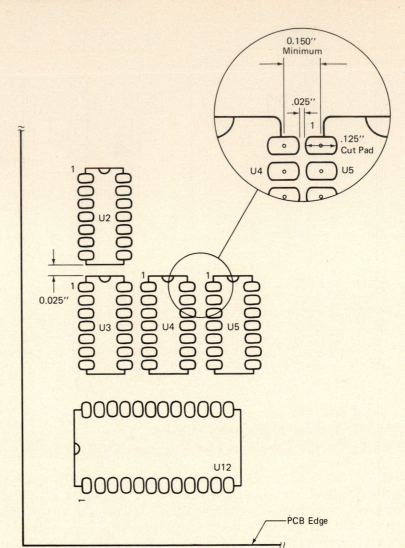

FIGURE 9–7   Spacing and orientation requirements of dual-in-line packages on dense double-sided PCBs.

10-mil path with two 10-mil air gaps (total 30 mils). Therefore, the minimum spacing to allow vertical path routing and still remain on grid is 0.200 in. This spacing allows three 10-mil paths with 10-mil spacings to be routed vertically between ICs. This amount of spacing between ICs is also necessary for automatic insertion of these devices and for access of an IC test probe.

9-6

Component Positioning for Automatic Insertion

When designing a PCB that is to be assembled manually, component positioning is not as critical as when automatic insertion of component leads is to be employed. For this application, parts placement is governed by specific design criteria. Designs that are most compatible with automatic assembly are those having a minimum variety of size and shape of components since they require less time to organize for the sequence of installation. As the variety increases, so does the cost of automatic parts assembly. It therefore behooves the circuit design engineer to standardize the components of a system so that a minimum variety will be involved to make the costs of automatic assembly feasible.

In addition to standardizing components as much as possible, there are other considerations that contribute to cost-effective automatic insertion. Components must be arranged on as few axes as possible in rows and in columns. Since PCBs are made of many components with different lead spacings, basic techniques for their arrangement must be established. As in other designs, pad centers must be accurately located on a precise grid system. The ideal arrangement for axial components is in a column having a parallel pattern with uniform lead spacings. Every effort should be made to standardize on-center lead spacings for axial or radial components so that the number of different spacings will be kept to a minimum. This also results in higher-density packages since less board space will be required.

Boards designed for automatic insertion are provided with two or more *tooling holes* from which all component insertion holes are dimensioned in increments of 0.025, 0.050, and 1.00 in. The size and location of the tooling holes are typically provided to the circuit design engineer from the manufacturing facility. Tooling holes do result in additional forbidden areas unless they are later used as mounting holes. Nevertheless, they are necessary to provide a means of securing the board to the insertion machine so that it remains immobile during this process.

The finished hole sizes that are designed for automatic insertion need to be made larger than those for manual assembly. In general, a drilled or punched hole made for automatic insertion is made 0.015 in. larger than the maximum lead diameter that it is expected to support. For rectangular IC leads, this same 0.015 in. is added to the diagonal of the rectangular dimension of the lead cross section. For example, an IC lead with a cross-sectional dimension of 0.010 × 0.020 in. has a diagonal measurement of 0.022 in. Therefore, a hole size of 0.037 in. (No. 63 drill) would be required. These larger holes may require larger pad diameters in order to meet the specified worst-case annular ring value. Because these larger pad diameters will require more space, it again emphasizes the need for standardizing lead spacings with minimum values in order to offset this additional space requirement.

The optimum arrangement of components for automatic insertion is shown in Fig. 9-8a. Figure 9-8b-g show compromise arrangements that result in successive increases of manufacturing costs, which range from *most economical* (Fig. 9-8b) to *unacceptable* (Fig. 9-8g).

The optimum orientation for multiple radial lead components, such as TO-99-style packages, is shown in Fig. 9-9a. The orientations shown in Fig. 9-9b-d result in greater assembly time and higher manufacturing costs.

The recommended arrangement, with minimum spacings, for dual-in-line packages using 0.125-in.-wide cut pads is shown in Fig. 9-10. The spacing between devices is determined to a large extent by

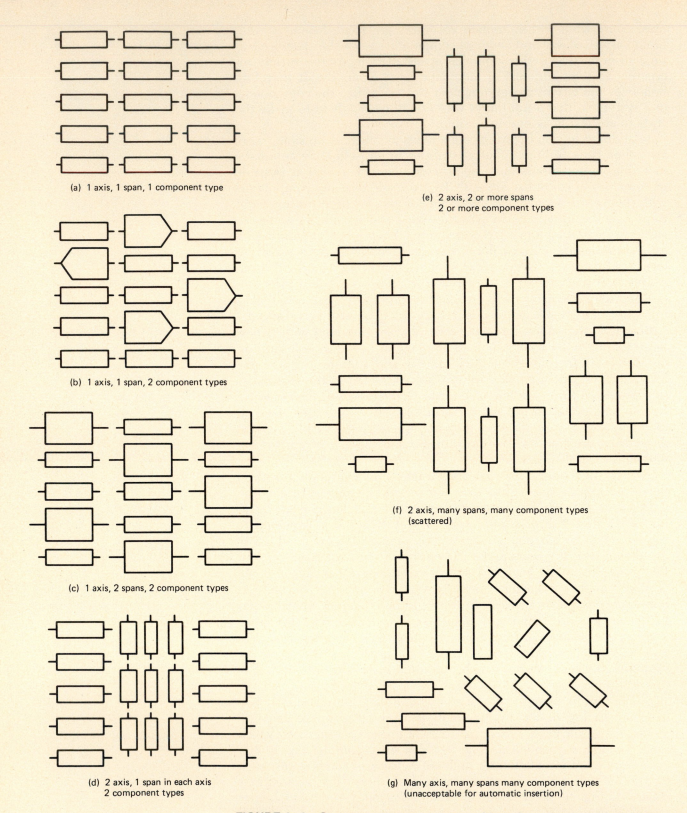

(a) 1 axis, 1 span, 1 component type

(b) 1 axis, 1 span, 2 component types

(c) 1 axis, 2 spans, 2 component types

(d) 2 axis, 1 span in each axis
    2 component types

(e) 2 axis, 2 or more spans
    2 or more component types

(f) 2 axis, many spans, many component types
    (scattered)

(g) Many axis, many spans many component types
    (unacceptable for automatic insertion)

FIGURE 9–8   Component arrangements with various degrees of acceptability for automatic insertion.

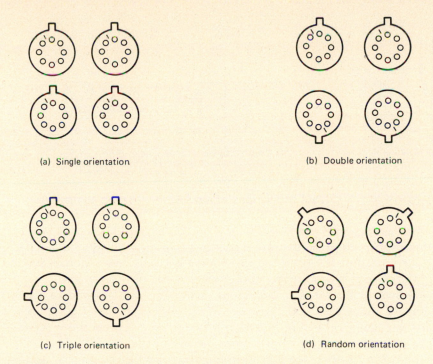

(a) Single orientation

(b) Double orientation

(c) Triple orientation

(d) Random orientation

FIGURE 9–9 Four different orientations for radial lead components.

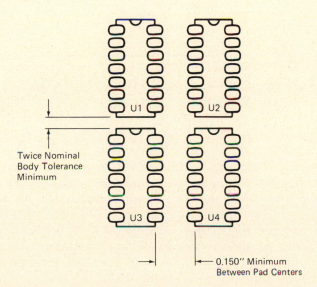

Twice Nominal
Body Tolerance
Minimum

0.150″ Minimum
Between Pad Centers

FIGURE 9–10 Recommended dual-in-line package orientation for automatic insertion.

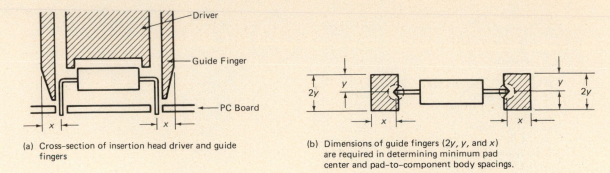

(a) Cross-section of insertion head driver and guide fingers

(b) Dimensions of guide fingers (2y, y, and x) are required in determining minimum pad center and pad-to-component body spacings.

FIGURE 9–11 Critical dimensions of automatic insertion equipment.

the number of paths that must be run vertically between device pads, their size, and the required insertion tool clearance. A cross section of the insertion head driver and guide fingers, or outside lead formers, is shown in Fig. 9–11a. Sufficient space must be allowed to accommodate the width and thickness of the fingers so that components are not damaged in the insertion process. The critical dimensions of the guide fingers are shown more clearly in Fig. 9–11b. The typical dimension for distance y is approximately 0.60 in. and for x, approximately 0.085 in.

The distance between pad centers or pad centers and component bodies also needs to be considered for automatic insertion of components. These spacings may become critical and are dependent on the order in which components are loaded onto the board. Some examples of how to determine pad and body clearances for a variety of axial component orientations with the sequence of installation are given in Appendix XX.

9-7

Conductor Path Widths and Spacings for High-Density Designs

A detailed discussion of conductor path width and spacing was presented in Sections 8–8 and 8–9. We will review that material and show how it applies to high-density designs. Recall that a minimum finished conductor width of 10 mils (0.010 in.) is capable of handling 0.75 A of current with a copper thickness of 1 oz/ft^2 and 1.3 A with a thickness of 2 oz/ft^2. Since this current-carrying capacity is typically in excess of the amount required by most signal paths, the 10-mil minimum width is normally recommended for reliable board manufacturing. Even though finished path widths of 8 mils or as small as 6 mils are possible under extremely controlled manufacturing processes, they result in reduced yield and, therefore, greater costs. Even for high-density layouts, we will continue to use the recommended 10-mil width for the reasons stated.

It should be emphasized that the finished conductor width on the PCB is smaller than that on the artwork due to (1) the initial thickness of copper used and (2) the reduction of the path width as a re-

TABLE 9–3 Finished Conductor Width as a Function of Artwork Tape Width Selection

Width of Conductor at 1X Scale Artwork	Conductor Width on Board	
	1 oz/ft^2	2 oz/ft^2
0.036	0.033	0.031
0.035	0.032	0.030
0.034	0.031	0.029
0.033	0.030	0.028
0.032	0.029	0.027
0.031	0.028	0.026
0.030	0.027	0.025
0.029	0.026	0.024
0.028	0.025	0.023
0.027	0.024	0.022
0.026	0.023	0.021
0.025	0.022	0.020
0.024	0.021	0.019
0.023	0.020	0.018
0.022	0.019	0.017
0.021	0.018	0.016
0.020	0.017	0.015
0.019	0.016	0.014
0.018	0.015	0.013
0.017	0.014	0.012
0.016	0.013	0.011
0.015	0.012	0.010
0.014	0.011	0.009
0.013	0.010	0.008
0.012	0.009	0.007
0.011	0.008	0.006
0.010	0.007	0.005
0.009	0.006	N/A*
0.008	0.005	N/A*

Not recommended

*N/A, not applicable.

sult of the etching process. A list of conductor widths comparing those on the artwork to those on the finished board for 1-oz and 2-oz copper thicknesses is given in Table 9–3. As an example of finished path width reduction, if a conductor on the artwork is made 0.020 in. wide, the resulting finished width after manufacturing is 0.017 in. wide for 1-oz/ft^2 copper and only 0.015 in. wide for 2-oz/ft^2 copper. It is apparent that the PCB designer must be given the copper thickness in order to determine the width of the path to use on the artwork. For a minimum of 0.010 in. conductor width, an artwork tape width of 0.013 in. (1:1 scale) for 1-oz copper and 0.015 in. for 2-oz copper must be selected. See Table 9–3. Using tape widths below the recommended minimum should be specified only with the approval of the engineer and within the capability of the manufacturer.

It can be seen from Table 9–3 that the artwork conductor width, due to the etching process, is typically reduced by 3 mils for 1-oz copper and by 5 mils for 2-oz copper. This is due to the fact that the etchant will attack not only the undesired copper but will also undercut a slight amount under the etchant-resist pattern. The formula for determining the minimum layout conductor path width taking into account the specified scale and the reduction due to manufacturing is:

$$\text{layout width (min.)} = \text{scale factor} \times (\text{finished path width} + \text{manufacturer's reduction}) \quad (9-1)$$

To determine the minimum layout width for a 0.010-in. finished path at 2X scale using 1-oz/ft^2 copper, we use Equation 9–1 as follows:

$$\text{layout width (min.)} = 2(0.010 + 0.003) = 0.026 \text{ in.}$$

For the same finished path width and scale using 2-oz copper, the equation would result in

$$\text{layout width (min.)} = 2(0.010 + 0.005) = 0.030 \text{ in.}$$

The discussion of minimum conductor width does not apply to power and ground conductors, which are much wider, in order to handle greater currents. For this application, the graph shown in Fig. 8–10 should be used.

The spacing between conductors was discussed in detail in Section 8–9. Recall that the minimum spacing of 0.025 in. required by military standards, shown in Fig. 8–12, is an extremely conservative value. High-density boards are currently being produced with 0.010-in. minimum spacings between conductors. For even higher density requirements, spacings of 0.008 in. to as low as 0.006 in. are not uncommon. In the high-density designs that follow in this book, we will specify a minimum spacing between conductors of 0.010 in. See Fig. 9–12.

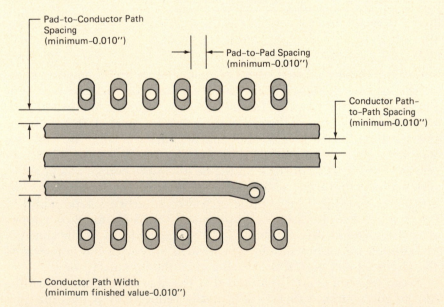

FIGURE 9–12 Minimum spacing between features on finished high density PCBs.

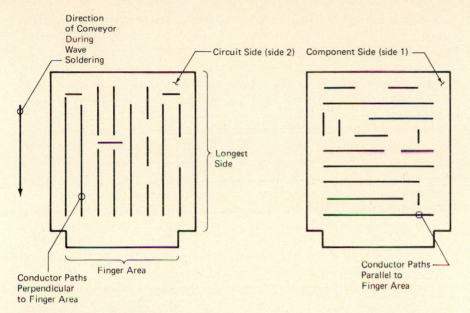

FIGURE 9–13 **Recommended conductor path orientation on high density PCBs.**

After the minimum conductor width and spacings have been determined for high-density applications, some attention must be given to the orientation and distribution of conductor paths on each side of the board. Refer to Fig. 9–13. The orientation of the conductor paths shown is recommended as good design practice for high-density layouts, especially for digital circuit designs, which will be discussed in the following chapter. Note that the majority of conductor paths on the component side (side 1) are positioned parallel to the *shortest* edge of the board (and the finger area if a part of the design). The majority of the paths on the circuit side (side 2) are routed either parallel to the *longest* side of the board edge or perpendicular to the finger area. In this scheme of conductor path routing, the majority of conductor paths on one side of the board are oriented perpendicular to those on the other side. Via holes are then used, as required, to change conducting paths from one side to the other.

Another advantage to this orientation of conductor paths for side 2 is to reduce soldering defects. After the board has been fabricated and all the components mounted, it is placed on a conveyor of an automatic wave soldering machine with the circuit side (side 2) facing down. It is then transported over a molten wave of solder which contacts the conductor pattern on this side and the ends of the protruding component leads, soldering them to their respective terminal pads. On all boards, but especially those with 10-mil conductor spacings, the board is placed on the conveyor so as to pass over the wave of solder in the same direction as the majority of conductors on that side. This is to minimize *solder bridging* between closely spaced conductors. Solder bridges are short circuits causing unwanted electrical connections as a result of solder solidifying between conductors.

Another concern of the PCB designer is to produce a uniform distribution of conductor paths on both sides of the board. The parallel–perpendicular relationship of paths, used particularly in high-density design, can lend itself to a more even side-to-side distribution of conductors.

Where double-sided boards are designed for plated-through holes, copper and then solder must be electroplated along the barrels of these holes. Efficient electroplating requires that the areas of copper on both sides of the board be as equal as possible. The calculation of the area of conductors on each side of the board can be a time-consuming process if done without the aid of automated equipment. It involves finding the area of all of the pads on one side, which is the same for the pads on the other side since they have the same number and size of pads. For path areas, the length multiplied by the width of each path must be calculated for each side. By adding the total pad and path areas for each side, an area comparison may be made. These calculations are not normally done this way, however. The whole process can be quickly and efficiently determined automatically with the use of a printed circuit board *area calculator*. This unit measures light transmission through either negatives or positives of the taped artworks and gives a digital readout of the total conductor area of each side. Uniform copper pattern areas on both sides of the board promote more uniform plating thicknesses of copper and solder in the hole barrels as well as more precise solder alloys.

As previously discussed, the layout of high-density boards is best done on a grid having 0.100-in. major divisions and 0.050-in. minor divisions. For the application of this grid system, refer to Fig. 9–14. For a 2:1 scale high-density design, two conductor paths have been drawn on grid at a distance of 0.050 in. apart and shown as thin solid lines. To show how conductor path spacings of less than 0.025 in. and approaching 0.010 in. are obtained, two 0.030-in. tape-width segments, labeled a and b, have been superimposed onto this grid system. This would be the required grid alignment position to result in reduced path spacing. At 2X scale on a 0.050-in. grid, a space of 0.020 in. results. When reduced to a 1:1 scale, this scale becomes 0.010 in. between two 0.015-in.-wide conductor paths. For designs that require exact 0.010-in. spaces and 0.010-in. conductor widths, it becomes necessary to use other aids rather than just the 0.050-in. grid system and the naked eye. The use of optical comparators with

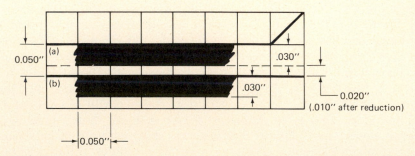

FIGURE 9–14 Use of a 0.050″ grid to achieve 1:1 scale 0.010″ spacing between conductors.

Conductor width	Code	
0.010″	Circuit Side	(Blue)
	Component Side	(Red)
0.025″	Circuit Side	(yellow)
	Component Side	(purple)
0.100″	Circuit Side	(Brown)
	Component Side	(Green)

— Draw with Appropriate Color

FIGURE 9–15   Conductor width color code legend.

scaled reticles and spacer tapes on the grid system are required. The use of these aids is discussed in Chapter 12.

It needs to be pointed out that there is a limitation on the degree of accuracy resulting from purely manual procedures. Where extremely tight spacings and high accuracy of tape placement are part of the design, it can be completed satisfactorily only with the use of computer-aided design (CAD) techniques, which are discussed in Chapter 15.

Because more than one conductor width is used on a typical design, a means of distinguishing between them must be employed. Further, since conductors will be laid out on both sides of the board, they must also be differentiated. You will recall that a color-coding system for differentiating conductor widths for single-sided boards was discussed in Chapter 8 and shown in Fig. 8–11a. We will expand on this system for applications to double-sided designs. This is shown in Fig. 9–15. Each conductor width (0.010, 0.025, and 0.100 in.) and the color used to represent that size on the conductor side and the component side are listed. This type of legend should be included as part of the completed layout drawing. Although any color may be selected for width identification, it is advisable to choose contrasting colors so that the determination of which side the conductor will be routed will be more readily observed. Suggested colors for the component side of the board are *red, blue,* and *green* and those on the circuit side are *yellow, purple,* and *brown.*

9-8

Packaging Feasibility Study

Since high-density packaging requires the placement of more components and circuitry into less space, the task of reducing package size results in more complicated boards and increased layout difficulties. Prior to initiating a layout sketch, the PCB designer may be confronted with problems concerning the size of the board and whether it will be a single-sided, double-sided, or multilayer design. Several possibilities exist. If the size of the finished board is predetermined, the designer must consider which type of layout is best suited for the amount of circuitry that is to be packaged in that given space. On the other hand, if the overall board size *and* the type of layout is specified, the designer must determine if the task of layout is plausible. This determination is the result of a *packaging feasibility study* which involves both *interconnection density* and *package density*.

Interconnection density refers to the number of conductor paths that can be placed on a board layer per given area. For example, it is apparent that it is possible to place more 10-mil conductors with 10-mil spacings in a 1-in.² area than it is to place 50-mil paths with 50-mil spacings in that same 1-in.² area. Interconnection density is typically specified in conductor width and spacing. To maximize the interconnection density, the designer must minimize conductor path widths and spacings and pad diameters.

Package density refers to the number of components mounted onto a given area or the number of lead access holes per given area. For example, it is a great deal easier to design a layout for a board having one resistor (2 holes) per square inch than it is to design a layout having five resistors (10 holes) per square inch. To increase package density, the designer must increase the number of components or holes per square inch of board space.

It can be seen that interconnection density and packaging density go hand in hand. If the designer initially decides that the design will incorporate minimum conductor path widths and spaces (maximum interconnection density), then the determination of how many components (or holes) can be packaged per square inch must be made. This will establish whether a single-sided design will suffice or if it will be necessary to provide a double-sided or even a multilayer design.

One method of *estimating* packaging feasibility is with the lead access *holes per square inch* (holes/in.²) technique. The designer first determines the usable surface area of the board in square inches by multiplying the length by the width. From this value, all areas of the board that will not be used for component mounting and conductor routing are subtracted. These unused areas include spaces around mounting holes and hardware, the spacing around the edge of the board, and cutouts and forbidden areas inside the board edges. The result of subtracting these unusable areas is the *total usable layout area* in square inches.

The next step in this feasibility analysis is to count all of the circuit interconnection points with the use of the circuit schematic. This is done as follows:

1. Count all two-terminal devices (resistors, diodes, capacitors, etc.) and multiply this count by *two*.
2. Count all three-terminal devices (transistors, UJTs, etc.) and multiply by *three*.
3. Count all 14-pin IC packages and multiply by 14, all 16-pin packages and multiply by 16, and so on.

4. Count all input/output points (i.e., external connections including power and ground).

   The total of the above is the number of interconnection points
(holes) to be made on the board. To determine the packaging density in units of holes per square inch, simply divide the total interconnection points by the usable board area. This is shown in the
following equation:

$$\text{package density (holes/in.}^2) = \frac{\text{total number of holes}}{\text{usable board area (in.}^2)} \qquad (9\text{--}2)$$

   To serve as an example of the calculation of package density, we
will use the single-sided design of the one-shot multivibrator circuit
described in Chapter 8. Refer to the dimensional drawing of Fig. 8–
6. The overall board dimensions are 3.50×5.00 in. Since the specifications require that no component be placed closer than 0.200 in.
of any board edge, the actual overall dimension for our calculation
becomes 3.10×4.60 in. There are a total of five areas on the drawing where no components are to be mounted. These are the four
areas around the mounting holes (each 0.600×0.600 in.) and the forbidden area (0.900×2.10 in.). The total usable board area can thus
be calculated as follows:

$$\begin{aligned} \text{usable area} &= 3.10 \times 4.60 - [4(0.600)(0.600) + (0.900)(2.10)] \\ &\simeq 10.9 \text{ in.}^2 \end{aligned}$$

   In order to calculate the total number of interconnection points,
refer to the circuit schematic shown in Fig. 8–1. We count that there
are eight resistors, two trim pots, six capacitors, four diodes, two 14-
pin DIP ICs, and five external connections. The total hole count is
as follows:

$$\begin{aligned}
8 \text{ resistors} \times 2 \text{ leads} &= 16 \\
2 \text{ potentiometers} \times 3 \text{ leads} &= 6 \\
6 \text{ capacitors} \times 2 \text{ leads} &= 12 \\
4 \text{ diodes} \times 2 \text{ leads} &= 8 \\
2 \text{ DIPs} \times 14 \text{ leads} &= 28 \\
5 \text{ external connections} &= \underline{5} \\
\text{total} &\quad 75 \text{ holes}
\end{aligned}$$

   The package density can now be calculated as follows:

$$\text{package density} = \frac{75 \text{ holes}}{10.9 \text{ in.}^2} \simeq 6.9 \text{ holes/in.}^2$$

To better understand the significance of this value, refer to Fig. 9–
16a, which is a plot of holes/in.^{2} versus recommended type of design, and Fig. 9–16b, which is a graph of package density versus board
type *or* layout difficulty versus packaging density in holes/in.^{2}. Figure 9–16a is used as follows. After calculating the package density

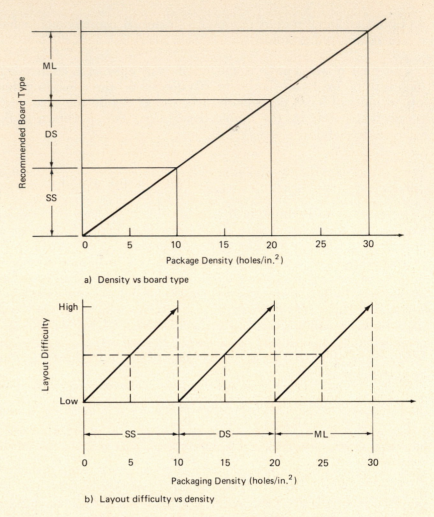

a) Density vs board type

b) Layout difficulty vs density

FIGURE 9–16 **Selection of the most appropriate type of layout as a function of package density and level of design difficulty.**

(holes/in.^{2}), it is located on the graph. As shown on this graph, an estimate of holes/in.^{2} below 10 can be considered suitable for a single-sided design. A package density of between 10 and 20 holes/in.^{2} usually requires a double-sided design. Densities of 20 to 30 holes/in.^{2} require a multilayer design.

Figure 9–16b shows that as package density increases, the degree of layout difficulty also increases as the holes/in.^{2} count approaches the upper limit of each board type. For example, a double-sided design with a package density of 11 holes/in.^{2} is normally much easier to lay out than would be a double-sided design with a much higher density of 19 holes/in.^{2}.

The graphs of Fig. 9–16 should be used only as a rough approximation for determining board type and difficulty of design since other factors also influence the nature of design. Some of these factors are the type of electrical circuit (analog, analog/digital, or digital) and special electrical and mechanical restrictions in the specifications. In general, however, the graphs of Fig. 9–16 can be used to determine (1) the type of design (single-sided, double-sided, or multilayer), (2) the feasibility of that level of design, and (3) the approximate degree of layout difficulty that can be expected.

Note in Fig. 9–16b that a horizontal projection, shown as a dashed line, extends across the graph from the midpoint of the layout difficulty scale and intersects a vertical dashed line at the midpoint of each of the packaging density groupings. Readings of package density of these vertical projections are 5 holes/in.^{2} for single-sided boards, 15 holes/in.^{2} for double-sided boards, and 25 holes/in.^{2} for multilayer designs. These projection points represent the ideal level of difficulty for each board type, that is, designs which can be laid out in a reasonable amount of time. Above this level, significantly more time would be required for each board type for completing a design.

In the next section we begin the high-density double-sided design of a dc motor control circuit. As a part of the preliminary planning, a packaging feasibility study is made in Section 9–10.

9-9

Analog Signal DC Motor Position Control Circuit

The design of the motor position control circuit shown in Fig. 9–17 will illustrate the problems encountered in the layout of a reasonably dense double-sided PCB.

The control circuit was selected because it has several features which have not been described previously. It is an analog circuit which contains five TO-style ICs. Four of these ICs are packaged in the 8-pin round-case style. The other IC has a 12-pin round-case style and is used as the power op-amp, IC5. This device requires a clip-on heat sink to dissipate the heat generated during its operation. This circuit also requires the packaging of seventeen $\frac{1}{2}$-W resistors, six 1-W resistors, four trim pots, eleven disc capacitors, and two electrolytic capacitors. The addition of six external connections and three dc power supply connections results in a PCB which must be designed to accommodate 45 discrete components and devices and 9 external connections.

Before beginning the layout of the motor control PCB, a brief description of the circuit will be presented. Refer to Fig. 9–17. Integrated circuit IC1 (LM301) is used as a differential amplifier. The input terminal IN(c) is supplied with the desired motor position information and the other input terminal IN(a) is used to receive feedback information from a potentiometer which has its control shaft geared to the motor drive shaft. This potentiometer is used to sense the actual position of the motor shaft. When information is supplied to the IN(c) terminal commanding the motor to turn, for example, in a clockwise direction (+ output to motor), the output terminal (pin 6) of IC1 applies the correct signal to the inverting input (pin 2) of the unity-gain amplifier IC4 (LM301). The output of IC4 (pin 6), in turn, provides the proper signal polarity to the inverting input (pin 5) of power amplifier IC5 (LH0041C). IC5 then delivers suffi-

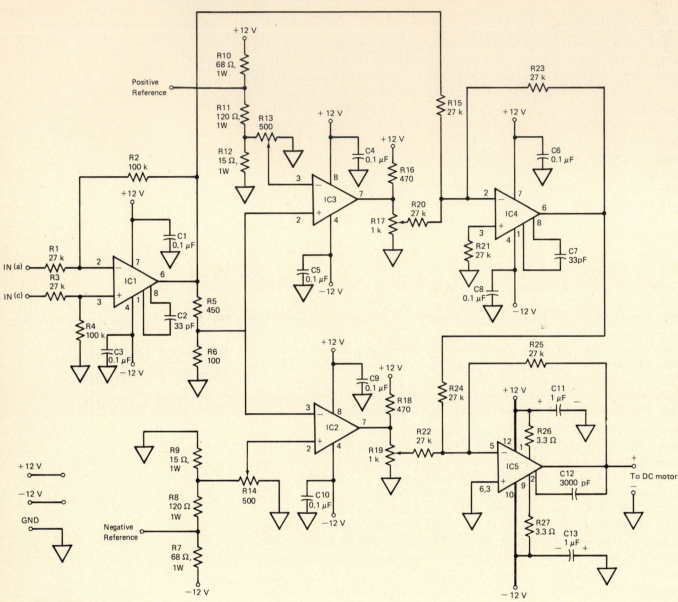

PARTS LIST

R1, R3, R15, R20, R21, R22, R23, R24, R25 — 27 kΩ, 1/2 W, 10%
R2, R4 — 100 kΩ, 1/2 W, 10%
R5 — 450 Ω, 1/2 W, 10%
R6 — 100 Ω, 1/2 W, 10%
R7, R10 — 68 Ω, 1 W
R8, R11 — 120 Ω, 1 W
R9, R12 — 15 Ω, 1 W
R13, R14 — 500 Ω trim potentiometers
R16, R18 — 470 Ω, 1/2 W, 10%
R17, R19 — 1 kΩ, trim potentiometers
R26, R27 — 3.3, 1/2 W, 10%
C1, C3, C4, C5, C6, C8, C9, C10 — 0.1 µF disc capacitor
C2, C7 — 33 pF disc capacitors
C11, C13 — 1 µF Tantalum Electrolytic Capacitors
C12 — 3000 pF disc capacitor
IC1, IC4 — LM301 Comparator Op-Amps
IC2, IC3 — LM 306H
IC5 — LH0041C

FIGURE 9–17   Schematic and parts list for the DC motor position control circuit. [From *Controlling Small DC Motors with Analog Signals*, by L. Sweer, T. Dwyer, and M. Critchfield, appearing in the August, 1977 issue of *Byte Magazine*. Copyright © 1977 Byte Publications, Inc. Used with the permission of Byte Publications, Inc.]

cient current to the motor with the correct polarity at its output terminals.

IC3 and IC4 (LM306H) are comparators which are wired as a *window detector*. This circuit is used to sense when the voltage level fed back from the potentiometer geared to the motor shaft is above or below upper or lower *window* or *threshold* voltage values. The upper threshold voltage is set by trim pot R13 and the lower voltage by R14. This window detector continuously monitors the position of the moving motor drive shaft. When the drive shaft has made the desired number of rotations and reaches the limit position in this direction, the appropriate voltage is set by the potentiometer at input terminal IN(c) and the output of IC1 goes to zero volts. This is sensed by the window detector circuit, which then delivers the proper polarity to IC4, feeding power amplifier IC5 and causing the output voltage to drop, which stops the motor rotation.

9-10

The Motor Control Layout Design Problem

Specifications for the layout of the motor control circuit will be described in the following design problem statement:

DESIGN PROBLEM
Design a printed circuit layout for the circuit shown in Fig. 9–17. The circuit is to be packaged within a 3.75×4.50 in. rectangular space with a 0.500-in. forbidden area along all four sides. Six 0.125-in.-diameter mounting holes are to be provided, one in each corner of the board, positioned 0.250 in. in from each board edge and one centrally located on the top and bottom of the 4.5-in. board dimension, 0.250 in. from the outside edge. No component body or conductor path (or pad) is to be placed closer than 0.025 in., at 1X scale, to the forbidden area.

All external connections are to be positioned along one board edge using turret terminals with 0.100-in. (maximum) shank diameters. The conductors shown as bold lines for IC5 on the circuit schematic must be capable of handling a minimum of 0.250 A of current. Sufficient space for a round-case-style clip-on heat sink with an outside diameter of 0.750 in. (maximum) is to be allowed for IC5. All power supply bus conductors originating at the external $+$, $-$, and ground supply terminals must also be capable of handling 0.250 A. The conductor widths for these connections are to be a minimum of 0.025 in.

The layout is to be designed for 0.059-in.-thick printed circuit stock having a copper foil thickness of 1.0 oz/ft^2, which, after plating, will be approximately a thickness of 2.0 oz/ft^2. The pad selection is to allow a 0.002-in. worst-case annular ring on the finished board. For this design, assume that the board will be assembled manually.

From the information provided in the problem statement, the first task is to determine (1) the package density in holes/in.^{2}; (2) if the

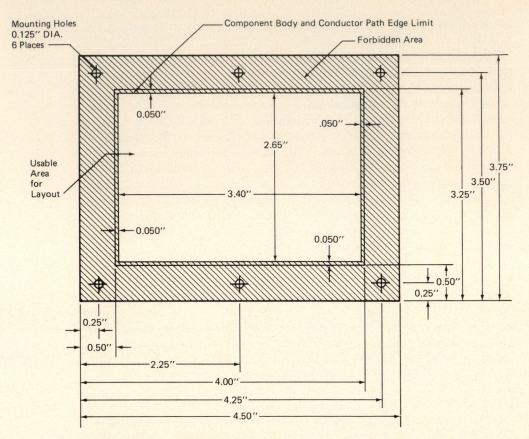

FIGURE 9–18 Dimensional information for the motor control circuit printed circuit board.

design is compatible to a single-sided, double-sided or multilayer board; and (3) the level of layout difficulty expected.

The usable area of the board will first be determined with reference to Fig. 9–18. This layout represents pictorially the dimensional information on the board size and shows the forbidden areas, which must be subtracted from the overall board size since they reduce the actual usable area. The total board area is 3.75 in. × 4.50 in. ≃ 16.9 in.^{2}. The total forbidden area, consisting of a 0.50-in. space along all four edges of the board, is found as follows:

$$\text{total forbidden area} = 2[4.5(0.5)] + 2[2.75(0.5)]$$
$$= 7.25 \text{ in.}^2$$

The amount of usable area for layout is determined by subtracting the total forbidden area from the total board surface area: $16.9 - 7.25 = 9.65$ in.^{2}.

The next step is to calculate the total number of component, device, and hardware holes required in the design. From the parts list information provided in Fig. 9–17, the total hole count is calculated as follows:

23 resistors (fixed) $\times$ 2 leads = 46
13 capacitors $\times$ 2 leads = 26
4 potentiometers $\times$ 3 leads = 12
4 ICs (8-pin) = 32
1 IC (12-pin) = 12
9 external connections = 9

total   137 holes

With the number of required holes calculated, the packaging density in holes/in.^{2} and the board type can now be determined. From Eq. 9–2, the package density is

$$\text{package density} = \frac{137 \text{ holes}}{9.65 \text{ in.}^2} \simeq 14.2 \text{ holes/in.}^2$$

Referring now to Fig. 9–16, we find that a package density of 14.2 holes/in.^{2} is approximately in the midrange of a double-sided design level. Within this level, an expected layout difficulty of about average is indicated. Remember that these findings are considered rough estimates. The calculations do not take into account larger components or special design requirements. For example, 1-W resistors are used in this design, and these require more board space than do $\frac{1}{2}$-W resistors. In addition, space must be provided for a clip-on heat sink to be used with IC5. These factors result in making the design more difficult than package density calculations initially indicate.

It is apparent that when considering all of the design specifications, a double-sided board with as high a density of parts as possible is required. Toward this end, we will select a 0.100-in. grid system with 0.050-in. minor graduations. This will allow conductors to be placed more easily with spacings of 0.010 in., which must be the criterion as a minimum conductor spacing specification.

The minimum conductor widths will be 0.010 in., which is the minimum recommended value for double-sided plated-through-hole designs. Since the majority of the circuit will be required to handle currents in the milliampere range, this choice of conductor width is more than adequate. To find the value that will result in a finished width of 0.010 in., refer to Table 9–3. The design problem statement specifies a copper thickness of 1 oz/ft^2. To obtain a 0.010-in. finished conductor width with this copper thickness, Table 9–3 shows that the layout must be done with a minimum taped artwork width of 0.013 in. at 1X scale, which, at 2X scale, will result in a width of 0.026 in.

The problem statement specifies that all power bus conductor paths for +12 Vdc, −12 Vdc, and ground (those connections shown as bold lines in the IC5 section of Fig. 9–17) be capable of handling a minimum of 0.250 A of current. It was shown in Section 8–8 that a finished conductor width of 0.010 in. will handle 1.3 A when plated to a thickness of approximately that of 2 oz/ft^2 of copper. Even though this is in excess of the 0.250 A required for this circuitry, the spec-

Conductor Width (inches)	Color Code	
0.010	Circuit Side	(Blue)
	Component Side	(Red)
0.025	Circuit Side	(Yellow)
	Component Side	(Purple)

FIGURE 9–19 Conductor width color code legend for the motor control circuit.

ifications require a minimum width of 0.025 in. for these connections. From the graph of Fig. 8–10, a 0.025-in. conductor for 2-oz/ft² copper foil can handle approximately 2.5 A of current. In order to result in a finished width of 0.025 in. on the 1-oz/ft copper foil that is to be etched, Table 9–3 tells us that a layout width of 0.028 in. is required.

Since the design is to be double-sided with two different conductor widths, a legend to distinguish between them is required as a part of the layout drawing. This is shown in Fig. 9–19. For the circuit side of the board, conductors drawn in blue are to be 0.010 in. wide and those drawn in yellow are to be 0.025 in. wide. The colors red and purple on the component side represent widths of 0.010 and 0.025 in., respectively.

The problem statement specifies that the pad diameters should be such that the finished annular ring specification will be, at worst case, 0.002 in. on the processed board. With an average packaging density to be laid out, all pad diameters should be as small as possible. Information on the selection of via hole sizes for boards with a thickness of 0.059 in. is provided in Section 9–1. All via holes should be drilled with a No. 69 drill bit (0.0292 in.) and provided with a pad diameter of 0.050 in. This is the smallest pad size that will result in a minimum annular ring of approximately 0.002 in. for the worst case.

Table 9–4 was constructed to aid in the determination of finished hole sizes and pad diameters for component and device leads. Much of the information in this table was obtained from Table 9–1. For each lead listed, the maximum diameter was obtained either by actual measurement or by manufacturers' specifications. A $\frac{1}{2}$-W resistor will be used to demonstrate how the data shown in Table 9–1 and other information were used in constructing Table 9–4. For a maximum lead diameter for the resistor of 0.032 in., we first read down the column of Table 9–1 labeled "0.002″ Annular Ring" until the next value of "Maximum Lead Diameter" which exceeds 0.032 in. is reached. This is found to be 0.034 in. The number immediately above the diagonal line is 0.051, which is placed in the "Finished Hole Size" column of Table 9–4. Projecting to the left from the 0.034-in. block to the column headed "Available Pad Diameters" in Table 9–1, the pad diameter is found to be 0.075 in. This is entered in Table 9–4 under the column labeled "Pad Diameter." The drill bit size is obtained from Appendix XIX, selecting bits with diameters that are approximately 0.003 in. larger than the finished hole size so as to allow for the plating of the copper and solder. Finally, the on-center hole spacings for all of the components and devices in the circuit are listed. For the resistors, Equation 8–3 is used. The spans (distance S) in Equation 8–3 were rounded off to the next highest grid, while dimension X was reduced to its minimum value to satisfy high-density design.

With the dimensional information tabulated in Table 9–4 and with consideration of the other design aspects discussed, the design of a double-sided layout for the motor speed control circuit may begin.

TABLE 9-4 Layout Information and Specifications for the Motor Control Circuit

Component or Device		Maximum Lead Diameter (in.)	Finished Hole Size (in.)	Drill Bit	Pad Diameter† (1X) (in.)	Lead Span or Hole Center Spacing (in.)
½-W resistors		0.032	0.051	No. 54	0.075	0.600
1-W resistors		0.042	0.061	1.65 mm	0.085	0.800
Disc capacitors	0.1 µF	0.025	0.046	1.25 mm	0.070	0.250
	33 pF					0.250
	3000 pF					0.200
Electrolytic capacitors		0.025	0.046	1.25 mm	0.070	0.200
Trim pots		0.025	0.046	1.25 mm	0.062	0.100
ICs	LM301	0.019	0.038	No. 59	0.062	0.350‡
	LM306H					0.350‡
	LH004 IC					0.350‡
External turret terminals		0.100	0.110	No. 33	0.130	N/A*
Via holes		N/A*	0.026	No. 69	0.050	N/A*

*N/A, not applicable.

†Recommended minimum pad diameter for 0.002-in. worst-case annular ring.

‡Lead circle diameter.

9-11

Double-Sided Component Layout Drawing for the Motor Control Circuit

The layout of the motor control circuit begins by first securing a sheet of precise 0.100/0.050-in. grid polyester onto a light table. A B-size sheet of good-quality vellum is then taped over the grid sheet. This B-size vellum is large enough to draw a 2:1 scale layout of a finished board with final dimensions of 3.75 × 4.50 in. At 2X scale, this will result in a drawing of 7.5 × 9.0 in., which leaves ample space for legends.

Refer to Fig. 9-18. With a straightedge and circle template, the 2:1 scale board border, corner brackets, and appropriate reduction scale are drawn. Mounting holes, located by targets, and the forbidden areas are then added to the drawing. See Fig. 9-20. Note in this figure that three special symbols have been placed outside the corner brackets. These symbols are called *register marks* or *targets*. In the fabrication of the board, they are used to align the patterns from one layer to another. In the preparation of the taped artwork master, which is discussed in Chapter 12, these targets are used to ensure alignment between individual terminal pads on each board layer. Even though the precise alignment of each side of double-sided taped artwork is essential, this alignment precision becomes even more

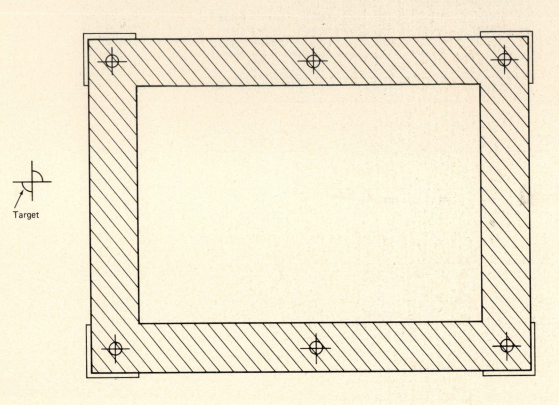

Target

— Reduce to 6.000 ± 0.005″ —

Pad Legend

*Symbol	Pad Diameter	Finished Hole Size	Plating	Remarks
○	0.050″	0.026″	Yes	Via Holes
○	0.062″	0.038″	Yes	Trim Pots and ICs
⊙	0.070″	0.046″	Yes	All Capacitors
⊗	0.075″	0.051″	Yes	1/2 Watt Resistors
○	0.085″	0.061″	Yes	1 Watt Resistors
○	0.130″	0.110″	Yes	External Turret Terminals
⊕	N/A	0.125″	No	Targets

*Color code pads for final hole size if required

Conductor Width Legend

Conductor Width	Color Code		
0.010″	Circuit Side	(Blue)	
	Component Side	(Red)	
0.025″	Circuit Side	(Yellow)	
	Component Side	(Purple)	

FIGURE 9–20   PC board outline and legends for the motor control circuit.

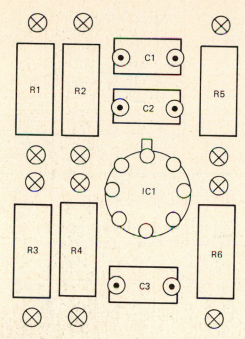

FIGURE 9–21 Initial component arrangement of first subcircuit.

demanding when working with multilayer boards. In addition to serving as a means of alignment, targets are also used to locate indexing or mounting holes. In this case, they would appear inside the corner brackets.

The legend for conductor width for both sides of the board, shown in Fig. 9–19, is positioned to the right of the board outline. In addition, a pad legend, using the information provided in Table 9–4, is also added to the layout. The pad legend shows the pad symbol, pad diameter (with color code, if required), finished hole size, which holes are to be plated and which unplated, and any other essential information. These legends become essential references for the construction of the taped artwork master.

After the preliminary layout, as just described, has been completed, the tentative placement of components and external connections can begin. Referring to the circuit schematic of Fig. 9–17, we see that it consists of essentially five subcircuits, each containing one IC package and several support components (resistors and capacitors). For example, IC1, together with the $\frac{1}{2}$-W resistors R1 through R6 and disc capacitors C1 through C3, form the first subcircuit. Concentrating on just this first subcircuit, the initial arrangement of its parts may be laid out as shown in Fig. 9–21, which closely follows the part positions as they appear in the schematic. Consideration is given to the correct component size, specified pad diameters and spacings, proper spacing of component bodies, and so on, all to the assigned 2:1 scale. Note that IC1 is positioned so that its pins are closest to the components to which each will be wired. This is to make the conductor path routing as easy as possible.

Continuing a similar layout of the parts of the subcircuits containing IC2 through IC5, the overall arrangement may appear initially

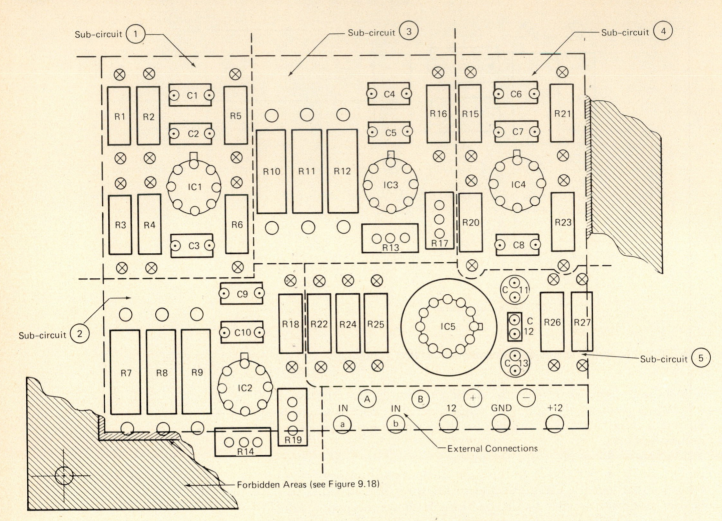

FIGURE 9–22 **Initial component layout of all five sub-circuits.**

as shown in Fig. 9–22. The dotted borders show the subcircuits. You will notice that, for two main reasons, this arrangement does not quite fit within the border of the forbidden area, shown in heavy dashed lines. First, the parts are too far apart. They can be positioned closer to utilize more efficiently the minimum allowable spacing of components and terminal pads (0.025 in. at 1X scale). Second, there appears to be a great deal of unused space above subcircuits one, three, and four and to the interior of the layout between subcircuits one, two, three, and five. Because this is a double-sided design and conductor paths can be routed on both sides of the board, greater latitude of parts rearrangement is possible than with a single-sided design.

To overcome this space problem, we will first move the $\frac{1}{2}$-W resistors R3, R4, and R6 in the IC1 subcircuit, the 1-W resistors R10, R11, and R12 in the IC3 subcircuit, and the $\frac{1}{2}$-W resistors R20 and R23 in the IC4 subcircuit along the top of the board as shown in Fig. 9–23. Notice that R24 from the IC5 subcircuit has also been moved to the top of the board to the right side of the new resistor grouping.

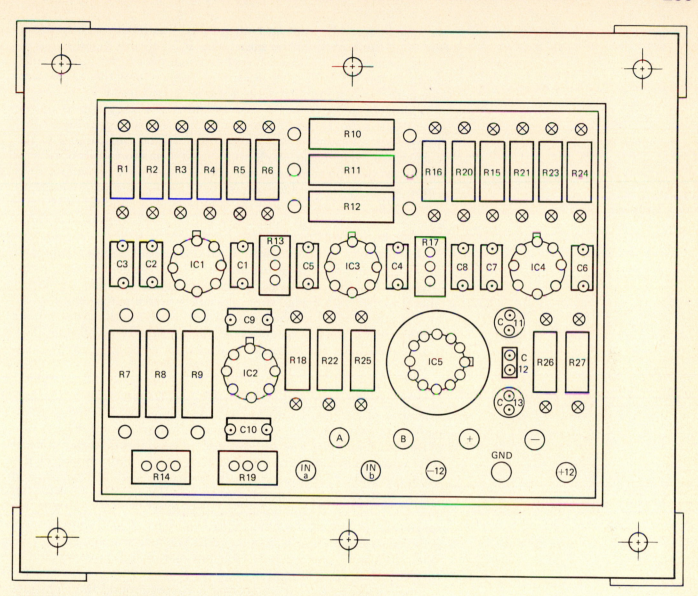

FIGURE 9-23 Final component layout after parts rearrangement.

This is acceptable since this resistor is also connected to pin 6 of IC4. The $\frac{1}{2}$-W resistors are spaced 0.050 in. apart to result in a 0.025-in. spacing at 1X scale. Also, the terminal pads and bodies of resistors R1, R10, and R24 are no closer than the minimum of 0.025 in. from the forbidden areas.

With the resistors aligned toward the top of the board, the other components and devices of the IC1, IC3, and IC4 subcircuits can now be balanced. Note in Fig. 9-23 that the rearrangement of those subcircuits in a row, running left to right across the board under the top row of rearranged resistors, has greatly reduced any unused board space. Also, it is seen that very little rearrangement of the parts in the IC2 and IC5 subcircuits is necessary for all of them to fit inside the usable area. By shifting trim pots R14 and R19 slightly to the left along the bottom of the board, the IC5 subcircuit and the staggered arrangement of the external connections can be moved sufficiently to the left so as to fit inside the usable area without crowding. Compare the bottom portions of Figs. 9–22 and 9–23.

With all of the parts positioned and labeled as they appear on the circuit schematic, the conductor path routing may begin. Recall that the majority of paths on the circuit side will be run parallel to the longest board dimension, while those on the component side will run parallel to the shortest side. See Fig. 9–13. The schematic should continuously be checked for correct path routing and path width. All signal paths represent a 0.010-in. minimum width and all power and ground paths (shown as bold lines in Fig. 9–17) represent a minimum width of 0.025 in. at 1:1 scale.

All short paths are drawn first. Each drawn path should be checked off on the schematic to ensure that all of the connections are properly made. As the next longer paths are routed, via holes are drawn where necessary to bridge over previously drawn paths. See Fig. 9–1a. Via holes should not be drawn inside a body outline (i.e., under a component). For the specified path color code for our design (Fig. 9–19), a via hole, for example, is necessary when it is not possible to route a red path without crossing a purple path.

The longer paths are finally routed to complete the layout. Whenever possible, power and ground paths are routed close to the outside edge of the board with individual component pads fed to these paths. For example, the −12-Vdc path is routed around the left side of the board and picks up the decoupling capacitor C10 and pin 4 of IC2. It is continued horizontally to pick up the bottom end of R7. From this pad, it is routed vertically to connect with the bottom end of C3. The path is then run horizontally to the right under ICs 1, 3, and 4 to pick up their decoupling capacitors C3, C5, and C8, respectively, as well as the pin 4 of each of these ICs. A short vertical path connects pin 10 of IC5 to the −12-Vdc external connection pad. With the use of one via hole, the top end of R27 is connected to complete the path routing to −12 Vdc.

The +12-Vdc external connection is routed around the right side of the board to the top of R26. With one via hole, it is connected to the top of C6 and then with another via, it is run horizontally above ICs 4, 3, and 1 to make all of the connections to the appropriate decoupling capacitors, IC pins, and resistors. A +12-Vdc path is also routed horizontally *below* IC4 to connect the top end of R26 to the positive lead of the electrolytic capacitor C11, and with the use of one via, this path is continued to pin 12 of IC5.

Because of the locations and number of ground points in this circuit, these connections are made through a network of paths on both sides of the board with the use of via holes. The same is true for routing inputs INa, INb, and A while one long and direct connection is made from input B to resistors R10 and R11.

A completed layout with all conductor paths routed is shown in Fig. 9–24. After all conductor path routing is complete, it should be carefully checked for accuracy and to see if any unnecessary via holes can be eliminated.

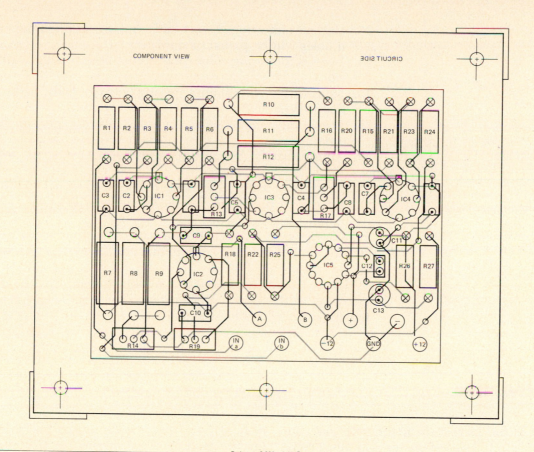

FIGURE 9–24 **Completed component and conductor pattern layout of DC motor position control circuit.**

The following checklist will serve as an aid in ensuring that the layout is complete and meets all circuit requirements:

1. Proper grid system used.
2. Mechanical dimensions of the board are correct.
3. No missing components.
4. Correct size and shape of components drawn to the appropriate scale.
5. All components and devices labeled as they appear on the circuit schematic.
6. Pads coded or keyed for multiple lead or polarized components.
7. No violation of minimum spacing requirements between component bodies.
8. No needless crowding of components.
9. Sufficient space to accommodate special hardware or mechanical adjustments.
10. All pad centers on grid.
11. Correct pad sizes.
12. Correct on-center pad spacings for all components.

13. Legend indicating correct pad symbols, finished hole sizes, and if holes will be plated or unplated.
14. Legend indicating color code system of conductor width on both sides.
15. No violation of minimum spacing between pads, paths, and pads and paths.
16. Proper clearance around via holes, board edges, and forbidden areas.
17. Approximately the same amount of path routing on both sides of the board.
18. Majority of paths on the circuit side running parallel to the longest side of the board and those on the component side running parallel to the shortest side.
19. Correct routing of conductor paths.
20. All unnecessary via holes eliminated.

When all minor corrections and adjustments have been made to satisfy all of the layout specifications, the final layout may then be transferred to the polyester sheet.

_____ EXERCISES _____

9–1 **(a)** What is the largest-diameter hole that can be drilled into a 0.075-in.-diameter pad and maintain a 0.005-in. worst-case annular ring? **(b)** What is the largest lead diameter that can be placed into that hole, after plating, and have a 0.010-in. clearance?

9–2 Determine the maximum number of 10-mil conductor paths with minimum air gaps of 10 mils that can pass between adjacent pads for both straight and offset arrangements for the TO-5, TO-99, and TO-100 packages shown in Fig. 9–4.

9–3 **(a)** Calculate the package density in holes per square inch for a double-sided board having the following description: 60 resistors, 12 capacitors, 2 diodes, 18 each 14-pin ICs, 9 each 16-pin ICs, 4 test points, and 75% usage of a 22-pin (each side) finger connector. **(b)** Using Fig. 9–16, determine the recommended board type and level of layout difficulty to be expected for the value of density calculated in part **(a)**.

9–4 **(a)** Construct a table similar to Table 9–4. Using the guidelines established in this chapter, calculate and record the values for each component in the multivibrator circuit shown in Fig. 8–1. **(b)** Using the information from part **(a)**, complete a pad legend and conductor width legend for the multivibrator circuit similar to those shown in Fig. 9–20. For this design, signal power and ground will all be 0.010-in. conductor widths and spacings.

9–5 Draw the 2X scale double-sided component and conductor pattern layout for Fig. 8–1 using the information obtained from Exercise 9–4 **(a)** and **(b)**. This double-sided design should be as dense as possible, with no mechanical restrictions as to board size or location of external connections. However, the finished layout should have all pertinent features, such as targets, corner brackets, and reduction scale, as well as pad and conductor legends (from Exercise 9–4).

10

Double-Sided Printed Circuit Board Design: Digital Circuits

_____ LEARNING OBJECTIVES _____

Upon completion of this chapter on the double-sided printed circuit design of digital circuits, the student should be able to:

1. Design and draw gold finger edge connector layouts.

2. Use circuit symmetry (when available) to position and code DIP packages on board designs.

3. Properly use decoupling capacitors on a layout.

4. Design power and ground distribution systems.

5. Perform a packaging feasibility study for digital circuit layouts.

6. Design double-sided digital circuit layouts using the guidelines presented in this chapter and in Chapters 8 and 9.

10-0

Introduction

The widespread use of digital electronics is the result of the technology which has developed powerful desktop and pocket computers. The semiconductor industry has packaged an astonishing amount of electronic circuitry onto a single chip, which has had its primary impact on the growth of digital circuits.

Digital circuits are generally packaged in the same form as other electronic circuit chips, with the dual-in-line package (DIP) predominating in the industry. The technology has advanced from 14- and 16-pin DIPs to 18- and 24-pin units, and more recently to 40- and 64-pin packages, which are used extensively in complex microprocessor chips. The DIPs are constructed so that their installation and pin interconnections are appropriate for printed circuit board applications.

Large digital systems, such as those in mainframe computers, are made up of many individual PCBs which are interconnected. Generally, the system is divided into a number of subcircuits, each of which is packaged onto a PCB. These individual boards are commonly referred to as *daughter boards*. Each of these boards, or cards, as they are often called, are fabricated with a series of gold-plated fingers on one end which plug into an edge card connector. The connector also has gold-plated spring tabs which make contact with the fingers. The group of edge connectors are, in turn, wired to a *backplane* or *mother board*. Signals terminating at the fingers of the individual cards are thus electrically interconnected to the other subcircuits through the mother board, which is typically of a multilayer design. It can be seen that the mother board performs the function of an extremely complex wiring harness.

The individual daughter boards are packaged as double-sided or multilayer boards and each has a well-defined block of circuitry. A typical daughter board might be a 64K RAM card (64,000-bit random access memory) which may serve as a memory board for a computer system.

It is beyond the intent or scope of this text to attempt to develop and describe an entire computer system. In this chapter we present the general considerations required to design a digital circuit daughter board. The style and form of printed circuit boards typically used in digital systems is presented first. The alignment and orientation of the ICs on the board as well as conductor path routing techniques for power, ground, and for decoupling capacitors are also discussed. A packaging feasibility study for double-sided digital circuits is demonstrated. This study involves finding the component density based on a method termed *equivalent DIP packages* rather than *holes per square inch*. The chapter concludes with the design of two practical digital circuits.

10-1

Printed Circuit Boards with Gold-Plated Fingers

One of the most common methods of making PCB interconnections is with the gold-plated finger edge connector design. This is shown in Fig. 10-1a. This type of design is used almost exclusively in digital board applications. The gold-plated edge terminations align and mate with contacts in the connector. These contacts provide spring tension to the fingers as they are inserted and seated into the connector. The contacts are also gold plated to provide optimum electrical contact and to prevent galvanic action, which occurs when two dissimilar metals are brought into direct contact with each other. The specifications for the gold plating provided by the PCB designer are typically a minimum of 30 μin. of 99.5+ pure gold (24 carat) plated over a minimum of 125 μin. of low-stress electroplated nickel.

External connections from the board are made through the edge connector contacts to the mother board. Note in Fig. 10-1a that the board is supported and aligned with the aid of *card guides* as it is inserted into the connector. Figure 10-1b and c show the *corner chamfers* and *edge chamfers* that have been fabricated onto the board.

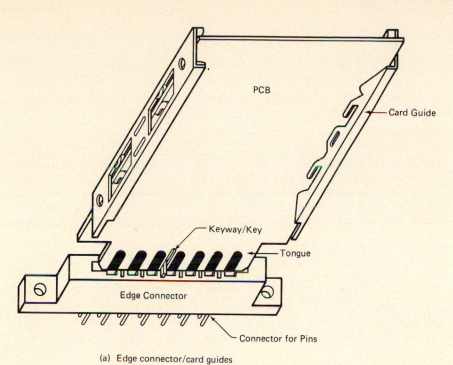

PCB

Card Guide

Keyway/Key

Tongue

Edge Connector

Connector for Pins

(a) Edge connector/card guides

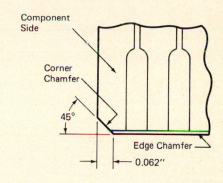

Component Side

Corner Chamfer

45°

Edge Chamfer

0.062''

(b) Detail for corner chamfer

FIGURE 10–1 Gold plated finger detail for edge connector designs.

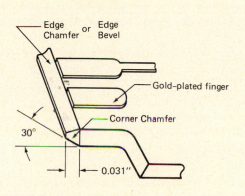

Edge Chamfer or Edge Bevel

Gold–plated finger

Corner Chamfer

30°

0.031''

(c) Detail for edge chamfer

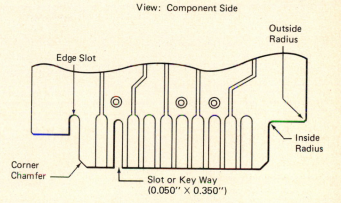

View: Component Side

Outside Radius

Edge Slot

Corner Chamfer

Slot or Key Way (0.050'' × 0.350'')

Inside Radius

(d) Detail for polarizing slot

(Refer to Section 5–6 for a detailed discussion of chamfers.) Corner chamfers facilitate the mating between the ends of the connector and the ends of the tongue (finger area). Edge chamfers at the leading edges of the tongue and fingers aid in board insertion.

To prevent incorrect electrical contacts when inserting the fingers into the connector, a *keyway* or *polarizing slot,* as mentioned in Section 5–6, is provided either in the finger area or at the end of the finger area. A keyway that is slotted into the finger area is shown in Fig. 10–1d. This is the most commonly employed method of keying. It is termed *between-contact* keying and results in no finger positions being lost. Some connectors are keyed *in contact,* which places the keying slot in one of the finger positions. In either method, the slot is located off-center on the tongue to allow for only one insertion position. When the board is properly aligned, the keyway mates with a *keying plug* in the connector ensuring correct board insertion. Refer again to Fig. 10–1a. *Between-contact* keying slots are dimensionally specified by width and depth (e.g., 0.050 × 0.350 in.). The tolerance of the keyway slot is not as critical as that of the tongue width. The slot must be sufficiently large to allow full insertion of the board into the connector. Because it is off-center, the slot can be fairly wide and still provide the correct insertion position. It is the close tolerance of the tongue width that ensures the correct mating between contacts and fingers.

For straight-edge cards with finger connections (those having no tongue) which plug into an open-ended line connector, the tolerance of the keyway is extremely critical. For this type of design, the slot must not only provide correct card polarization, but also must align mating contacts and fingers. Because of the exacting tolerances required for this mating hardware, fingers and slots must always be dimensioned from a 0,0 datum. As was shown in Section 5–7, this datum may or may not be a tooling hole. Slots should never be dimensioned from the edges of fingers or from a board edge. The 0,0 datum should be used as an exact zero reference point for all critical dimensions.

The detailed dimensional drawing of a typical PCB with gold-plated fingers and edge slot is shown in Fig. 10–2a. Because the fingers are designed to be on *both* sides of the board, the view shown in the drawing must be identified. Note that the drawing in Fig. 10–2a represents the component view. The reference point (0,0 datum) is located just outside the board in the lower left corner. Two primary datum lines, labeled x and y, are used to define the leftmost and bottom-most edges of the board. The overall board dimensions are 6.000 by 4.900 in., with tongue dimensions of 0.500 by 3.400 in. Three outside corners are broken with a 0.031-in. radius (R) and one inside corner is formed with a 0.0625-in. radius. Two 0.0625-in. × 45-degree corner chamfers are shown at the tongue ends to aid in board insertion. At the upper right corner of the board, note that another chamfer having dimensions of 0.125 in. × 45-degrees is shown. This is to accommodate a *circuit board ejector*, which is used to provide lever action against the rack to facilitate board insertion and removal from the edge connector. See Fig. 10–2b. On wide cards, two ejectors are recommended to prevent binding against the guides and unnecessary stresses on the connector.

Note that an *edge slot* is provided on the board shown in Fig. 10–2a for positive keying. The height of the slot from the leading edge of the fingers is 0.375 in. Its width is obtained by using the dimensions shown from the 0,0 datum to each of the slot edges. Thus it is seen that the slot width is 0.750–0.625 or 0.125 in. The location of the center of the slot is found by taking one-half of the width and adding it to the smaller edge dimension (0.625 in.) or subtracting it from the larger dimension (0.750 in.).

Refer to the finger area of the layout of Fig. 10–2a. Note that the tongue begins 0.750 in. from the y-axis datum and extends to a distance of 3.400 in. The first finger center is 0.812 in. from the y datum and the center-to-center distance between the first and last fingers is 3.275 in. The fingers are 0.093 in. in width and have a 0.062-in.

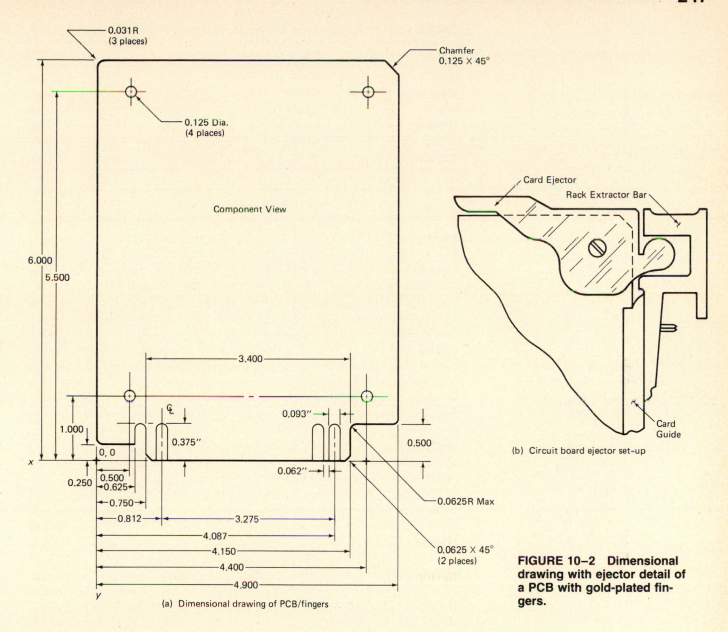

(a) Dimensional drawing of PCB/fingers

(b) Circuit board ejector set-up

FIGURE 10–2 Dimensional drawing with ejector detail of a PCB with gold-plated fingers.

spacing between them. The distance between finger centers is determined as follows. The finger legend shows that this board has 22 pins (fingers) per side with a finger center-to-center span of 3.275 in. For 22 fingers, there are 21 spaces. Thus the span is divided by the number of fingers minus 1 ($N-1$). This calculation is as follows:

$$\text{center-to-center finger spacing} = \frac{3.275}{22\text{-}1} = \frac{3.275}{21}$$
$$= 0.1559 \text{ or } 0.156 \text{ in.}$$

This distance is one of the most common finger spacings used in PCB designs. The dimensions of several other finger spacings used in the industry are shown in Table 10–1.

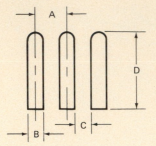

TABLE 10–1 Insertion-Type Connector Pattern Dimensions (in.)

A	B	C	D
0.050	0.031	0.019	0.250
0.100*	0.062	0.038	0.500
0.125*	0.062	0.062	0.500
0.150	0.080	0.070	0.500
0.156*	0.093	0.062	0.500
0.200	0.125	0.075	1.000

*Most common connector style.

When the specifications require that fingers be plated (gold over nickel), the PCB designer needs to include *finger extensions* and a *bus bar* in the layout. These are shown in Fig. 10–3. The finger extensions are used to electrically connect all of the fingers to a single bus bar for purposes of plating. After the plating has been completed, the board is routed (cut) to shape. The finger extensions and bus bar are removed when the tongue portion is routed to its finished dimensions. See Fig. 8–9.

The number of fingers specified on a board depend on the number of individual input and output connections required in the design and also the capacity of the connectors to be used. Connectors are available with all of the standard center-to-center spacings for the finger dimensions shown in Table 10–1 and with contacts on only one side for single-sided boards or on both sides for double-sided designs. The number of individual contacts available (total on both sides) ranges from 6 to as many as 120. The 44- and 60-pin-style connectors are most common to the industry.

Several methods are used to identify fingers and connector contacts on a layout drawing. Two methods of identification are shown in Fig. 10–4. For designs requiring up to 44 pins for double-sided boards (22 pins per side), the alpha/numeric system shown in Fig. 10–4a is used. The alphabet, omitting the letters G, I, O, and Q for clarity, are used on the circuit side of the board. For the component side, the consecutive numbers 1 through 22 are shown. When working with larger connectors, such as those with 60 pins (30 per side),

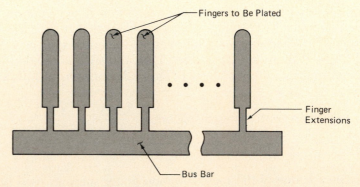

FIGURE 10–3 Finger extensions and bus bar for plating.

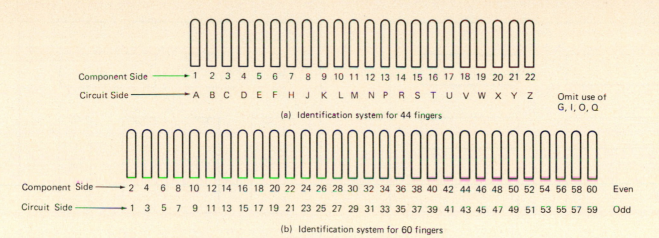

(a) Identification system for 44 fingers

(b) Identification system for 60 fingers

FIGURE 10–4   Coding systems for contact and finger identifications.

a common system of identification is to number the pins on the circuit side with odd numbers (1 to 59) and those on the component side with even numbers (2 to 60). See Fig. 10–4b.

10-2

Layout Considerations for Digital IC Packages

One of the advantages of designing the layout of digital circuits compared to that of analog circuits is that most digital circuitry is packaged in DIP-style cases which require very few, if any, support components. It is thus often possible to obtain a good balance and symmetry of parts distribution on the board.

A digital circuit containing 10 DIP packages on a rectangular board with fingers along one edge is shown in Fig. 10–5a. The arrangement of the ICs is in two rows of five DIP packages each with their long axes parallel to the shorter sides of the board. Note the uniform distribution and balance of parts that result from this layout.

Another layout of the same 10 ICs on the same shape and size board is shown in Fig. 10–5b. In this design, ICs 1 through 8 are mounted vertically and parallel to the longest sides of the board. The two remaining ICs, 9 and 10, are positioned horizontally at the top of the card in the space created by starting the first row of ICs (1 through 4) closer to the fingers.

The determination of which of the two layouts in Fig. 10–5 is more suitable is dependent on ease of conductor routing. The circuit schematic is reviewed for interconnections between ICs and between ICs and fingers, for power and ground buses, and for decoupling capacitor wiring. To maximize path routing flexibility, the number of available connector pins and the IC count per board, it is typical to position the ICs in the horizontal arrangement and parallel to the long sides of the board with the fingers also along one of the long sides. However, specifications do not always permit this optimum layout and alternative arrangements, such as those shown in Fig. 10–5, are employed.

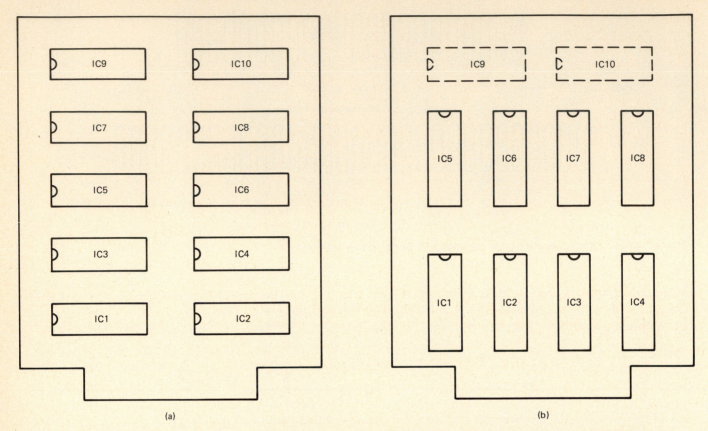

(a)

(b)

FIGURE 10–5   Two typical IC arrangements for digital PCB designs.

The conventional numbering order of ICs on a board begins at the bottom left and continues left to right and bottom to top. This numbering arrangement is for the purpose of IC positioning and orientation on the board only and is not to be confused with the IC designations on the schematic. For example, an IC designated as IC4 on a schematic may have several interconnections to fingers. In this case, that IC may be located on the board in position IC1 or IC2 in Fig. 10–5b. Again, an IC on the same schematic may be designated as IC7 and may interconnect only with ICs other than power and ground. There would be no reason for this IC to be positioned close to the edge connector, and thus it may appear in board position IC9 or IC10 in Fig. 10–5b. Basic layout guidelines for digital ICs are (1) those ICs having the most interconnections between them are positioned closely together, and (2) ICs having the most interconnections to fingers should be positioned closest to the edge of the card.

10-3

Decoupling Capacitors

Before considering the detail of various power and ground distribution systems used on double-sided board designs, the requirements of decoupling capacitors will be discussed. Although they are not always shown on the circuit schematic, decoupling capacitors are required in digital circuits to keep high- and low-frequency noise and other undesirable signals entering or being generated on the power supply conductor lines from being transmitted to any of the IC chips in the circuit. Their function is to shunt or bypass these unwanted signals around the chip to ground before they can enter the chip circuitry. Disc capacitors are the type most commonly used to filter out high-frequency noise from each digital package. The reactance of these capacitors (Xc) at the noise frequency is such that these unwanted signals see a smaller impedance path to ground through the capacitors than they see looking into the chip. The purpose of connecting individual decoupling capacitors to each IC is to provide a low-impedance path between the power supply pin of the case to the ground terminal of the same case. Decoupling capacitors must be placed as close as possible to the individual chips. They are generally nonpolarized and have relatively low capacitance with values in the order of 0.1 μF and lower.

Refer to Fig. 10–6a, which shows a single IC chip positioned on a PCB. Note that a +5-V bus is routed from a finger and is positioned above the IC case. The ground bus, which is routed from another finger, is routed below the case. It is seen that the IC is correctly powered with pin 7 connected to ground (GND) and pin 14 connected to +5 V. Observe that capacitor C1 is placed across the +5 V-to-ground connections at their entry points on the board. This capacitor is called a *bus decoupler* and its purpose is to provide a low-impedance path to bypass low-frequency noise before it can be transmitted to any branch lines that power the ICs. Polarized electrolytic capacitors (tantalum) having relatively large capacitance values of 10 μF and higher are used as bus decouplers. They are placed as close as possible to the power input finger connection and the ground connection.

We will now examine both low-frequency and high-frequency noise. In Fig. 10–6a, let i_{n1} represent low-frequency noise which is generated on the +5-V supply line. Capacitor C1 will provide a low-impedance path to ground and not allow the noise to be transferred to the ICs. High-frequency noise which is not bypassed to ground at the input power connections is labeled i_{n2}. As shown in Fig 10–6a, this noise will be transmitted to pin 14 of the IC and through the chips circuitry, causing potential operational problems before it eventually returns to ground. To prevent these unwanted signals from entering the chip, a decoupling capacitor has been added to the power supply terminals of the IC package. This is shown in Fig. 10–6b. The decoupling capacitor provides a protective bypass path to ground for noise signal i_{n2} and prevents it from causing internal chip problems.

In order to accommodate the decoupling capacitor, note in Fig. 10–6b that a ground path had to be extended from pin 7 to the ground lead of C2. The disadvantage of this wiring arrangement is that it adds a great deal more difficulty to the overall wiring. Another decoupling layout which is similar to that of Fig. 10–6b but results in less wiring difficulty is shown in Fig. 10–7. Note that the +5-V bus is positioned on the component side of the board and the ground bus on the circuit side. Again, low-frequency noise signals (i_{n1}) are shunted to ground through C1. High-frequency noise signals (i_{n2}) will

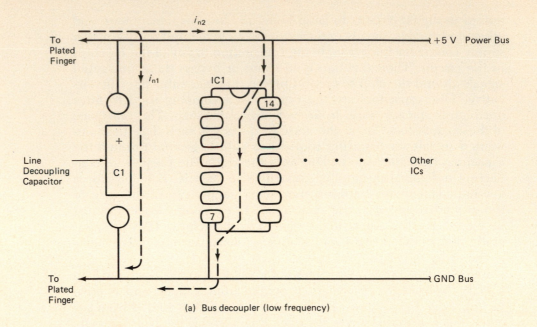

(a) Bus decoupler (low frequency)

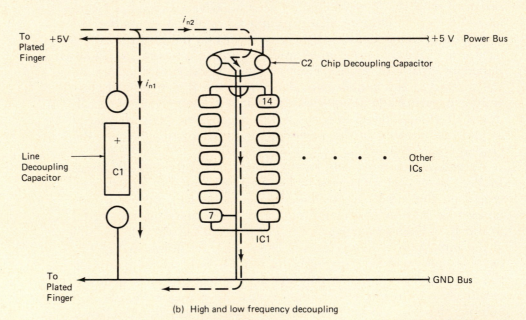

(b) High and low frequency decoupling

FIGURE 10–6   Single chip decoupling directly off main power and ground buses.

enter the IC at pin 14. To avoid this, a decoupling capacitor is shown properly wired in Fig. 10–7b. It provides a low-impedance path for i_{n2} around the IC to ground, thus preventing it from entering pin 14.

Still another wiring scheme for decoupling capacitors is shown in Fig. 10–8. Note here that decouplers C1 and C2 are positioned between the branch line of the 5-V bus (which connects power pins 14 of IC1 and IC2) and pins 7 which are connected to ground. Noise signals i_{n2} generated in the +5-V power line will thus be diverted

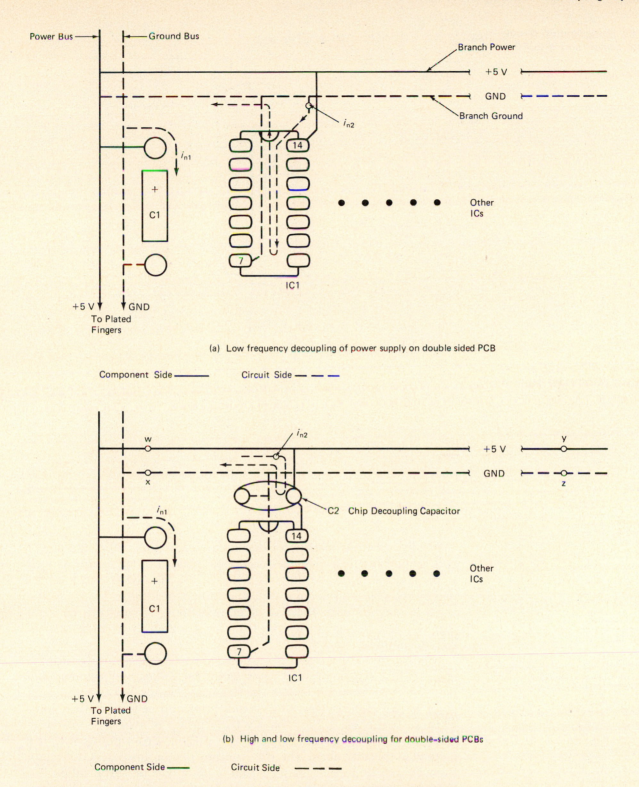

(a) Low frequency decoupling of power supply on double sided PCB

Component Side ———   Circuit Side — — —

(b) High and low frequency decoupling for double-sided PCBs

Component Side ———   Circuit Side — — —

FIGURE 10-7   Single chip decoupling off branch power and ground paths.

through C1 to ground. Any other noise signals i_{nx}, generated in the power line between pins 14 of each IC, will be shunted to ground and away from the power pin of IC2.

FIGURE 10–8 Multiple chip decoupling off main power and ground buses.

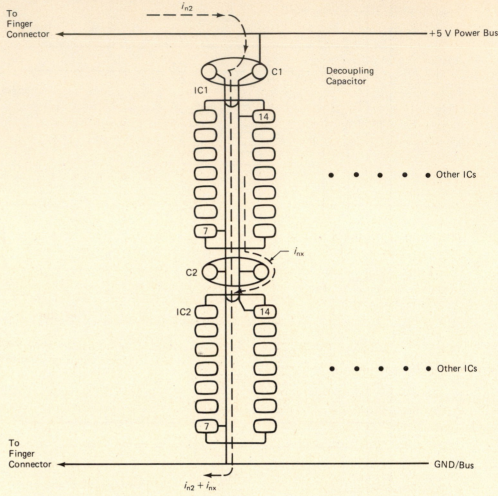

FIGURE 10–9 Multiple chip decoupling off branch power and ground paths.

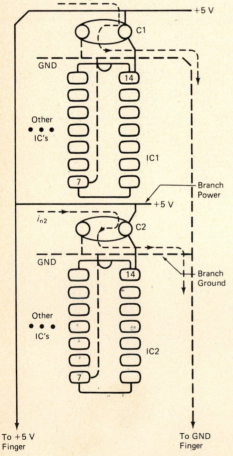

Finally, a wiring scheme for decoupling capacitors where the power and ground bus lines are run on opposite sides of the board as well as on opposite sides of the ICs is shown in Fig. 10–9. Each of the power and ground pins of the ICs is connected independently to the appropriate bus with the use of branch lines. It can thus be seen that IC1 and IC2 are protected from noise signal i_{n2} by their respective decoupling capacitors.

It is not generally necessary to provide a decoupling capacitor for each IC in a system. In fact, most circuit specifications require only one capacitor for a set of ICs. In this case, they need to be located close to the main power and ground lines and at the starting point of these branches as they supply their group of ICs. For example, capacitor C2 in Fig. 10–7b would be placed across points w and x instead of directly above the IC. Another decoupling capacitor could then be positioned at points y and z after two or three other ICs have been connected to these lines. Another common circuit specification is to place two decoupling capacitors (one for high-frequency noise and the other for low-frequency noise) directly across the main power and ground lines at the fingers. On boards that accommodate many ICs (100 or more), both schemes are often required, that is, across the main power and ground lines at the fingers *and* in the branch lines feeding groups of ICs. For these types of designs, the circuit design engineer and the PCB designer need to work together to obtain the best arrangement of all decoupling capacitors to achieve optimum circuit performance.

The two primary considerations in the placement of decoupling capacitors are: (1) connect them directly from the IC power pin to a ground bus or branch, and (2) never position them in such a manner as to route noise generated on the power lines from one chip to another.

10-4

Power and Ground Distribution Systems

Power and ground conductors are typically routed from their appropriate edge connector pins around the perimeter of the board. The design for the path widths of these main conductors as well as the number of pins required is determined by the total amount of current required by the circuit. Refer to Fig. 10–10a. The main power and ground conductors are labeled A and E. The branches (conductors B, C, D, F, G, and H) are reduced in width since their current-carrying requirements are much less than conductors A and E. For example, conductor B is required to handle only the current drawn by ICs 1, 2, 3, and 4. Pins 14 of these ICs are connected to branch B similar to the connections of the other power branches shown.

The main power supply path (conductor A) of Fig. 10–10a originates at pin 1 of the edge card connector and is routed along the left edge of the board. Branches B, C, and D are routed horizontally to the right of branch A and have smaller widths. The main ground bus (conductor E) originates at pin 22 and is routed along the right side of the board with branches F, G, and H, also of reduced widths, providing the ground connections (pin 7) of the individual ICs. Although this power and ground distribution system provides for ease of connection to the IC pins, there are several trade-offs required to improve the design. Note that the horizontal and outermost vertical spaces are congested with conductor paths. This type of layout requires greater separation between the horizontal rows of vertically aligned ICs in order to accommodate the pads for decoupling capacitors. If we space the rows of ICs farther apart, this defeats the objective of high-density packaging. A trade-off to reduce this problem is to place the decoupling capacitors close beside each IC in the vertical spaces. Because the power and ground buses originate from fingers at opposite ends of the tongue, bus decoupling cannot be done as efficiently as in other designs discussed in Section 10–3. However, with unobstructed access to fingers on one side of the board, via holes can be used for routing paths to fingers in order to bridge the ground bus (E) where required.

The layout of Fig. 10–10a represents one PCB power distribution system which can be designed for a digital circuit. For clarity in the illustrations, decoupling capacitors are omitted. In a finished layout, however, they must be included as part of the total design. In this section, the +5-V pin on all ICs shown will be pin 14 and ground will be pin 7.

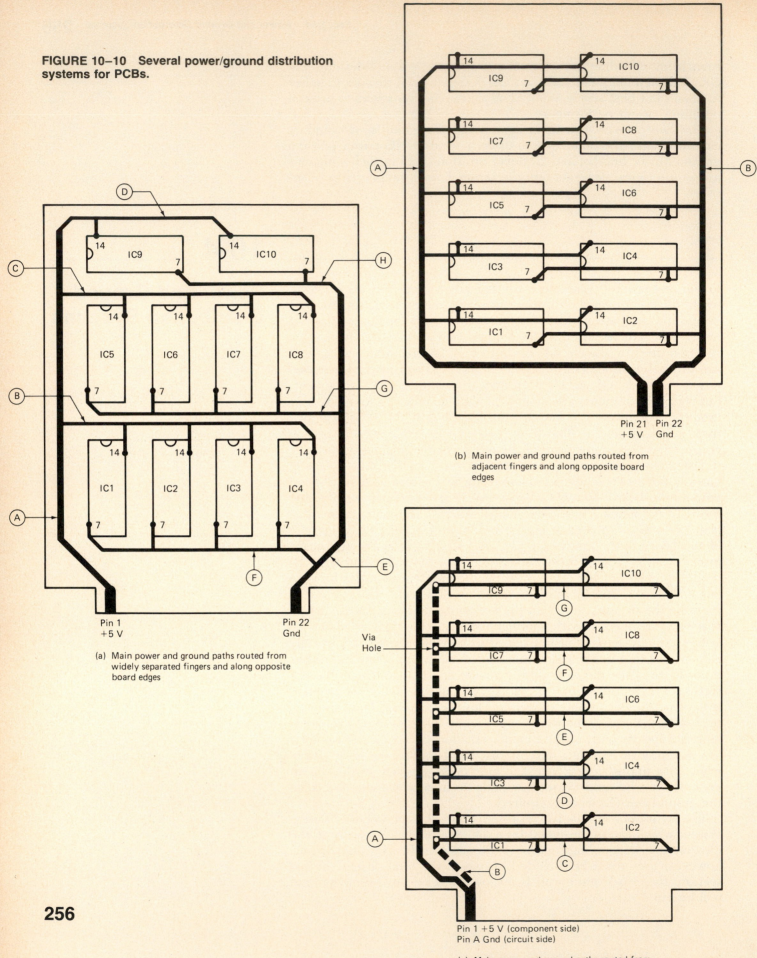

FIGURE 10-10 Several power/ground distribution systems for PCBs.

(a) Main power and ground paths routed from widely separated fingers and along opposite board edges

Pin 1
+5 V

Pin 22
Gnd

(b) Main power and ground paths routed from adjacent fingers and along opposite board edges

Pin 21
+5 V

Pin 22
Gnd

(c) Main power and ground paths routed from closely spaced fingers on opposite sides of the board

Pin 1 +5 V (component side)
Pin A Gnd (circuit side)

Via
Hole

Another power/ground distribution system is shown in Fig. 10–10b. The main power and ground conductors, labeled A and B, respectively, are again routed along the edges of the board, but in this design the power bus runs across the lower portion above the fingers. This is the result of using adjacent fingers for power and ground. Connections to pins 14 (power) and 7 (ground) of each IC row are made by routing the branches from the main lines horizontally and under the ICs. This method of routing results in more unobstructed space being available between the ICs for running signal paths and locating via holes. In addition, decoupling capacitors can be conveniently placed at the ends of the ICs. This arrangement also allows the ICs to be placed closer together, both vertically and horizontally. Because the main power bus runs horizontally above the fingers, via holes are required for routing paths on that side of the board to the fingers. By designing the power and ground to originate at adjacent fingers, bus decoupling at these fingers is efficiently achieved.

The last power and ground distribution system to be discussed is shown in Fig. 10–10c. The main power bus (A) is positioned on the component side of the board, originating at pin 1, and runs along the left side. The main ground bus (B, shown as bold dashed lines) originates from pin A and is also run along the left side of the board, but on the circuit side directly below the power bus. The power and ground branches are routed under the ICs from their respective trunks, similar to the arrangement of Fig. 10–10b. Note that the ground branches (C, D, E, F, and G) are routed on the component side of the board with the power branches. Connections to the main ground bus are made with via holes. This arrangement leaves unobstructed vertical routing for signal paths on the circuit side of the board to the fingers. Note the open vertical space at the right side of the board. This design makes it convenient to place the decouplers at the ends of the ICs. It also affords a great deal of flexibility in signal path routing on the component side and the placement of via holes. The unobstructed vertical space on the right side of the board is also open to all horizontal spaces between the ICs, which simplifies signal path routing to the fingers. Bus decoupling can be efficiently accomplished not only at the fingers (Fig. 10–7) but also at the branches as they run off the main bus lines.

In any power and ground distribution system design, sufficient space must be provided for any groups of long closely spaced (0.010 in.) signal paths which are common in digital circuit layouts. These paths are a potential source of board faults in the fabrication process. The problem arises from the difficulty encountered in the complete removal of the plating resist film between the paths prior to etching. Any resist not removed results in incomplete etching between paths, which causes bridges of copper to remain. These bridges short out adjacent signal paths. For this reason, the power and ground distribution system on any design should provide the maximum space for signal paths and via holes to ensure that the minimum air gap between long closely spaced paths will be 0.010 in. and greater, where possible.

As with other PCB design considerations, it has been shown that power and ground bus routing as well as decoupling capacitor placement involve trade-offs in order to maximize the available board space.

10-5

Packaging Feasibility Study for Digital Designs

Recall from Chapter 9 that a procedure for evaluating the level of package density for analog circuits was presented. This evaluation aided in the determination of what type of printed circuit board design should be employed (i.e., single-sided, double-sided, or multi-layer design). This packaging feasibility study also showed that for a given usable board area, as the number of holes/in.^{2} increases, so does the density of the parts. As the density increases, the level of layout difficulty became greater. Rough guidelines were established which recommended that single-sided boards have a maximum of 10 holes/in.^{2}, and that double-sided designs be considered for packaging densities of 10 to 20 holes/in.^{2}, with 15 holes/in.^{2} considered a typical value. Multilayer boards would be considered for densities above 20 and toward 30 holes/in.^{2}

A similar packaging feasibility study is applied to digital designs. However, the method of determination and the units will be different from those presented in Chapter 9. Because digital ICs are typically enclosed in dual-in-line-packages (DIPs), the packaging density is defined in units of the number of *DIPs per square inch (DIPs/in.^{2})*. Since DIP packages vary in size from 14-pin to 64-pin, it becomes necessary to develop the concept of an *equivalent DIP package* which will represent this range of sizes. Typically, the *16-pin* DIP is considered one equivalent dual-in-line package or *EDIP*. The number of EDIPs in a design is determined by dividing the total number of lead holes (to include the number of fingers used) by the number of holes for one EDIP (16):

$$\text{EDIPs} = \frac{\text{number of holes}}{16} \qquad (10\text{--}1)$$

The packaging density is now given as

$$\frac{\text{packaging density}}{(\text{EDIPs/in.}^2)} = \frac{\text{number of EDIPs}}{\text{in.}^2 \text{ of usable board area}} \qquad (10\text{--}2)$$

For the application of Equations 10–1 and 10–2, we will consider the following packaging problem.

DESIGN PROBLEM

On a 4.5×4-in. double-sided printed circuit board having 15 in.^{2} of usable layout area, the following are to be mounted: ten 16-pin DIPs, two 24-pin DIPS, and six decoupling capacitors. The design will be fabricated on a board having 36 fingers (18 per side), 15 of which will be used for external connection points.

We will begin the determination of packaging density by finding the number of holes that will be required in the design, including the count of fingers used for making external connections:

$$
\begin{array}{llr}
10 \times 16\text{-pin ICs} & = & 160 \\
2 \times 24\text{-pin ICs} & = & 48 \\
6 \times 2 \text{ lead capacitors} & = & 12 \\
15 \text{ external connections} & = & \underline{15} \\
& \text{total} & 235 \text{ holes}
\end{array}
$$

The number of EDIPs in the design are

$$
\text{EDIPs} = \frac{235 \text{ holes}}{16} \simeq 14.7
$$

Finally, the packaging density is determined using Equation 10–2 as follows:

$$
\text{packaging density} = \frac{14.7 \text{ EDIPs}}{15 \text{ in.}^2} = 0.98 \quad \text{or} \quad \simeq 1 \text{ EDIP/in.}^2
$$

This example illustrates a typical packaging density for a double-sided digital design. The 1.0-EDIP/in.^{2} density is considered a valid compromise for dense packaging to include 10-mil conductor path widths and spacings, 50-mil via hole pad diameters and minimum pad diameters for plated-through component lead holes, as discussed in Chapter 9.

At a packaging density level of 1 EDIP/in.^{2}, it is estimated that an experienced PCB designer can produce a layout at the rate of one EDIP every $1\frac{1}{2}$ hours. For our example design problem having 14.7 EDIPs, it would take approximately 22 hours to complete the layout. Of course, as the packaging density decreases below 1 EDIP/in.^{2}, double-sided layout becomes less difficult and requires less time to complete. Double-sided designs may be employed for densities up to approximately 1.5 EDIPs/in.^{2}. This is regarded as the point at which consideration needs to be given to the use of multilayer boards. Clearly, the time required to complete these layouts will be markedly increased.

One advantage of digital circuit layouts that is rarely seen in analog circuits is that there is a high degree of built-in symmetry of parts. This is due to the fact that the vast majority of the parts used are ICs which have the same size and shape. The result is that the designer is able to produce much denser layouts on double-sided plated-through-hole printed circuit boards.

10-6

Design of a Digital Memory Board

To demonstrate the design of digital circuits, we will begin with an extremely symmetrical card with 4K of *random access memory* (RAM). In the next section, a more complex circuit consisting of primarily digital circuitry but including a few analog devices will be designed.

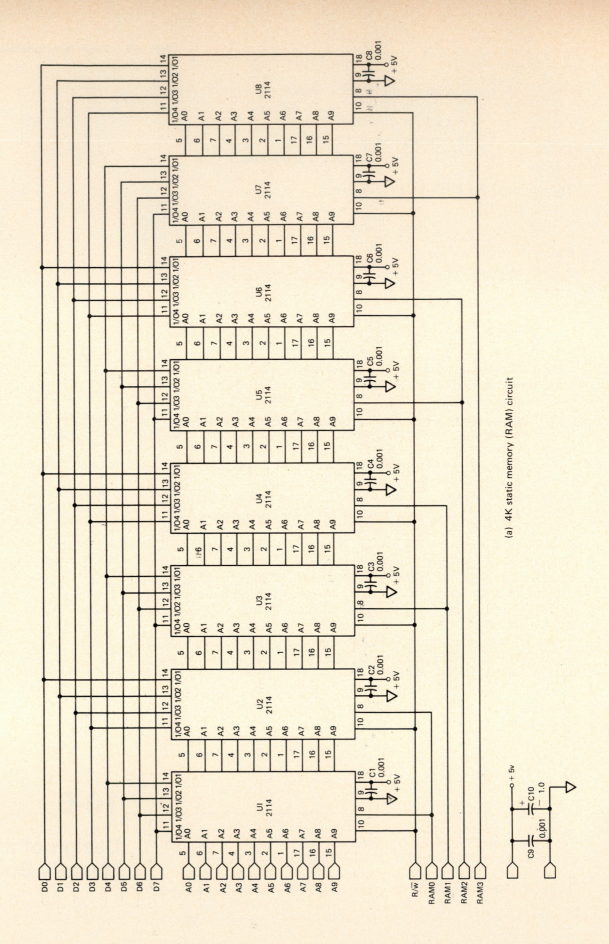

(a) 4K static memory (RAM) circuit

260

Side	1	2	3	4	5	6	7	8	9	10	11	12	13	14	15	16	17	18
Component Side			A0	A1	A2	A3	A4	A5	A6	A7	A8	A9					+5 Vdc	
Circuit Side	A	B	C	D	E	F	H	J	K	L	M	N	P	R	S	T	U	V
			D4	D5	D6	D7	R/W	RAM 0	RAM 1	RAM 2	RAM 3		D0	D1	D2	D3		GND

(c) Finger identification for the 4K static memory board

(b) Dimensional drawing for the memory board

FIGURE 10–11 Electrical and mechanical information for the RAM memory PCB.

261

The circuit schematic for the memory card is shown in Fig. 10–11a. Note that except for the decoupling capacitors, there are no support components. Because of this, the circuit can be designed onto an extremely dense double-sided board with a high EDIP/in.^{2} value.

As shown in Fig. 10–11a, the +5-Vdc power line and the ground bus are to be immediately decoupled with a 1-μF electrolytic capacitor in parallel with a 0.001-μF disc capacitor as they enter the board. A power distribution system will be adopted so as to supply +5 Vdc to pins 18 and ground connections to pins 9 of each of the 18-pin DIP packages. Each of the ICs (U1 through U8) are to be provided with a 0.001-μF decoupling capacitor. Ten *address bus lines* (A0 through A9) are connected to each IC and interconnected to fingers 3 through 12 on the component side of the board. Further, all *data bus lines* (D0 through D7) are to be connected to each IC and interconnected to the circuit side of the board to fingers C through F for bus line D4 through D7 and fingers P through T for D0 through D3. All of the finger connections for both the component and the circuit side of the board are listed in the table in Fig. 10–11c.

Note in Fig. 10–11a that every other chip (U1, U3, U5 and U7) is connected to the four data lines labeled D4 through D7. If we position the IC packages with commonly grouped data lines together (side by side), this will greatly simplify the routing of data lines to their respective pins (11, 12, 13, and 14).

The R/$\overline{\text{W}}$ (Read/Write) line is routed to all of the ICs at pin 10 and is connected to finger H on the circuit side of the board. Finally, note that pins 8 of an odd/even pair (1–2, 3–4, 5–6, 7–8) of chips are wired together and then to one of the four RAM lines. This is to form an 8-bit digital word having a capacity of 1K. The entire circuit will thus have 4K of 8-bit static memory.

Before beginning the board layout, we first need to decide on the overall orientation of the IC packages. From the circuit diagram, observe that many of the parallel pin connections of each chip are to be routed to fingers at the edge of the card. For this reason, we will position the ICs with their long sides parallel with the edge connector. This will provide a more direct approach of many signal paths to the fingers on the circuit side of the board, while power and ground bus lines can be routed on the component side.

The circuit will be laid out on the board outline shown dimensioned in Fig. 10–11b. The usable layout area, labeled *Component and Conductor Path Limit*, is approximately 9 in.^{2}. From the table of Fig. 10–11c, it is seen that 27 out of the 36 fingers are to be used. (The coding of individual fingers in this table is done in accordance with the system established in Fig. 10–4; that is, numbers are used to identify fingers on the component side of the board and letters used to identify those on the circuit side.) In addition, a total of 10 decoupling capacitors will be included which require 2 leads × 10, or 20 holes. The hole count for the ICs is 8 × 18 pins or 144 holes. Thus the total hole count for the design is 191.

To determine the EDIP count for this layout, we use Equation 10–1:

$$\text{EDIPs} = \frac{191 \text{ holes}}{16} \simeq 11.9$$

Using Equation 10–2, we can now calculate the package density:

$$\text{package density} = \frac{11.9 \text{ EDIPs}}{9 \text{ in.}/^2} \simeq 1.32 \text{ EDIPs/in.}^2$$

This figure represents a fairly dense package which approaches the limit of double-sided design. The main reason that such a dense package would be considered for double-sided design is due to the almost perfect symmetry of parts to be positioned on the board. This results in very large reductions in space requirements for path routing, allowing a much denser package to be designed than in the typical 1.0 EDIP/in.^{2} density figure. Of course, this layout would require rigid adherence to having both conductor paths and air gaps at the minimum widths of 10 mils. In addition, the IC packages will be spaced at a distance of 0.1 in. from each other. The decoupling capacitors will be positioned at the ends of each chip close to pin 1.

To achieve the high-density requirements of this digital circuit, it is essential that the 10-mil signal paths be routed between adjacent pins of the IC. This would require the spacing between adjacent pins to be 0.050 in. at 1:1 scale. For the pins, *oval* pads measuring 0.050 in. wide × 0.125 in. long at a 1:1 scale will be required. With these reduced pad sizes, a smaller drill hole size of 0.029 in. is needed to result in a 2-mil annular ring. These trade-offs are a requirement in all highly dense packages.

10-7

Component Layout Drawing for the Memory Board

The drawing of the layout for the static memory card will be done in three phases: (1) component placement with power and ground distribution system, (2) signal path routing, and (3) layout refinement.

The first phase of the memory card is shown partially designed in Fig. 10–12. Note that the odd-numbered ICs (U1, U3, U5, and U7) are positioned one above the other forming a column on the left side of the board. This position places the long side of each IC parallel with the fingers. In similar fashion, the even-numbered ICs (U2, U4, U6, and U8) are aligned along the right side of the board. Recall from Section 10–2 that the IC numbering system on a board begins at the bottom left and continues left to right and bottom to top. It works out in Fig. 10–12 that this conventional numbering system happens to correspond to the IC designations shown in the wiring diagram.

The high-frequency chip decouplers (C1 through C8) are positioned at the ends of each IC. The high- and low-frequency main power decouplers (C9 and C10) are connected in parallel at the 17U

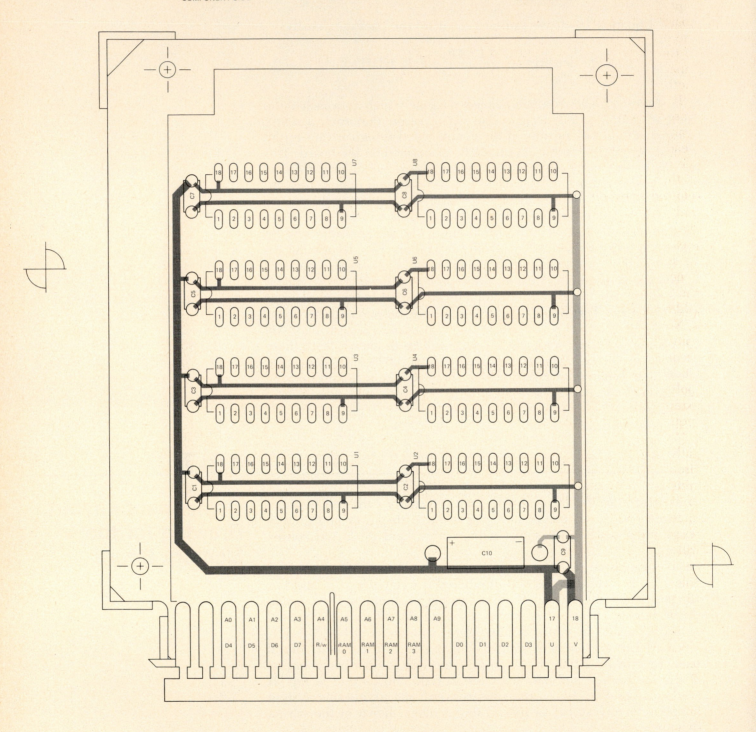

REDUCE TO 4.000 ± 0.005"

FIGURE 10–12   Layout Phase I of static memory board showing all ICs and decoupling capacitors with power and ground bus routing.

and 18V fingers (see Fig. 10-4). The power and ground distribution system is routed similar to the technique shown in Fig. 10–10b, except that the main ground bus is routed on the circuit side of the board and its branch lines placed on the component side with the use of via holes. This will allow unobstructed access for signal paths to the fingers on the circuit side. In addition, the chip decoupling capacitors are easily positioned in this distribution system using the method shown in Fig. 10–8.

It may initially appear in Fig. 10–12 that the ICs are positioned closer together than necessary. However, note in the wiring diagram of Fig. 10–11a that U1 through U8 have common Address and Read/Write connections. For this reason, space is allotted between the top of the board and ICs U7 and U8 for the convenient routing of the signal paths between the two columns of ICs. Space is also allotted between U1 and U2 and the fingers for decoupling power and ground as well as for convenience of routing signal paths from both columns of ICs. Generally, there is a need for via holes in this area. Again, the initial layout appears that the two IC columns could be spaced closer together. This vertical space between the columns is necessary, however, for the convenient routing of signal paths for the random access memory (RAM) connections between pairs of odd/even numbered ICs and fingers.

Note that the layout of Fig. 10–12 includes the labeling of all components, IC pins, and fingers. These are necessary before the second phase of our design, signal path routing, may begin. Also included are the mechanical requirements of the board, such as the tooling holes, card ejector mounting hole, corner brackets, finger extensions with bus bar, and the reduction scale. This information is required for the preparation of the taped artworks.

With the first phase of the layout completed, the routing of signal paths may begin. All of these paths will initially be positioned on the circuit side of the board and perpendicular to the branch power and ground paths on the component side. Since signal paths will be routed to fingers, let us first discuss the criteria for the positioning of each finger on the board. Finger connections should be placed as close as possible to the IC pin connections to which they will be tied. This will result in a minimum of crossovers which will require via holes for making the connections. Remember that the excessive use of via holes drives up the production costs of the board. Refer to Fig. 10–13. Note the designation of each finger is such that the routing to the appropriate pin connections is as close and direct as possible.

Signal path routing can best be demonstrated by considering each of the following steps: (1) connecting common Address and Read/Write pins of each IC in both columns; (2) linking these between columns; (3) connecting to the appropriate fingers; and (4) routing the Data lines and RAM lines. These steps are not necessarily accomplished in the order given, but a better understanding of path routing will be realized if each of these steps is considered separately.

We will begin by connecting all of the pins 1 in each vertical IC column. These paths will pass between pins 17 and 18 of each chip as they make their way to each pin 1. See Fig. 10–13. All of the pins 2 are then connected together, each path being routed between pins

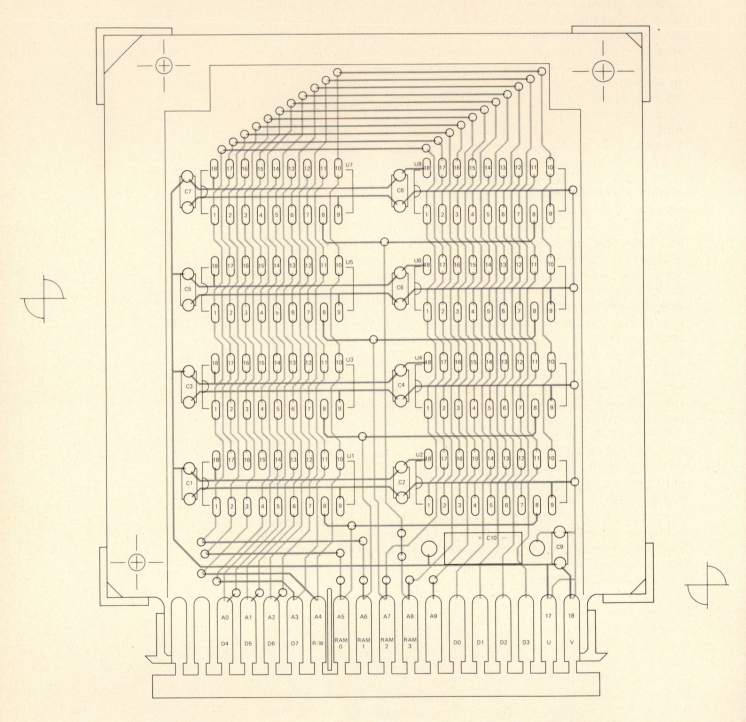

Reduce to 4.000 ± 0.005″

FIGURE 10–13 Layout Phase II of static memory board with the addition of signal paths.

16 and 17. This procedure is continued until all related pin numbers of each IC are connected according to the wiring diagram.

The next series of paths that need to be routed are those which link the Address and Read/Write connections in the two IC columns. These connections are conveniently made with the use of two series of via holes and 11 closely spaced horizontal signal paths which are positioned on the component side in the space provided at the top of the board. The initial layout of these paths is also shown in Fig. 10–13.

With reference to the table of Fig. 10–11c, connections from the paths routed to the appropriate fingers are made. The closest IC pins to the designated fingers are routed first. As these paths begin to block unconnected fingers, bridging paths with the use of one or more via holes is required. Remember that the vertical and horizontal orientation of conductor paths between both sides of the board needs to be maintained. If a signal path is designated to be connected to a finger on the component side of the board, possibly only one via hole will be required. Refer again to Fig. 10–13. An example of making a finger connection using a single via hole is the routing of the path from pin 15 of U2 to finger A9. Again, note that the connection from pin 1 of U1 to finger A6 is made by using three via holes. In the initial path routing phase, the use of as many via holes as appear necessary are included. It is in the refinement stage of the design that adjustments of path routing are made in an effort to reduce the number of via holes required.

The remaining signal paths to route are the Data lines and the RAM lines. The Data line connections for D0 through D7 are made vertically on the circuit side of the board to connect to the appropriate pins in both IC columns. Note that a linkage between IC columns for these lines is not required since fingers D0 through D3 are connected only to the even-numbered ICs and D4 through D7 to the odd-numbered ICs. In addition, no via holes are required for making these connections due to the positions of the fingers and ICs which results in an unobstructed path route to the respective pins.

The RAM lines are routed vertically on the circuit side of the board in the space provided between the IC columns. The connections to the appropriate IC pins from fingers RAM0 through RAM3 are made with the use of via holes. The paths are routed directly downward to these centrally positioned fingers. All of the paths can make direct connection to the fingers with the exception of RAM3, which requires two via holes for connection. The resulting drawing, which includes the initial signal path routing added to the component layout and power and ground distribution system, is shown in Fig. 10–13.

The final phase for the design of the memory card is the refinement of the layout. No major modifications, such as the repositioning of components, is made in this phase. The initial layout of Fig. 10–13 is reviewed for the purpose of making simple adjustments in signal path routing with the objective of achieving a more efficient use of via holes to result in an improved overall layout. Remember that any path designated as a blue color (circuit side) can be changed to one designated as a red color (component side) as long as it does not cross another red path or result in any circuit specification to be violated. The reverse of this is also true for changing red color paths to those having blue color.

In the refinement phase, caution must be exercised when simplifying path routing, eliminating unnecessary via holes, or in routing a path from one side to another that severe reductions in the air gaps do not result. These reductions will be prone to solder bridging and may cause manufacturing problems. Where air gaps are reduced below the minimum of 0.010 in., a *solder mask* becomes a requirement to prevent solder bridging. (Solder masks are discussed in detail in Chapter 14.) The mask is resistant to the solder generated by wave soldering systems. It covers essentially the entire conductor pattern with the exception of the pads. Although this is an effective means of eliminating bridging problems, it increases the overall cost of the board since it requires additional artworks and steps of processing. For this reason, any reduction of air gaps resulting from changing path positions should make the use of a solder mask optional rather than required.

Recall that the paths appearing on the circuit side of the board are routed in the direction of motion across the solder wave. For this reason, care must be exercised when considering the changing of paths or groups of tightly spaced paths from the component side to the circuit side. This problem becomes more significant for long parallel paths whose spacings become less than 10 mils and whose direction does not conform to that of the predominant paths on that side. These cautions do not apply to paths moved from the circuit side to the component side, however, because the component side is not subjected to wave soldering, which eliminates the bridging problem. Thus the relocation of paths to the component side, even if their direction is perpendicular to that of the existing paths, is acceptable practice as long as the specifications for minimum spacings are not violated.

The first memory card layout improvement is to eliminate one of the three via holes used to route the Address line from pin 1 of U1 to finger A6. As shown in Fig. 10–14, only two via holes are required for this connection since vertical path direction is acceptable on the component side of the board.

Second, a major design refinement is at the upper portion of the board. Careful inspection of the conductor pattern in this area suggests that by making a number of simple changes, the entire series of via holes on the right side can be eliminated. This reduction by 11 via holes is extremely advantageous in terms of lowering production costs. The path adjustments are made as follows. The power and ground paths from C7 to pin 18 of U8, which also connects to C8, are moved from the component side to the circuit side. This will also require the changing of sides for many of the signal paths leading upward from U5 and U7. With the branch power and ground lines now located on the circuit side, the signal paths leading upward to the left series of via holes from pins 1 through 7 of U7 are repositioned to the component side of the board. This is also true for the paths connecting pins 11 through 17 of U5 to the same pin numbers of U7.

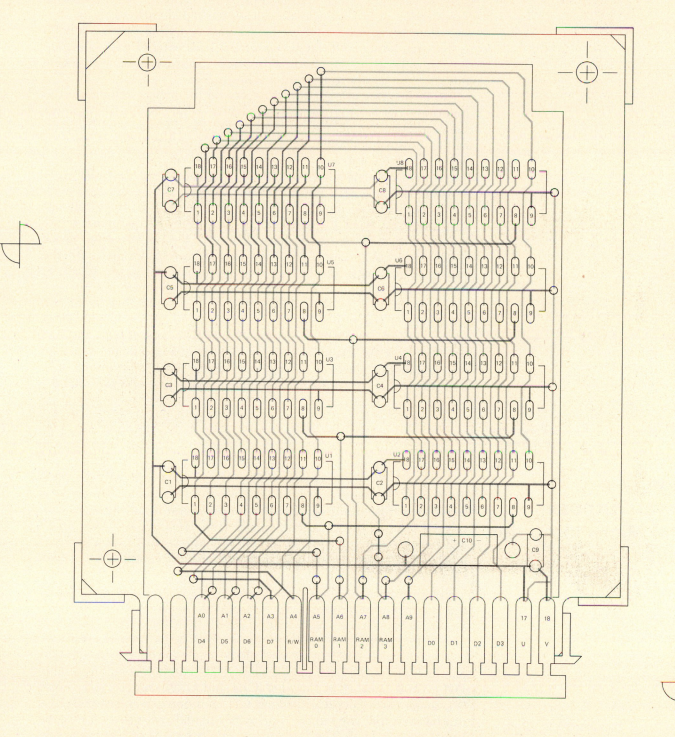

COMPONENT SIDE   CIRCUIT SIDE

Reduce to 4.000 ± 0.005"

269

FIGURE 10–14   Phase III completes the static memory board layout. Note the refinements of signal path locations which have resulted in the use of fewer via holes.

The horizontal signal paths at the top of the board are next moved from the component side to the circuit side with the use of the left series of via holes. In comparing Figs. 10–13 and 10–14, this change places these paths on the same side as the signal paths. Leading upward from U8 to this point, we have eliminated 10 of the 11 via holes in the right series. Although this places long parallel paths perpendicular to those on the circuit side, their spacing is sufficiently wide (0.050 in.) so as not to require a solder mask.

The final series of path adjustments will be made to eliminate the uppermost via hole in the right grouping, which is used to connect the Read/Write signal path to pins 10 of U5, U7, and U8. The path leading upward from pin 10 of U5 will first be moved from the circuit side to the component side, causing it to bridge the branch power and ground lines which are under U7. To overcome the problem of shorting this path with the RAM3 line to pin 8 of U7, the path leading to pin 8 will be moved to the circuit side. The path leading upward from pin 10 of U7 will next be moved to the component side and run directly to the uppermost via hole in the left grouping, which places it on the circuit side. This path is now in parallel with the other 10 signal paths on that side and is run directly to pin 10 of U8. This eliminates the total top right grouping of via holes.

In comparing Figs. 10–13 and 10–14, it is found that by careful inspection of the initial layout and adjustment of conductor paths, we have eliminated a total of 12 via holes. This amounts to a reduction of 26% of the via holes initially laid out in Fig. 10–13. Although this refinement stage of design may appear time consuming, it is well worthwhile in terms of producing a more cost-effective board.

10-8

Design of a General-Purpose Digital Microcomputer Circuit

The circuit schematic shown in Fig. 10–15, together with its parts list, is that of a general-purpose microcomputer system. The level of layout that this circuit will present to the beginning designer is considered to be at the maximum complexity. The design of this circuit will be on a double-sided PCB and will utilize all of the pertinent information, procedures, and selection criteria established in previous sections of this book.

Before discussing the design of the component layout drawing for the microcomputer system, we will provide an overview of the functioning of this circuit. Note in Fig. 10–15 that the system consists of nine integrated circuits in various sizes of DIPs, two resistor networks (R101 and R102) in molded plastic single-in-line packages (SIPs), one molded switch pack having four SPST switches in an 8-pin mini-DIP (SWA), two 1N914 diodes (CR1 and CR2), one 1-MHz crystal (Y), seven capacitors (C1 through C7), and one $\frac{1}{4}$-W resistor (R1). It can be seen that the circuit is primarily digital, with several discrete components added. The integrated circuits consist of a microprocessor (μP) chip (U1), a 7400 Quad NAND (U5), memory chips (U2, U3, and U4), an input/output (I/O) device (U7), and two RS–232C interface chips (U8 and U9).

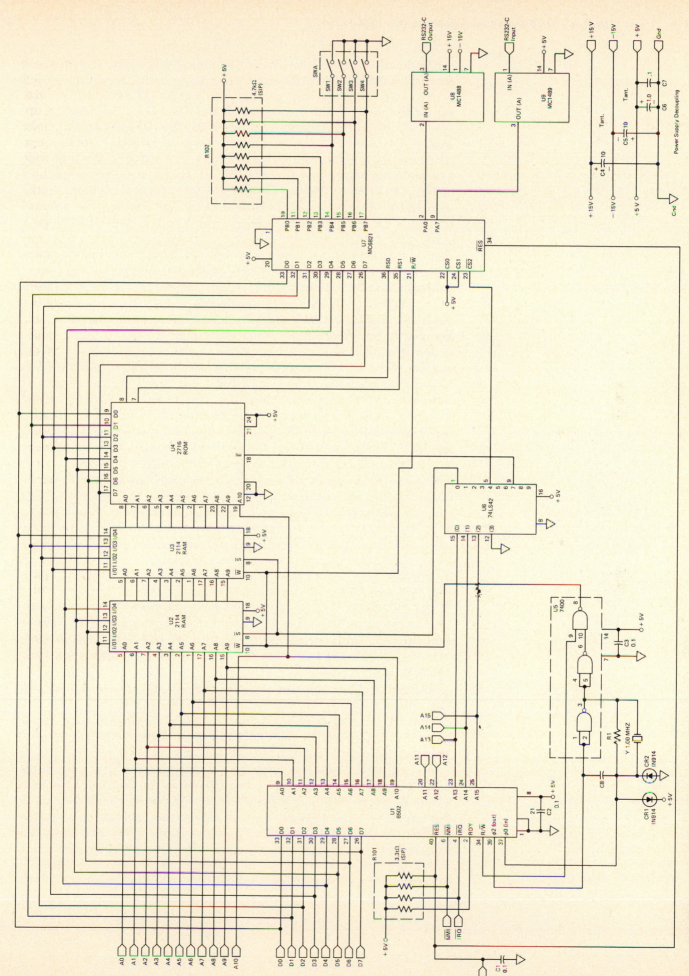

FIGURE 10–15 General purpose microcomputer system.

The heart of the system is the 6502 microprocessor. It communicates with the other devices using the Address bus lines labeled A0 to A15 and the Data bus lines labeled D0 to D7. To facilitate the use of the system, the Address and Data bus lines must be terminated as external connections on the PCB. The Read/Write line (pin 34 of U1 labeled R/$\bar{\text{W}}$) is at a high voltage level (logic 1) when the μP is *receiving* data from either memory or an I/O device and is at a low voltage level (logic 0) when the μP is *sending* data to memory or to an I/O device. All read and write operations are synchronized with the on-board crystal-controlled clock circuitry by the ϕ2 line (pin 39) of the μP. The clock circuitry consists of the 7400 Quad NAND (U5), diodes CR1 and CR2, capacitors C2 and C3, resistor R1, and the 1-MHz crystal (Y). The crystal allows the μP to read or write data to memory or to an I/O device every 100 microseconds.

The on-board memory circuitry consists of two RAM chips and one ROM chip (U2, U3, and U4 respectively). Chips U2 and U3 are 2114 devices and serve as the *random access memory* of the system. They are able to hold a program or maintain data only when the +5 Vdc power is applied to the system. If the power is turned off, the information in these chips is lost. For this reason, RAM is also often referred to as *temporary memory*. Chip U4 is the system's *permanent memory* or ROM (read-only memory). This chip holds programs and data that allow the system to operate as soon as power is applied. For example, the ROM chip is able to store a program that allows the μP to communicate with an external keyboard and cathode ray tube (CRT) display through the RS–232C interface terminals.

Device U7 is a MC6821 chip which performs two functions. First, it is used to transmit and receive data serially, and second, it sets the rate at which these data are transmitted or received. When power is applied initially, the microprocessor executes a program in ROM which instructs the μP to retrieve the data from switches SW1 to SW4 (SWA). The different binary switch combinations are also used by another program in ROM to fix the rate at which data are to be sent and received over the RS–232C lines.

The RS–232C voltage levels are different from TTL (transistor-transistor logic) voltage levels. For this reason, voltage-level *shifters* are required. This is accomplished by the MC1488 (U8) and MC1489 (U9) chips. The MC1488 chip is a TTL-to-RS-232C shifter and the MC1489 is an RS-232C-to-TTL shifter. Chip U8 must thus be powered by a $\pm$15-Vdc source. All other chips and circuitry are at a 5-Vdc voltage level.

The 74LS42 chip (U6) is a 4-to-10 decoder. Its function is to allow the μP to communicate with only one part of the system at a time: either to RAM (U2 and U3), ROM (U4), or I/O (U7).

We will now begin the discussion of the layout of the microcomputer circuit. The schematic of Fig. 10–15 shows that it is primarily a digital circuit with only a few discrete components added to allow the circuit to be functional. We will design the package on a daughter board having gold-plated finger connections. This will allow the external connections, shown with the symbol $\Longleftarrow$ on the schematic, of all address lines (A0 to A15) and all data lines (D0 to D7) to terminate on one edge of the board. In addition, all +5 V, +15

1	2	3	4	5	6	7	8	9	10	11	12	13	14	15	16	17	18	19	20	21	22	
R S 2 3 2 C	I N P U T		A10	A9	A8	A7	A6	A5	A4	A3	A2	A1	A0					S L O T	+5v			Component Side
A	B	C	D	E	F	H	J	K	L	M	N	P	R	S	T	U	V	W	X	Y	Z	
R S 2 3 2 C	O U T P U T	A11	A12	A13	A14	A15	D7	D6	D5	D4	D3	D2	D1	D0	$\overline{NMI}$	$\overline{IRQ}$	$\overline{RES}$	S L O T	G N D ⏚	+15v	−15v	Circuit Side

FIGURE 10–16 Finger identification for the 6502 microcomputer system shown in Figure 10–15.

V, −15 V, and ground connections can be terminated at the fingers along with the connections for both the input and output lines of the RS–232C terminal points. Note also in Fig. 10–15 that several circuit monitoring points, labeled $\overline{NMI}$, $\overline{IRQ}$, and $\overline{RES}$, have been brought out for circuit testing. Finally, one finger is to be removed to provide the keyway slot for card polarization.

The finger location and identification for the microcomputer circuit is shown in Fig. 10–16. Note that a 44-contact finger connector (22 fingers per side) is specified. A total of 35 contacts are required, including the two removed for the keyway slot. Observe the symmetry of the groups which are established for the Address and Data bus line finger connections. This layout is commonly referred to as a *standard bus* configuration. Note also that power and ground finger connections are grouped on one end of the connector while the RS–232C connections are located at the opposite end. Since the fingers are to be gold plated, finger extensions with a bus bar, such as that shown in Fig. 10–3, will be required.

With the configuration and orientation of the connector established, we will now consider the various holes and pads required for this design. Refer to Fig. 10–17a. From the schematic of the microcomputer circuit, it can be seen that the majority of the terminal pads must accommodate DIP, SIP, and mini-DIP pins. We will use the oval-style pad, having dimensions of 0.050×0.125 in., with a finished plated hole size of 0.033 in. This hole size is typical for all of these pins since their cross-sectional dimensions are approximately 0.023 in. Where via holes are required, the same recommended dimensions as used on previous designs are shown (i.e., 0.050-in. pads with 0.026-in. holes). All remaining terminal pads are to be 0.070 in. in diameter with a 0.046-in. finished plated hole except for the crystal, which requires 0.085-in. pads and 0.061-in. finished plated holes.

As previously discussed, nonplated holes are shown with targets and are to be drilled 0.125 in. in diameter. At least one of these nonplated holes should be positioned at the corner of the board to

FIGURE 10–17 Layout legends for the general purpose mirocomputer circuit.

(a) Pad legend

*Symbol	Pad diameter	Finished hole size	Plating	Remarks
○	0.050″	0.026″	Yes	Via Holes
○	0.070″	0.046″	Yes	All R, CR, C
⊙	0.085″	0.061″	Yes	Crystal
⬭	0.050 × 0.125″	0.033″	Yes	All IC, SIP & m-DIP*
⊕	N/A	0.125″	No	Targets

*mini-DIP

(b) Conductor width legend

Conductor width	Color code		Function
0.010″	Circuit Side ————— (Blue)		Signal +15 V and −15 V
	Component Side ————— (Red)		
0.050″	Circuit Side ————— (Yellow)		+5 V and Ground
	Component Side ————— (Purple)		

allow the assembly of a circuit board ejector, which is shown in Fig. 10–2b.

The conductor width legend is shown in Fig. 10–17b. Note that only 0.010- and 0.050-in. tape widths are used for the total microcomputer package. The +5-V and ground bus path widths are specified as 0.050 in. and all signal paths are to be 0.010 in. The +15-V and −15-V power lines are also specified as 0.010 in. since they are to handle only minimal currents. To maximize component density, a minimum air gap of 0.010 in. is permissible in this design.

While laying out the PCB design, remember that wherever possible, all paths on the circuit side and component side should run perpendicular and horizontal, respectively, to the finger area. Refer to Fig. 9–13, which shows this type of layout.

As a final consideration in designing the PCB for the microcomputer circuit, note from the schematic of Fig. 10–15 that the +5-V and the ±15-V connections are to be decoupled at their initial entry points on the board. The +5 V-to-GND decoupling is to be done with a 1-μF (tantalum) and a 0.1-μF (disc) capacitor combination. The ±15 V-to-GND decoupling will be done with 10-μF tantalum eletrolytic capacitors. Note the required polarity reversal for the −15-V supply decoupling capacitor (i.e., −15 V-to-GND).

The schematic also shows that only two of the ICs are to be individually decoupled with 0.1-μF disc capacitors. These are the microprocessor (U1) and the clock chip (U5).

Recalling all of the techniques and layout criteria presented in this chapter and with reference to Figs. 10–15 through 10–17, the component and conductor pattern layout for the microprocessor board is completed. This is shown in Fig. 10–18.

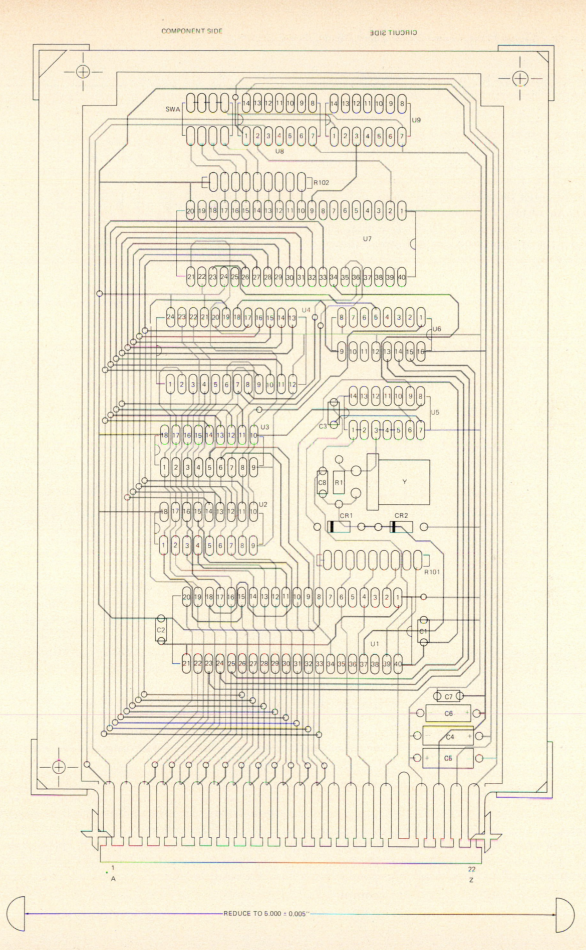

COMPONENT SIDE

CIRCUIT SIDE

REDUCE TO 5.000 ± 0.005"

275

FIGURE 10–18   Completed component and conductor pattern layout for the microprocessor board.

EXERCISES

10–1 Calculate the board density in EDIPs for the microprocessor board shown in Fig. 10–18. Use Fig. 10–15 to determine the number of required lead access holes. Use Fig. 10–16 to count the number of connections to fingers (both sides). Remember that this number must be added to the value of lead access holes to obtain the total value to be used as *number of holes* in Equation 10–1. Use Fig. 10–19 to calculate the dimensioned usable area, shown as shaded.

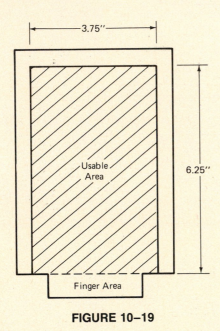

FIGURE 10–19

10–2 In calculating for Exercise 10–1, it will be seen that the board density is not considered to be in the moderate range of 1.5 EDIPs/in.² for the microprocessor layout. Again refer to Fig. 10–19. The density of the design required is to be changed to 1.5 EDIPs/in.². Keeping the 3.75-in. dimension the same, by how much does the 6.25-in. dimension have to be reduced to obtain the 1.5 EDIPs/in.² density?

10–3 Refer to Fig. 10–20, which gives a symmetrical layout of ICs. Draw the double-sided layout for the conductor path routing and position the main bus and branch decoupling capacitors. The power bus is

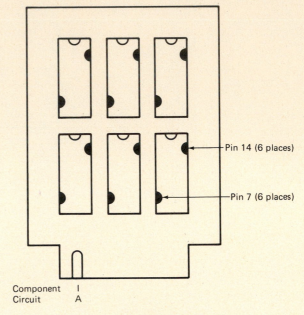

FIGURE 10–20

to be connected to pin 1 and routed on the component side, while the ground bus is to be connected to pin A and routed on the circuit side. Use pin 7 as the ground connection and pin 14 as the power connection on all chips.

10–4 Draw the double-sided component and conductor pattern layout for the digital clock display circuit shown in Fig. 10–21a. The overall size of the printed circuit board is to be 1.00 in. by 3.60 in., with the external connections positioned as shown in Fig. 10–21b. Each LED display is a 10-pin DIP package with pin connections as shown in Fig. 10–21c. This layout is to be made at 4X scale with a 0.050-in. forbidden area around the entire periphery of the board. Group the ICs in seconds, minutes, and hours, positioning as shown in Fig. 10–21a.

10–5 Using the information given in Exercises 10–1 and 10–2 as well as Figs. 10–15 through 10–18 as a guide, redesign the general-purpose microprocessor board to meet the 1.5-EDIP specification given in Exercise 10–2.

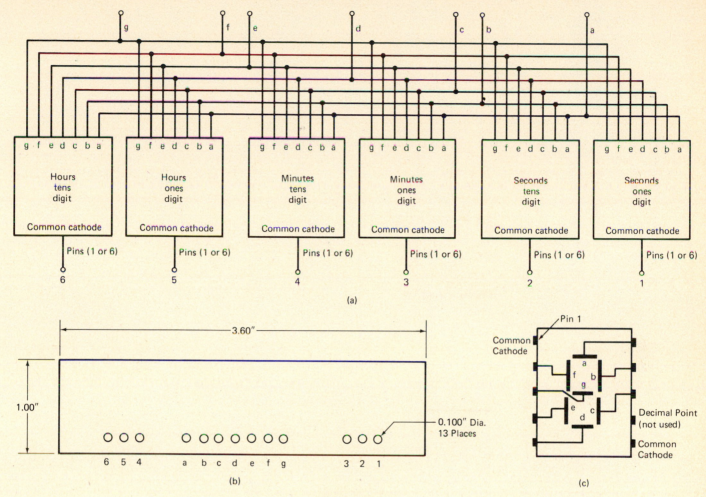

(a)

(b)

(c)

FIGURE 10–21

11

The Design of Multilayer Printed Circuit Boards

Upon completion of this chapter on the design of multilayer printed circuit boards, the student should be able to:

1. Understand how to show layer-to-layer interconnections on a drawing.

2. Visualize all layers simultaneously in a composite drawing.

3. Draw connections between inner layers and external fingers.

4. Properly provide antipads and thermal relief pads.

5. Design four-layer multilayer boards using the guidelines presented in this chapter and in Chapters 8, 9, and 10.

_____ **11-0** _____

Introduction

In Chapters 9 and 10 we discussed double-sided PCB designs and the problems and restrictions involved in their layout. Prior to beginning the layout of a board, some of the criteria which needed to be determined included the selection of conductor widths and spacings, center-to-center lead spans, and pad diameters. In addition, a design feasibility study resulted in the determination of package density. This aided in the decision of what level of design (single-sided, double-sided, or multilayer) was required. In this chapter we treat dense packages that require a multilayer design. Although it is true that multilayer designs will generally allow for a denser packaged circuit, we will see that this is not the only consideration in the selection process. In general, however, density levels greater than 25 to 30 holes/in.^{2} for analog circuits and 1.5 to 1.8 EDIPs/in.^{2} for digital circuits should be considered to be laid out on a multilayer board (MLB).

Another advantage of using a multilayer design over a double-sided board is easier routing of conductor paths on the outer layers. Taking a four-layer board as an example, one inner layer is used to distribute power to the circuit and the other inner layer serves as ground or common reference connections. In a double-sided layout, space would have to be allowed for power and ground interconnections. With these designed onto the inner layers of a multilayer board, the outer layers need include just signal paths, which allows for easier routing of denser packages.

Digital circuits designed onto a multilayer board with a large copper ground plane results in a reduction of some forms of noise which is commonly generated in double-sided boards. Thus fewer decoupling capacitors are required to effectively protect the ICs.

Multilayer boards also have the advantage of the option of using one of the inner layers to dissipate large amounts of heat which may be generated from on-board components. Heat dissipation is becoming an increasingly difficult problem as higher-density designs are laid out on PCBs. As the EDIPs/in.^{2} value increases, so does the generated amount of heat per square inch increase. If the increase is high enough, elaborate and expensive methods of heat removal need to be incorporated into the design to optimize circuit performance.

One must also consider the disadvantages of a multilayer board compared to a double-sided design. Most important, multilayer boards are more costly to manufacture than a typical double-sided board having plated-through holes. This cost factor is two to three times higher. In addition, a multilayer board requires a great deal more time to lay out and to manufacture. As a general rule, unless specific advantages unique to a multilayer design can justify its use, consideration should be given to a dense double-sided board which will not affect cost and production time as severely.

When a multilayer design is required, specific rules and methods of illustration are necessary for the layout designer to convey the correct information to the manufacturer. In this chapter we expand on the introductory treatment of multilayer boards presented in Chapter 6. A four-layer power/ground board will be used to demonstrate the layout of a multilayer board. The information presented will allow more complex designs to be laid out. It will be shown how electrical connections are made to the inner layers as well as how to insulate between component leads and inner layers where isolation is required as they pass through the board. The use of *thermal relief pads* for electrical connections and *antipads* for isolation is discussed. Finally, a simple IC timing circuit will be used to demonstrate a complete four-layer power/ground design. This circuit was selected only to show methods of layout and in no way suggests that a multilayer board is required for this design. The four-layer power/ground configuration is perhaps the least difficult to design, and the least expensive to manufacture when compared to more complex multilayer boards. These advantages make this the most popular of all the multilayer boards.

11-1

Layer-to-Layer Interconnections in a Multilayer Board

A cross section of a four-layer board, cutting across a plated-through hole, is shown in Fig. 11–1a. The top layer is the *component side* and is labeled *layer 1*. On this layer, a typical component lead hole and pad connected to a conductor path is shown. The bottom layer is the *circuit side*, labeled *layer 4*. A pad is also shown on this layer. Note that the pads shown in layers 1 and 4 are oriented in a similar manner to that of a double-sided board; that is, the two outer layers are electrically connected together by a copper-plated barrel surrounding the hole wall which is formed during the manufacturing process. In the view shown in Fig. 11–1a, it is not evident whether there are any conductor paths connected to the pad of layer 4. This fact is not pertinent to our discussion at this point.

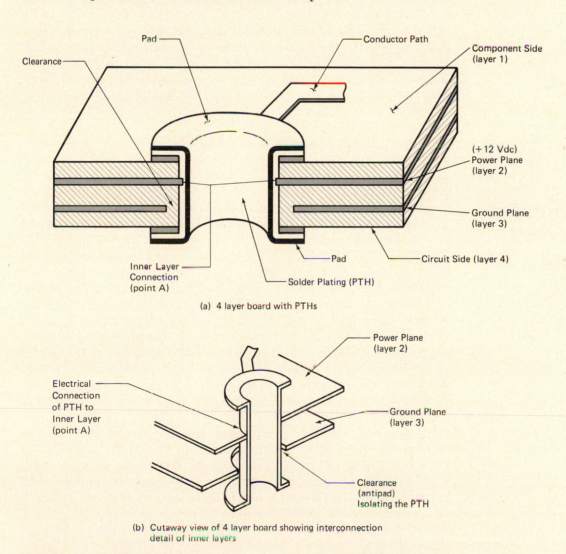

(a) 4 layer board with PTHs

(b) Cutaway view of 4 layer board showing interconnection detail of inner layers

FIGURE 11–1   Cross-sectional views of layer-to-layer interconnections in a multilayer PCB.

Note that references made to Fig. 11–1a so far do not differ greatly from those of a double-sided board with a plated-through hole. We will now consider the inner layers of the cross section shown in Fig. 11–1a. The *+12-Vdc power plane*, labeled *layer 2*, is shown electrically connected to the outer layer pads by means of the plated-through hole. This connection is shown as point A, which is the edge of the copper plane of layer 2. It was exposed during the drilling operation prior to the plating of the hole. Note that with +12-Vdc power connected to layer 2, it will also be electrically connected to the pads on layers 1 and 4 and to any conductor paths that may be connected to these pads.

The *ground plane* shown in Fig. 11–1a is labeled *layer 3*. Note that no electrical connection exists between this layer and the barrel of the plated-through hole. The clearance shown to result in this electrical isolation is made with an *antipad*, which is discussed in the next section.

The cross-sectional view shown in Fig. 11–1b is the same as that shown in Fig. 11–1a, but with *all of the insulating board material removed*, exposing only the individual layers of copper. The view more clearly illustrates the association of the pads, antipads, plated-through hole, and the power and ground planes of this four-layer configuration.

11-2

Layer Representation on the Component Layout Drawing

From our discussion in the preceding section on layer-to-layer interconnections on a processed multilayer board, it is seen that there are four separate and distinct layers of circuitry. Conventionally, the outer layers are labeled layer 1 (component side) and layer 4 (circuit side). The inner layers are labeled layer 2 for the power plane and layer 3 for the ground plane. Thus the "hot" terminal of the power supply (either + or −) is electrically connected to layer 2. This will result in every lead access hole which contacts layer 2 (see Fig. 11–1a and b) being electrically connected to dc power. Similarly, layer 3 must be electrically connected to the GND terminal of the power supply. Accordingly, every lead access hole contacting layer 3 is electrically connected to ground. (This is not shown but would simply be the reversal of layers 2 and 3 connections to the plated-through hole as shown in Fig. 11–1b.)

While laying out a four-layer board, the designer must be able to visualize four layers of conductive planes simultaneously as a composite drawing. This is not as difficult as it may appear initially. Each layer will be constructed separately but will ultimately require layer-to-layer registration. There is one of three options of making electrical connection combinations for every plated lead access hole. These options are shown in Fig. 11–2a and are designated as *case A, case B*, and *case C*. For each layout possibility, each of the four layers will be shown independently. The symbolic representation for the layout of copper pads for each of the following three cases is shown in Fig. 11–2a.

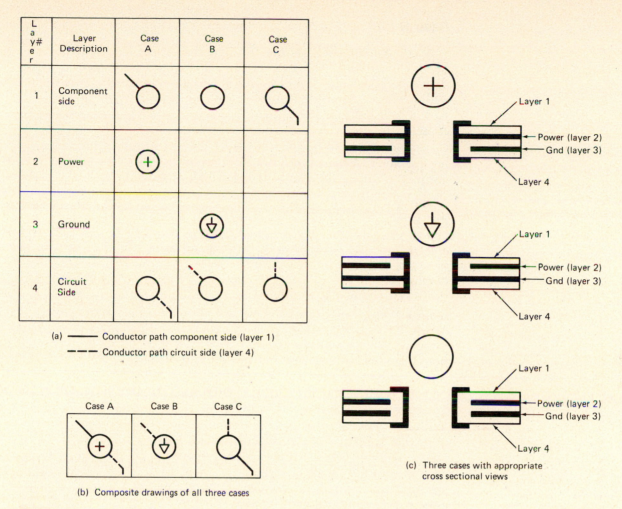

L a y # e r	Layer Description	Case A	Case B	Case C
1	Component side			
2	Power			
3	Ground			
4	Circuit Side			

(a) ——— Conductor path component side (layer 1)

     - - - Conductor path circuit side (layer 4)

Case A	Case B	Case C

(b) Composite drawings of all three cases

Layer 1
Power (layer 2)
Gnd (layer 3)
Layer 4

Layer 1
Power (layer 2)
Gnd (layer 3)
Layer 4

Layer 1
Power (layer 2)
Gnd (layer 3)
Layer 4

(c) Three cases with appropriate cross sectional views

FIGURE 11–2 Coding pads for the three possible layer-to-layer interconnections.

Case A

A component lead passing through an access hole is to be connected to +dc (power). In the layout, a pad would be drawn on both outer layers 1 and 4. This is a requirement of all layouts. (Conductor paths may or may not be required on both the component side and the circuit side. Recall that paths on the component and circuit sides are shown in this book as solid and dashed lines, respectively. In an actual drawing, they would be drawn as solid color-coded lines to indicate the appropriate side of the board.) To electrically connect to power, layer 2 is provided with a round pad symbol which represents a thermal relief pad and is drawn with a + symbol inside the circle. This indicates that this pad makes electrical connection to layer 2, which is the power plane of the layout. Because no electrical power connection is to be made to the ground plane, no pad is shown on layer 3. The absence of a pad indicates that clearance (antipad) is required between the ground plane and the plated-through hole.

Case B

A component lead passing through the access hole is to be connected to ground. See Fig. 11–2a. This would be shown as a pad drawn on both outer layers (1 and 4) and appropriate conductor paths added where required. The electrical connection to ground is shown with a round thermal relief pad drawn on layer 3 and the *ground* or *common reference* symbol added within the circle. Since no electrical connection is to be made to layer 2, no pad or symbol is shown on this layer. Again, the absence of a pad indicates the need for an antipad to isolate the plated-through hole from layer 2.

Case C

A component lead passing through an access hole is not to be electrically connected to either the power or ground planes but is to connect directly from layer 1 to layer 4. The symbol for this connection is shown in Fig. 11–2a. This type of connection is typical in a double-sided board design. Pads are shown on both outer layers (1 and 4) but no pads or symbols are drawn on either inner layer. This indicates the requirement of antipads to isolate the plated-through hole from both the power and the ground planes.

Note in all of the three cases presented, no pad or symbol is required when no connection is to be made to an inner copper plane. With the absence of a pad, it is understood that an antipad is required for isolation of the plated-through hole.

The symbols listed in Fig. 11–2a are shown as if they were drawn on four separate sheets of vellum (i.e., one sheet for each layer). They are shown in composite form in Fig. 11–2b, which is how they would appear on a single sheet of vellum to represent all four layers.

Using the convention of symbols shown in Fig. 11–2a, the composite drawings of Fig. 11–2b are interpreted as follows:

Case A: Path connections—layers 1 and 4. Thermal relief pad—layer 2. Antipad—layer 3.

Case B: Path connection—layer 4. Thermal relief pad—layer 3. Antipad—layer 2

Case C: Path connections—layers 1 and 4. Antipads—layers 2 and 3.

For each hole to be drilled into the multilayer board, the designer simply relates it to one of the three cases above to determine how it will be drawn on the conductor pattern layout. Each hole symbol will thus show which plated-through hole will connect to the appropriate inner layer to satisfy the power and ground connections as dictated by the circuit schematic. Remember that for representing plated-through holes that are to be connected only to layers 1 and 4, the round pad drawn will have no symbol associated with it, which indicates no inner layer connections. The cross-sectional views representing the three cases just described are shown in Fig. 11–2c.

The symbols used to represent layer representation on the component layout drawing may be more conveniently read if provided with a coded system. For example, it may be much more convenient to represent the three cases of connections shown in Fig. 11–2b with different geometric shapes (e.g., round, square, and rectangular). Another form of coding is by using a different color for each of the three cases. Regardless of the system of coding used, a legend should accompany the component layout drawing so that there will be no confusion in interpretation.

In all cases in the layout of multilayer boards, the power plane and the ground plane must be electrically isolated. One of the first tests that a board is subjected to after it has been manufactured is to ensure that this isolation has not been violated. This test is referred to as *continuity test* or *ring out*. An ohmmeter is used to ascertain if there is infinite resistance (open circuit) between the power plane and the ground plane, which indicates that no electrical connection exists. This test is performed by first attaching one lead of the ohmmeter to a plated-through hole (or finger edge connector) which connects the power inner layer to a point on one outer layer. The other meter lead is attached to another plated-through hole, or finger, which connects the ground layer to a point on one of the outer layers. Any finite value of resistance indicated on the ohmmeter renders the board defective.

Figure 11–3a shows how electrical connections between inner layers and external fingers are drawn on the component layout. This

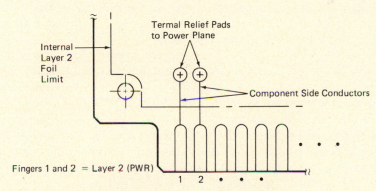

(a) External finger pins 1 and 2 are connected with one or more conductors on the component side through via holes to the internal power layer

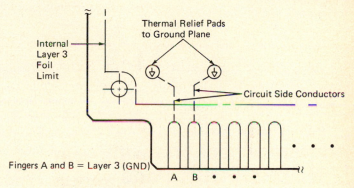

(b) External finger pins A and B are connected, with one or more conductors on the circuit side through via holes to the internal ground layer

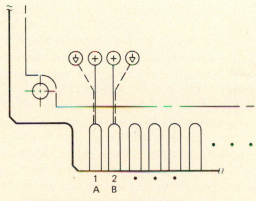

(c) Four layer composite view of finger to inner power and ground planes connections

FIGURE 11–3 Finger connections to internal power and ground layers.

figure shows the internal power plane (layer 2) electrically connected to the fingers labeled 1 and 2 on the component side of the board. Two via holes are shown inside the power plane edge limit and drawn with the appropriate symbol. Two conductor paths (shown as solid lines) are drawn from the pads to the tops of fingers 1 and 2. In low-current applications, it may only be necessary to connect one finger to the power plane using only one via hole. Where current requirements are high, it is common practice to use several fingers and via holes.

Connections from the ground plane (layer 3) to fingers A and B are shown in Fig. 11–3b. Again, two via holes with the appropriate symbols are drawn above the fingers and inside the ground plane edge limit. Two conductor paths (shown as dashed lines) are then drawn on the circuit side to the tops of fingers A and B.

Note that Fig. 11–3a, showing layers 1 and 2, and Fig. 11–3b, which shows layers 3 and 4, were drawn separately for the sake of clarity of explanation in connecting fingers to inner layers. On a component layout drawing, this would be drawn as a composite view showing all four layers on one sheet of paper. This composite view is shown in Fig. 11–3c, which is simply the result of superimposing Fig. 11–3a on top of Fig. 11–3b.

The number of fingers connected to inner layers for any design is determined by the amount of current that is to be supplied to the board. Of course, the number of fingers connected to the power plane should be the same as those connected to the ground plane.

11-3

Clearance with Antipads and Interconnections with Thermal Relief Pads

It has been shown that a clearance or insulation gap is required on all inner layers around plated-through holes that are not to be connected to power or ground planes. For each plated hole, at least one, or at most two, clearance pads (or *antipads*) will be required on the inner-layer artworks. One form of these antipads is shown in Fig. 11–4a. (These are available in preformed shapes to aid in preparing the artwork master, which is discussed in detail in Chapter 12.) The copper inner layer of the foil plane is electrically isolated from the inner copper pad (through which the lead drill hole will pass) by a circular insulating gap, shown as a shaded area. This gap can vary in width from 0.010 to 0.025 in. It can be seen that where an antipad is used, the plated-through hole will pass through the inner pad but will be electrically isolated from the foil plane. A table of typical values of inner pad diameters with accompanying maximum finished (after plating) hole diameters is shown in Fig. 11–4b. Antipads are selected on the basis of the finished hole size. Their diameter should be a minimum of 0.040 in. larger than the hole size. This is to provide a degree of additional safety in the manufacturing phase in cases of misregistration of the inner layers. Even with a 0.040 in. clearance, if the misregistration is severe, the result could be unwanted electrical connections throughout the board. Remember that an antipad is required at each lead access or via hole on each inner layer where *no* electrical connection is to be made.

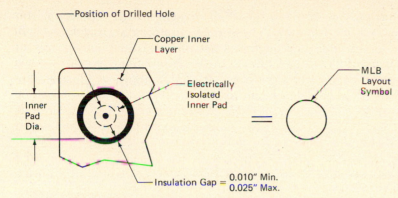

(a) Clearance pad or antipad for isolation of lead access holes on inner layer

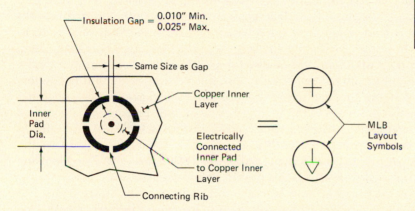

Inner Pad Dia	Max. Finished Hole Size
0.075″	0.035″
0.093″	0.053″
0.100″	0.060″
0.125″	0.075″
0.150″	0.110″

(b) Antipad diameter vs maximum finished hole size

(c) Thermal relief pads for electrical connection of lead access holes to inner layer

FIGURE 11–4   Antipad and thermal relief pad configurations.

To make electrical connections to an inner layer represented by pads having a *plus* or *ground* symbol, a special pad called a *thermal relief pad* or *spider pad* is required. One form of thermal relief pad is shown in Fig. 11–4c. (These are also available in preformed shapes.) This pad is composed of an inner pad diameter which is electrically connected to the copper foil plane of the inner layer by *connecting ribs*. These ribs have the same width as the surrounding insulation gap and vary from 0.010 to 0.025 in. The design of this style of pad also provides high-resistance paths to heat flow from the inner pad to the foil plane. During the soldering process, a lead passing through a plated hole and the inner pad heats quickly and alloys properly with the solder inside the hole. This is the result of not allowing the heat being generated at the lead to be rapidly sinked from the plated hole and pad to the large copper plane because of the high thermal resistance provided by the connecting ribs. Thus maximum heat is maintained at the lead/hole interface and results in sound soldered connections. The inner pad diameter of the thermal relief pad should be at least 0.040 in. larger than the finished plated hole size. Remember that a thermal relief pad is placed at each lead access hole on each layer where an electrical connection is to be made.

It needs to be emphasized that the antipad and the thermal relief pad configurations shown on the left side of Fig. 11–4a and c are illustrated only to provide the PCB designer with a better understanding of their function. These configurations are produced with the use of drafting aids on taped artwork masters, which are discussed in the following chapter. On the component layout drawing, they are shown as circles, with or without symbols, as illustrated on the right side of Fig. 11–4a and c.

11-4

Setup Considerations for Multilayer Boards

Many of the points discussed in the layout of densely packaged double-sided PCBs also apply to that of multilayer boards. For example, the determination of the lead span dimensions between pad centers for axial lead components are based on the same criteria. In addition, those procedures for the selection of outer layer pad diameters, component spacing, and conductor width and spacing also apply to the layout of multilayer boards. For this reason, they will not be repeated here. In developing the multilayer component layout drawing in the following section, these criteria will be applied.

In addition to those layout considerations which are common to both double-sided and multilayer designs, there are several others which are unique to multilayer boards. These are the following: the grid and scale, defining the usable board area on each layer, layer-to-layer registration, and legends and codings used on drawings.

Positional accuracy to result in precise layer-to-layer registration in multilayer boards is more critical than in the layout of double-sided designs. In extremely high multilayer packaging densities greater than 25 holes/in.^{2} or 1.5 EDIPs/in.^{2}, registration is much more challenging to the designer. It thus becomes necessary to use a 0.050-in. grid and an enlarged drawing four times the actual size (4:1 scale). This scale will make the overall size of the drawings and artworks much larger than those normally required for single- or double-sided boards. However, the size of the artworks is kept within practical limits by the fact that the inner layers allow more circuitry to be packaged into a smaller area. This reduces the overall size of the finished board. In terms of accuracy, any positional errors are greatly reduced when using a 4:1 scale and a 0.050-in. grid when compared to a 2:1 scale with a 0.10-in. grid. When photoreduced to their actual size, positional errors will be reduced by $\frac{1}{4}$ when using a 4:1 scale instead of $\frac{1}{2}$ with a 2:1 scale. Recall from Section 8–6 that at 2X scale using a 0.10-in. grid, the placement of a pad can be off position by as much as 0.015 in. This error translates to a reduction of 0.0075 in. when photoreduced. When using a 4:1 scale and a 0.050-in. grid, this 0.015-in. error is reduced to only 0.00375 in. when produced to actual size.

Corner brackets drawn on layers 1 and 4 again define the edges of the finished board. For the inner layers, a typical limitation is that the edges of the copper foil planes are no closer than 0.10 in. (1X scale) to the finished board edges. This is to ensure total sealing of the inner layers and also to prevent layer delamination. In addition, inner foil clearance normally must be provided around all mounting

FIGURE 11-5 Mechanical specifications for multilayer PC Board design.

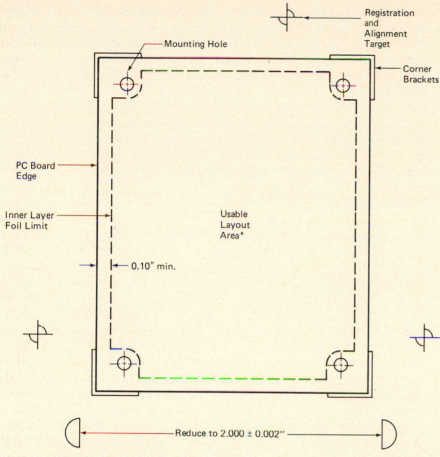

Registration and Alignment Target

Mounting Hole

Corner Brackets

PC Board Edge

Inner Layer Foil Limit

Usable Layout Area*

0.10" min.

Reduce to 2.000 ± 0.002"

*Specification may indicate usable layout area for components may be less than foil area

Conductor Path Legend

Conductor Width	Code
0.010"	Component side*————
	Circuit side*– – – –
0.025"	Component side N/A
	Circuit side N/A

*Color code signal paths if required

Pad Legend (MLB)

*Symbol	Pad Diameter		Finished Hole Diameter	Plating	Remarks
	Outer Layer	Inner Layer (inner pad dia.)			
◯	0.050"	0.075"	0.026"	Yes	Via. Only
◯	0.062"	0.093"	0.038"	Yes	
◯	0.075"	0.093"	0.051"	Yes	
◯	0.130"	0.150"	0.110"	Yes	External Turret Terminals
⊕	N/A	N/A	0.125"	No	Target (tooling holes)

*If required color code for pad size

Layer Legend

Layer #	Layer Orientation	Connection Symbol	Final Foil Thickness
1	Component side	◯	2 oz/ft²
2	Power	⊕	2 oz/ft²
3	GND	⊗	2 oz/ft²
4	Circuit side	◯	2 oz/ft²

and hardware assembly holes. To define the limits of the foil planes and the required mounting hole clearance for the inner layers, an outline of the inner copper border is drawn on the layout. This is shown as *inner layer foil limit* in Fig. 11-5.

For precise layer-to-layer registration of the four individual boards, a minimum of three registration and alignment targets are required. They are positioned on the layout outside the finished board outline in the same staggered arrangement as that used for the registration of double-sided boards. These are also shown in Fig. 11–5 and discussed further in Section 12–2.

Of prime concern in any layout is the necessity of conveying all required information to eliminate the possibility of misinterpretation. For multilayer designs, there are three types of information required on the component layout drawing. Similar to double-sided layouts, pad and conductor path legends are shown. The third legend, which is unique to multilayer designs, is one that codes all the layers by number, name, connections (symbols, colors, or pad shapes), and the final foil thickness of each. The appropriate legends are shown together with the basic outline of a rectangular multilayer board in Fig. 11–5. Note that the board outline includes corner brackets, borders, mounting holes, targets, reduction scale, usable layout area, and the inner layer foil limit shown as dashed lines. This outline will be used in the next section to illustrate the techniques used to draw a multilayer component layout design.

11-5

Four-Layer Multilayer Design

To demonstrate the basic layout techniques for designing a multilayer board, we will use the simple schematic of a relaxation oscillator shown in Fig. 11–6. This circuit employs a 555 IC timer chip packaged in a TO-99 style 8-pin case. Power is applied between the +12-Vdc terminal and the ground terminals of the IC. Current flows through R1 and then the diode CR1, which is properly biased for conduction. This current charges capacitor C1. When the voltage across C1 reaches the predefined trigger level of the 555 timer, the

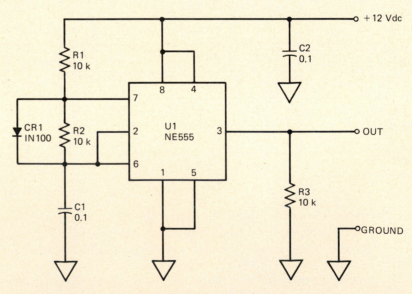

All resistors are 1/2 watt ± 10% unless otherwise specified.
All capacitors are in microfarads unless otherwise specified.

FIGURE 11–6   Circuit schematic of a simple NE555 IC timer/oscillator.

capacitor quickly discharges through resistor R2 and into pin 7 of the IC. This charge/discharge cycle of capacitor C1 applied to the IC produces a square-wave output across the load resistor R3 that is connected to pin 3. The output waveform is measured across the OUT and GROUND terminals.

Using the layout practices established in previous chapters, we will design a multilayer board for the circuit of Fig. 11–6 using the board outline and accompanying legends shown in Fig. 11–5. As usual, the layout begins with the parts placement, using the schematic as a guide for initial component positioning. Attention is given to achieving a uniform distribution and balance of parts throughout the usable board area. This component layout is shown in Fig. 11–7a. Note that all parts are provided with a pad for each lead and that all parts are coded with both their appropriate designators and values.

At this point, we proceed with the routing of conductor paths as though the layout were for a double-sided design. Signal paths having 0.010-in. widths as per the conductor path legend of Fig. 11–5 will be routed along the circuit and component sides (layers 1 and 4). Note in the conductor path legend that a conductor width of 0.025 in. is not applicable (N/A) in this design. Recall that in double-sided designs, this conductor width was specified for power and ground paths. Because we will be making power and ground connections only to inner layers 2 and 3, respectively, no conductor paths will be required for these connections. Thus the only routing of conductors that is required for a multilayer board is that for signal paths on layers 1 and 4. These are shown in Fig. 11–7b. Note that no outer-layer conductor paths are routed to the external terminal pad labeled +12 Vdc or to any component lead requiring a connection to +12 Vdc. Similarly, no outer-layer paths are routed to the external ground terminal pad or to any component lead requiring a connection to ground. In short, no conductor path is used to make any power and ground connections.

The final phase of our design will be to code the appropriate component and external terminal pads for inner layer power and ground connections using the layer legend of Fig. 11–5. Refering to the circuit schematic of Fig. 11–6, we begin by making all the power connections. Pins 8 and 4 of U1 must be connected to power, as are the top ends of resistor R1, capacitor C2, and the external turret terminal labeled +12 Vdc. The pads associated with these points will be connected to the inner power layer (layer 2). Using the symbol +, all of these pads are coded to show that they are to be connected to layer 2 as well as to both outside layers.

In like fashion, ground connections are made to layer 3. The schematic shows that pins 1 and 5 of U1 as well as the external ground pad and the bottom ends of capacitors C1 and C2 and the resistor R3 are to be connected to ground. These points are coded on the drawing with a ground symbol inside of each pad. The completed component layout drawing for the multilayer board is shown in Fig. 11–7c.

Recall that all pads shown on the outer layers (1 and 4) which have no plus or ground symbols are to have no electrical connection to either power or ground. This means that pairs of antipads will be used on layers 2 and 3 with these uncoded pads. The artworks re-

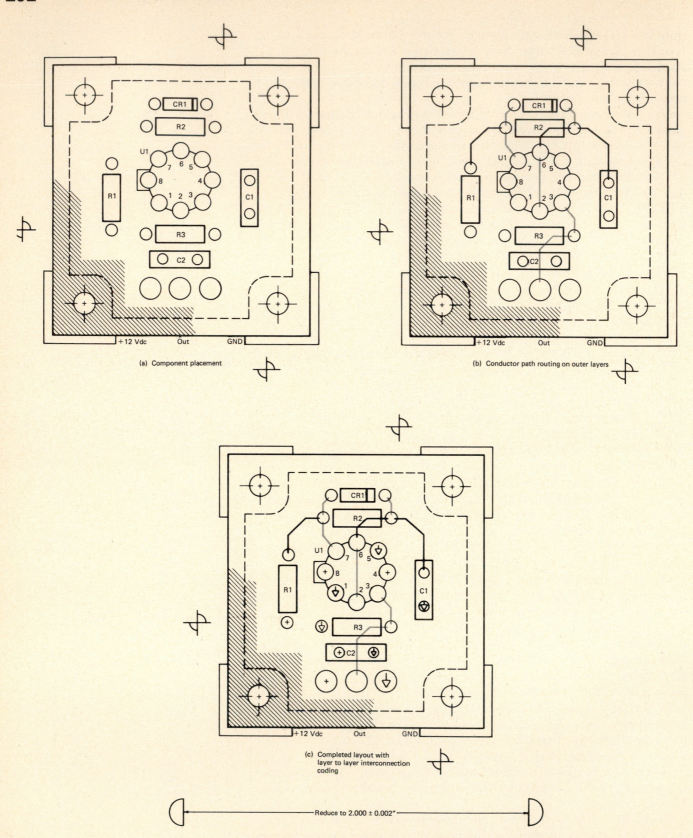

(a) Component placement

(b) Conductor path routing on outer layers

(c) Completed layout with
layer to layer interconnection
coding

Reduce to 2.000 ± 0.002"

FIGURE 11–7 Sequence of multilayer board design.

quired for the inner layers, including antipads, for the composite layout drawing of Fig. 11–7c will be developed in the following chapter.

This chapter completes the treatment of printed circuit component and conductor pattern design. The following chapter details procedures and techniques for the generation of manually taped artwork masters from component layout drawings.

EXERCISES

11–1 For the circuit of Fig. 10–11a, construct MLB pad, conductor path, and layer legends similar to those shown in Fig. 11–5.

11–2 Use the dimensional drawing shown in Fig. 10–11b to design a 2X scale multilayer component and conductor layout for the 4K static memory board shown in Fig. 10–11a. Also use the legends prepared in Exercise 11–1, the finger identification information given in Fig. 10–11c, and as a guide, the double-sided layout shown in Fig. 10–14. Emphasis is to be placed on layer coding and conductor path routing simplification through the use of inner power and ground layers with antipads and thermal relief pads.

11–3 Repeat Exercise 11–1 for the general-purpose microcomputer system of Fig. 10–15.

11–4 Draw a 4X scale MLB layout for the general-purpose microcomputer system of Fig. 10–15. Use the legends prepared in Exercise 11–3, the finger identification information in Fig. 10–16, and as a guide, the double-sided layout shown in Fig. 10–18. Emphasis is to be placed on determining the most dense package possible.

11–5 Calculate the package density, in equivalent DIPs/in.^{2}, of the design completed in Exercise 11–4.

12

Taped Artwork Masters

Upon completion of this chapter on preparing taped artwork masters, the student should be able to:

1. Assemble the appropriate tools, materials, and equipment for preparing taped artwork masters.

2. Precisely align each of the artwork layers.

3. Select and properly position terminal pads and preforms.

4. Precisely tape conductor paths.

5. Add additional literal and numerical information to the taped artwork master.

6. Prepare artwork masters for single-sided, double-sided, and multilayer designs.

7. Understand the need to improve the distribution of copper areas on both sides of a double-sided board or the outer layers of a multilayer board.

8. Prepare complete documentation packages.

12-0

Introduction

The previous five chapters have been devoted to the design of single-sided, double-sided, and multilayer printed circuit boards. Recall that for purposes of accuracy and precision, these layouts were drawn to an enlarged scale, typically two or four times larger than the actual finished board size.

With the component layout drawings completed, the next phase for producing a PCB is the preparation of the *taped artwork masters*. These artworks are made to the same scale that was used for the component layout drawing. They are produced on clear polyester sheets using specially prepared printed circuit drafting aids. These aids are adhesive-backed elements that are positioned onto the polyester sheet to form a positive scaled image of the conductor pattern designed in the layout drawing. When completed, the artwork master is photographically reduced with the use of a precision copy camera. The film is then processed using standard developing techniques to produce 1:1 scale negatives of the artworks. These negatives are contact printed to make 1:1 scale positives, called *phototools*, that are used in one of the fabrication phases of the board.

Because each layer of a PCB is designed with a different pattern, an artwork master is required for each. For a single-sided board, only one artwork master of the circuit side is needed. Double-sided boards require two artwork masters, one for each of the conductor patterns on each side. For multilayer designs, an artwork master is required for each of the outer layers in addition to one for each of the inner layers. Some PCB specifications also require *component marking masks* and/or *solder masks* to be produced. These special types of artworks will be considered separately in the following two chapters.

The artworks show the exact detail, position, and orientation of all terminal pads and conductor patterns that will be processed from the laminated copper foil. For this reason, the artwork masters must be laid out with extreme accuracy. In this chapter, information is presented on how to manually prepare a precise taped artwork master required for all board types previously discussed. Included are the tools, materials, and equipment required in addition to the preparation of a *documentation package*, typically required for the manufacturing phases of the board.

12-1

Tools, Materials, and Equipment Required for Preparation of the Taped Artwork Master

Even though a taped artwork master can be produced on any hard, flat surface, such as a table or drafting table, it is highly recommended that a suitably sized light table be used. The illuminated surface area of the light table should be at least as large as the artwork to be prepared and preferably larger. It should be equipped with a high-quality light-diffusing plastic panel and plate glass work surface. It is important that the panel distribute the light uniformly over the entire work area. In addition, it should have fluorescent lamps and fans, if necessary, to prevent the generation of heat, which would increase the temperature of the glass and also of the polyester sheet.

If the component layout were drawn on plain vellum, an accurate underlying grid system would be required. The taped artwork master is made on a sheet of clear polyester film having a thickness of 0.007 in. This must be of the stable-based type as described in Chapters 1 and 8. Again, if the layout was drawn on a grid sheet, only a clear sheet of polyster would be required.

The typical clear plate glass work surface may be replaced with a precise glass grid which provides maximum accuracy and dimensional stability. This type of grid is least effected by temperature and humidity changes. However, owing to their cost ($\simeq$ \$100 per square foot), this type of work surface would be necessary only for designs that require the ultimate in layout precision. The use of the polyester film affords a sufficient degree of accuracy and economy, which generally justifies its use in the design of artwork masters.

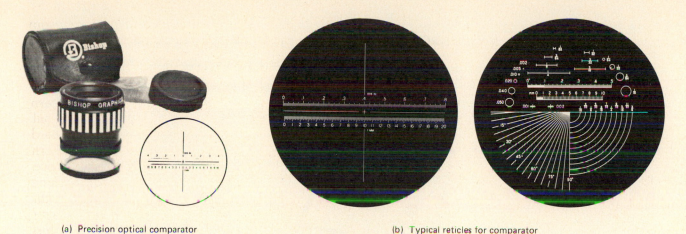

(a) Precision optical comparator

(b) Typical reticles for comparator

FIGURE 12–1 Optical comparator used for printed circuit layout. [Provided courtesy of Bishop Graphics, Inc., Westlake Village, California 91359]

For some portions of the artwork layout, it may be required to place pads, preforms, or tapes more accurately than is generally possible to do with the naked eye. To aid in this precise placement and inspection, a wide selection of precision magnifiers *(optical comparators)* such as the one shown in Fig. 12–1a are available. They come in magnifications of 5X, 8X, 10X, 15X, 22X, and 30X, with 10X interchangeable reticles. The reticles are precision-printed on round pieces of flat, distortion-free glass. Several types of reticle scales and configurations are available, two of which are shown in Fig. 12–1b. The 10X comparator is the most popular. It is full-focusing, with an achromatic 35-mm lens system. The coating on the magnifier acts as a neutral density filter which allows true color perception of the image being viewed.

To use a comparator, it is simply placed onto the polyester sheet over the area to be inspected. The viewing light is provided by the light table and enters through the clear 360-degree protective plastic apron around the base of the comparator. Larger magnifications reduce the field of view, that is, the amount of area visible without having to move the comparator. The field of view varies from approximately 1.2 in. for 5X magnification to 0.3 in. for a 30X comparator. The 10X comparator provides a field of view of approximately 1.0 in.

Comparators provide excellent resolution and accuracy of measurement. They are used to aid in the exacting measurements required for on-grid registration of pads and targets in addition to the spacing between pads and between paths and pads on taped artworks. On fabricated boards, they may be used to inspect pad and hole diameters, conductor widths, and plating and soldering quality.

In the preparation of an artwork master, it is often required to make accurate measurements or to precisely position pads or targets over distances which are outside the comparator's field of view. To meet this need, accurate polyester scales are available which are clear and flexible. See Fig. 12–2a. These scales are manufactured from 7.5-mil

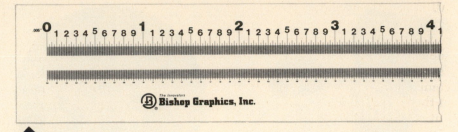

(a) 7.5 mil precision polyester scale for 1:1 scale measurement

(b) Precision scale being used with comparator

FIGURE 12–2 Precision polyester scales.

polyester film similar to that used for grids. Scales are available in 18, 24, and 36 in., in addition to 500 and 1000 mm. They are marked for precise measurements at 1:1, 2:1, and 4:1 scales. Each scale has two distinct images, one to be read with the naked eye and the other to be read with a 10X comparator for higher-precision layouts. The 1:1 scales have increments every 0.005 in. with accent lines every 0.025, 0.050, 0.100, and 1.000 in. Every 0.100-in. increment is numbered. The 2:1 scales have increments every 0.002 in. (actual 0.004 in.), with accent lines every 0.010, 0.050, 0.100, 0.500, and 1.000 in. This scale is numbered every 0.050 in. The 4:1 scales have increments every 0.001 in. (actual 0.004 in.), with accents every 0.005, 0.025, 0.100, 0.500, and 1.000 in., and numbered every 0.025 in. The metric scales have increments every 0.1 mm, with accents every 0.5, 1, and 10 mm. These scales are numbered every 1 mm. The regular increment line widths for all these scales are 0.001 in., with accent widths of 0.002 in.

The accuracy of the polyester scales at 70°F and 50% RH is ±0.002 in. (0.051 mm) and ±0.005 in. (0.0127 mm) between adjacent lines.

All images are printed on the bottom surface to eliminate parallax error due to the thickness of the scales.

To use the scale, the left index is aligned with one feature (e.g., the center of a target), and the distance to the location of the second feature is initially determined by eye. For the precise distance, a comparator with a clear reticle is positioned over the scale location at the second feature. It is essential that the left index not be moved out of its exact position when setting the comparator in the field of view and adjusting the focus. See Fig. 12–2b.

The final tools that will be required are a selection of scalpels to aid in the positioning of the drafting aids and the cutting of the tape. Several common blade types and handle designs are shown in Fig. 12–3. Handles are available in swivel and fixed designs, both finding application in the positioning and alignment of drafting aids and tapes onto the polyester sheet. A wide selection of the printed circuit drafting aids used in artworks is discussed in subsequent sections.

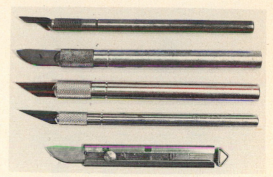

FIGURE 12–3 Various blade and handle styles used in making taped artworks.

12-2

Initial Setup for Taped Artwork Masters

The grid system used for the component layout must be clearly visible in order to serve as an accurate guide to prepare the artwork. If the component layout was not drawn directly on a grid sheet, one must be secured to the light table and then overlaid with the component layout drawing. The drawing needs to be carefully positioned to ensure that the pad centers are registered on grid. The 10X comparator is extremely helpful in this alignment.

With the drawing secured to the light table, a sheet of clear 0.007-in. polyester film is positioned over the drawing and taped to the table. The sheet of film should be several inches larger than the overall board outline as defined by the corner brackets on the drawing. This setup, shown in Fig. 12–4, is ready for making taped artwork.

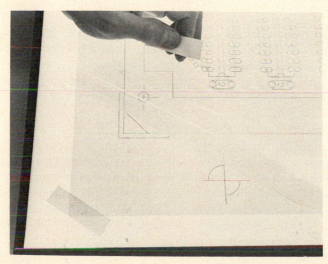

FIGURE 12–4 Typical setup for making a taped artwork master on a light table.

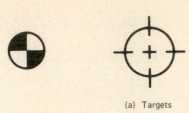

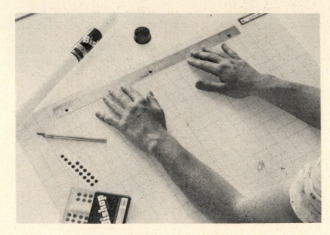

(a) Targets

(All Stainless Steel)

(b) Carlson pin

FIGURE 12–5 Graphic symbols and systems to obtain layer-to-layer registration. [(b) and (c) provided courtesy of Bishop Graphics, Inc., Westlake Village, California 91359]

(c) Pin registration bar

(d) Symbols used to locate datum and reference points and identify special holes or terminal areas

(e) Universal target symbol

For double-sided and multilayer boards, additional clear polyester sheets, one for each layer, are positioned onto the initial setup just described and secured to the table. Precise layer-to-layer registration is essential to ensure the exact locations of pads, preforms, and tapes. The procedure for maintaining this registration is first to secure one sheet of polyester film onto the component layout drawing. A *target*, such as one of the type shown in Fig. 12–5a, is positioned over the *registration marks* (see Section 9–11) on the underlying component layout drawing. Targets are precision patterns which have a pressure-sensitive adhesive backing and are supplied on a liner from which they are easily removed with a scalpel. They adhere to the tip of the scalpel blade and are easily positioned and placed onto the polyester sheet. Three registration targets are typically required on each sheet and are located outside the component border of the design.

Where the design is double-sided or multilayer, a second sheet of polyester film is placed over the first and another set of three targets are precisely aligned over those on the first sheet. The 10X comparator should be used to obtain perfect alignment of the targets. This procedure is repeated for all board layers to be produced.

Another technique for achieving precise layer-to-layer registration is with the use of *Carlson pins*, which are shown in Fig. 12–5b. With this method, the polyester sheets are first stacked together with their edges aligned. A single 0.250-in. diameter hole is punched near one edge of the stack using a paper punch, and a Carlson pin is inserted into the hole. Another hole, spaced at a uniform distance from the same edge as the first hole, is made and a second pin is inserted. The stack of film with the pins is then placed onto the light table and the pins are securely taped in place. With this arrangement, any layer of film can be easily removed without losing the preset registration. Thus replacement of any sheet can be quickly and accurately made. For convenience and to ensure buckle-free overlay systems, *pin registration bars*, made of spring-tempered stainless steel, and prepunched films are also available. See Fig. 12–5c. Note that the A-size film has one round hole and the other is a universal slotted hole. They are approximately $7\frac{1}{2}$ in. apart and along one edge. Sizes B through E have one round center hole and one to three slotted holes spaced from 6 to $7\frac{1}{2}$ in. apart to either side of center. The larger the sheet of film, the more slotted holes there are along the edge. Film sizes C through E have a slotted *tail pin* hole which is centrally positioned close to the opposite long edge of the sheet. The direction of this slot is perpendicular to those along the top edge. The top slotted holes provide tight hole-to-pin tolerance at the top and bottom edges of the elongated metal pins. They also allow a small degree of expansion and contraction along the length of the sheet to prevent buckling. The round center hole and pin establish a fixed point from which registration radiates in all directions across the sheets. The bottom tail pin maintains the vertical registration which is required when working with larger sheets. Again, a small amount of expansion and contraction in the vertical direction is allowed to prevent buckling. This system thus allows each sheet of film to react to changes in temperature and humidity while maintaining precise registration. It needs to be emphasized that even though the use of Carlson pins aid in precise film registrations, the use of targets is still required for use in the fabrication of the board.

With the layer-to-layer registration completed, the next consideration will be to indicate on each sheet of polyester film the exact position from which all mechanical dimensions will be made. Special pressure-sensitive symbols, shown in Fig. 12–5d, are used for this purpose. Typically, one of these symbols is positioned on the film over the datum point, which may be inside or outside the board outline on the component layout drawing. If the reference datum is a tooling hole (lower left corner of the drawing), one pattern is centered about it. To locate the y axis, a second symbol is accurately positioned with a scale and a comparator. This symbol is placed at another tooling hole, if on the drawing, or positioned approximately 1 in. beyond the board outline if another tooling hole is not shown. Similarly, the x axis is located with the use of a third symbol. These are shown in Fig. 12–8.

For precise locations of datum points, a *universal target* symbol, shown in Fig. 12–5e, is commonly used. The lines of the underlying grid system are clearly visible through the four arms of this target. This allows the target to be aligned with the grid lines passing through the datum points to an accuracy of ±0.001 in. A selection of reference symbols and their sizes is shown in Appendix XXI.

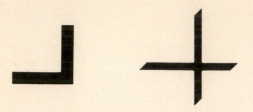

FIGURE 12–6   Corner bracket symbols.

To define the board edges, pressure-sensitive corner bracket symbols, shown in Fig. 12–6, are used. Recall that they are not part of the finished board. For this reason, they are located on the pad master so that their inside edges are perfectly aligned with the outside edges of the board as shown on the underlying layout drawing. The size of the brackets used depends on the scale of the artwork.

The last feature to be added to the initial setup of the artwork master prior to pad positioning and tape routing is the inclusion of a reduction scale. This scale is shown on the pad master to inform the photographic department of the amount of reduction required in order to result in a finished 1:1 scale. This reduction scale is shown in Fig. 12–7. Note that the two pressure-sensitive half-circles have the same shape as those drawn on the component layout as shown in Fig. 8–8. The straight edges of the half-circles are aligned exactly with those on the drawing with the same spacing between them. The precise scaled distance should be made with the use of a plastic scale and a comparator. Two pressure-sensitive arrowhead symbols are next positioned with their points just touching the vertical edges of the half-circles. Finally, dry transfer lettering (described in Section 2–5) is used to produce the words *Reduced to* followed by the 1:1 scale distance in inches and the ± tolerance, also in inches (refer to Sec-

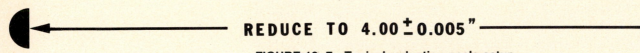

FIGURE 12–7   Typical reduction scale setup.

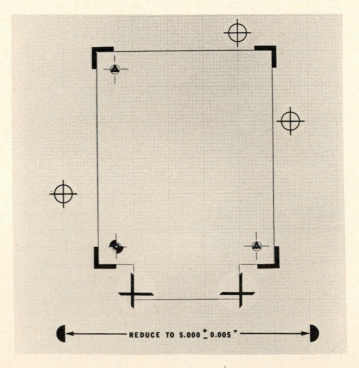

FIGURE 12–8   Initial setup of features for a typical taped artwork master.

tion 8–6 and Fig. 8–9c). The tolerance values used depend on the scale and the degree of accuracy required in the application of the artworks. A typically specified tolerance is ±0.005 in. The higher the degree of tolerance required, the more expensive the camera needed to obtain the specified photoreduction. The initial setup with the features described in this section for a typical taped artwork master is shown in Fig. 12–8.

12-3

Terminal Pads and Preforms

The terminal pads and via holes drawn on the component layout drawing must be reproduced exactly on the taped artwork master. Pressure-sensitive drafting aids called *pads* are used to define these features on the artwork. These pads are die-cut from 0.005-in. stock and are accurate in diameter to within ±0.002 in. They are available with outside diameters (OD) ranging from 0.050 in. to as large as 1.500 in. They come as solid circles or with a center hole. See Fig. 12–9a. Those with holes are used for unsupported pads and the solid configurations are used for supported holes on the board. For many of the OD pad diameters, there are a variety of inside diameters (ID) available. When selecting a pad that is to have a center hole (e.g., single-sided unsupported pads), the following guidelines may be used to ensure compatibility with hole sizes to be formed in the board and the pad's ID dimension. The popular sizes of center hole ID dimensions for pads used on 2X and 4X scale artwork are 0.031 and 0.062 in., respectively. These hole sizes are large enough to allow the center of the pads to be easily aligned with the grid intercepts which locate the center of the hole. After etching, the 2X scale center holes will be approximately 0.016 in. This size, when drilled for even the smallest lead size normally encountered, will result in there being copper right up to the edge of the drilled hole and around its entire circumference. Available pad sizes with center hole ID di-

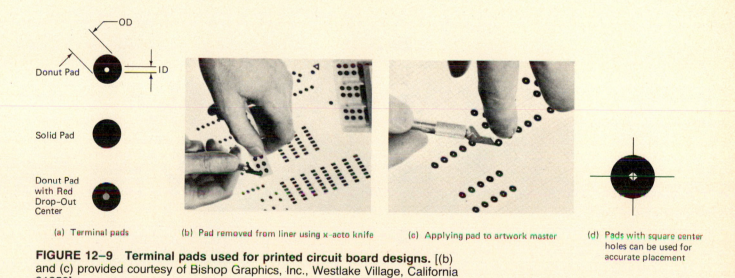

(a) Terminal pads   (b) Pad removed from liner using x-acto knife   (c) Applying pad to artwork master   (d) Pads with square center holes can be used for accurate placement

FIGURE 12–9 Terminal pads used for printed circuit board designs. [(b) and (c) provided courtesy of Bishop Graphics, Inc., Westlake Village, California 91359]

mensions and solid pads are listed in Appendices XXII and XXIII, respectively.

The procedure for removing a pad from its liner and transferring it to the artwork master is shown in Fig. 12–9b and c. The point of the scalpel is gently inserted under the pad and peeled from the liner, care being taken not to damage its edge. The scalpel blade now acts as a holding tool to aid in positioning the pad over the desired grid intercepts. When perfect alignment is obtained, the pad is firmly pressed onto the artwork and the blade withdrawn. For those pads with holes, the grid lines are viewed directly for alignment and centering.

For plated-through-hole pads, solid pads are used on the artwork master. For the alignment of these pads, their outer edges are positioned with the use of the circumference of the terminal pad drawn on the componenet layout. Because this method does not lend itself to reliable layout accuracy, it should be avoided and a *red eye pad* used instead. This drafting aid, shown in Fig. 12–9a, has a transparanet red center hole. The grid intercepts are viewed through the red center to aid in precise alignment. When photographed with standard orthochromatic film, the red centers will appear as black and opaque, thereby producing solid terminal pads as required for plated-through holes. A limited number of the pads listed in Appendix XXIII are available with red centers. Those which are available are listed in Appendix XXIV.

Another type of round terminal pad used when optimum accuracy of placement is a requirement is the *square center-hole* pad shown in Fig. 12–9d. As shown, accurate grid intercept placement is achieved by aligning the square hole corners with the underlying grid lines. See Appendix XXV.

When a smoother interface between a pad and its associated conductor path is required, a *teardrop* pad, shown in Fig. 12–10a, is commonly used. Teardrop pads are in two styles. One has radius fil-

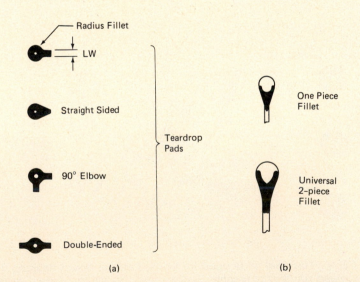

FIGURE 12–10 Single- and double-entry teardrop pads and fillets.

lets and the other has straight-sided fillets. The latter type are used only when a straight-angled interface between the terminal pads and the conductor paths is preferred. These pads are available with one conductor path entry point or with two entry points, either 90 degrees apart (elbow) or 180 degrees apart (double-entry). These are also shown in Fig. 12–10. A selection of teardrop-style pads with available sizes is listed in Appendix XXVI. Great care must be exercised in the placement of teardrop-style pads so that the associated taped conductor path will be aligned with the entry point of the pad. The width of the conductor path should be the same as the entry dimension (shown as LW in Fig. 12–10a) of the pad. Because there is a limited selection of pad diameters and entry widths, flexibility in their use is somewhat limited. This is overcome with the use of either *one-piece* fillets or *universal two-piece* fillets, shown in Fig. 12–10b. These aids are used in conjunction with the round pads. After the round pad is positioned, the fillets are placed over it with their tails extending in the anticipated direction of the conductor path entry point. The conductor tape is then overlapped onto both the fillet and the pad. See Fig. 12–10b. Available sizes of fillets are listed in Appendix XXVII.

Oval-style pads, shown in Fig. 12–11, are available with a lead access hole which is centered for normal lead entry or offset for lead clinching. Oval pads are also supplied with two holes for entry of two leads onto a single pad. This style is also shown in Fig. 12–11. The use of these styles for individual pads is not recommended for high-density designs. The reasons for this are given in the following section. A listing of available oval-style pads is given in Appendix XXVIII.

In addition to individual pads, *multipad preforms* of device patterns are also available for various round-pin packages, the popular dual-in-line packages, and flat-pack designs. All of these patterns come in a variety of shapes and sizes to meet the needs of a wide range of taped artwork applications. These preforms are opaque black-inked patterns on a 0.002-in. dimensionally stable polyester film carrier and have a backing of pressure-sensitive adhesive. The patterns are packaged on a strip of paper or release liner, which affords ease of removal with a scalpel. To remove a preform from the liner, the strip is held flat in one hand and the scalpel blade is gently slipped under one edge of the pattern. See Fig. 12–12a. The pattern is then peeled from the liner by a slow upward motion of the scalpel. The backing should not be touched with the fingers since body oil will degrade the adhesive qualities. With the preform completely removed from the liner, it is positioned onto the pad master with the aid of the scalpel blade. When properly aligned, the pattern is firmly pressed against the polyester film, burnishing it lightly with the index finger. See Fig. 12–12b. Once positioned onto the pad master, care must be taken to avoid scratching any of the inked patterns.

Multipad preforms save a great deal of layout time in the preparation of an artwork master. In addition, they ensure accurate positioning between the individual pads in the pattern.

The round multipad patterns are made for TO-style cases, such as the TO-5 and TO-18 styles having 3, 4, and 6 pins, in addition to

Centered Lead Hole   Offset Lead Hole   Double-Hole Lead Entries

FIGURE 12–11   Oval pads.

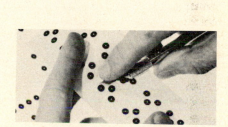

(a) Removing preform from liner with knife blade

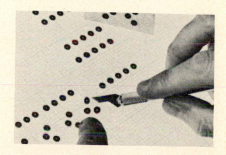

(b) Positioning preform and pressing onto polyester film

FIGURE 12–12   Procedure for applying small preforms to taped artwork master. [Provided courtesy of Bishop Graphics, Inc., Westlake Village, California 91359]

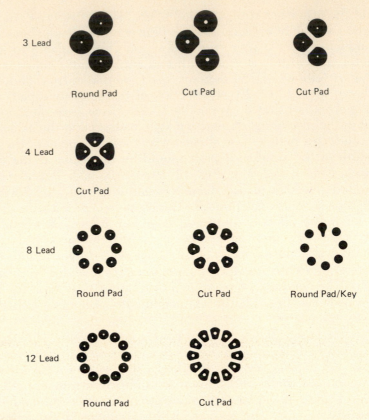

FIGURE 12–13   Selection of TO-style preforms.

the TO-99 and TO-100 cases for 8- and 10-pin ICs as well as those with 12 pins. As shown in Fig. 12–13, the individual pads on a particular pattern may be round or teardrop-shaped. In some patterns, one of the pads may have a different shape from the others for purposes of keying to aid in device assembly.

As discussed in Section 9–4 and shown in Fig. 9–4, the leads of the TO-style packages may be specified to enter the board straight through or with an offset to increase the device's pad-to-pad clearance. The selection of the appropriate pattern is made on the basis of how the leads are to enter the board.

Some available preforms and pad styles for TO packages are shown in Appendix XXIX. Included is dimensional information regarding hole sizes, pad shapes and sizes, clearances between pads, and lead circle sizes.

A preformed pattern for a flat-pack case style is shown in Fig. 12–14a. The pattern is designed for surface mounting of the case with its leads formed as shown in Fig. 12–14b. These preforms are available with pattern fingers having on-center spacings of 0.050 in. and with various spacings between rows of fingers to accommodate several flat-pack styles. Dimensional information for these patterns is given in Appendix XXX.

One of the most popular device cases is the DIP, which is used in both analog and digital circuits. Because of its wide use, many preform styles are made to satisfy a variety of applications. Several

Surface Mounted In-Line Type

Through-Hole Mounted Staggered Type

(a) Flat pack patterns

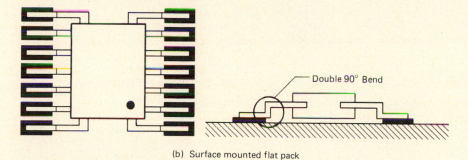

Double 90° Bend

(b) Surface mounted flat pack

FIGURE 12–14 **Preforms for flat pack style cases and surface mounting configuration.**

of these patterns are shown in Fig. 12–15. As can be seen, designs are available with round, rectangular, cut, oval, and elliptical pad shapes. In addition, they may have conductor feed-through sections or a keying pad for the quick location of pin 1. The wide variety of pattern styles are made to meet the demands for high-density packaging. Dimensional information for DIP patterns is given in Appendix XXXI.

The selection of the most appropriate pattern depends on the application and specifications, such as minimum conductor spacing and annular ring. Patterns with 8 to 40 pins are available, but not in all styles. However, preforms can be cut, aligned, and spliced to result in a design requiring a larger number of pins.

On single-sided boards, maximum bond strength of each copper pad is absolutely essential. For this reason, rectangular, round, or cut-style pads (without conductor feed-through sections) are used. The density of single-sided designs normally does not require that conductor paths be routed between adjacent pins. In any case, this should be avoided since it will necessitate a reduction of the unsupported pad sizes, which would have an adverse effect on the bond strength of the pads.

On double-sided plated-through-hole designs, where the pad bond strength is dependent not only on the pad size but also on the plated hole wall, oval and cut pads with reduced widths may be used. Patterns with conductor feed-through sections should be used where the layout requires critical centering of paths between adjacent pads.

Oval pads are used when paths must pass between adjacent pads and maximum pad diameter is required. The oval shape allows the conductor path to be routed between pads with a larger clearance.

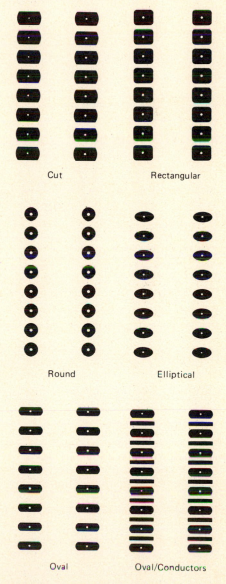

Cut   Rectangular

Round   Elliptical

Oval   Oval/Conductors

FIGURE 12–15 **Various styles of preforms for dual-in-line packages.**

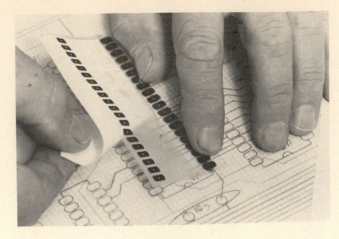

(a) Initial alignment using release sheets as an aid

(b) Removing release sheets

FIGURE 12–16 Best method of applying large preforms.

If the cut pad pattern is used in this application, corner clearance would be reduced. The narrow cut pad would cause a reduction in the clearance space area.

The overall size of preforms can become considerable, especially when a 40-pin package is laid out to a 4:1 scale. To position these large patterns accurately, a technique similar to that used for laying down large pieces of adhesive-backed vinyl (contact paper) is effective. The underlying grid is used as a guide. The pattern is removed from its liner and replaced onto a set of used release liners, allowing any one row of pads on one side of the preform to be exposed. See Fig. 12–16a. The pattern can then be aligned and realigned without sticking until the exposed pads are exactly located on grid. The exposed row of pads is then pressed against the artwork. Holding the pattern in place, the release liner sections are next rolled out from under the preform. This is shown in Fig. 12–16b. When the first section of liner is removed, the second row of pads is pressed onto the artwork. The remaining liner is gently raised and peeled away as the preform contacts with the artwork. The entire pattern is finally burnished with the finger, care being taken to avoid scratching any of the inked surface.

When the design requires a DIP having more than 40 pads (20 per row) or when a large number of individual pads must be positioned with even spacings of, for example, 50 or 100 mils (1:1 scale), prespaced pads, shown in Fig. 12–17, may be used. They are available on pressure-sensitive polyester strips, approximately 8 in. long, with a thickness of 0.002 in. Depending on the pad size, there are as many as 80 pads on each strip. There are a variety of pad styles on these strips for 1:1, 2:1, and 4:1 designs.

For continuous in-line and staggered connector patterns for connector plugs that are staked or soldered directly to a board, preforms on 18-in. strips are available. Several of these patterns are shown in Fig. 12–18.

The final preform to be considered is the insertion-type edge connector pattern, shown in Fig. 12–19. Note that the evenly spaced

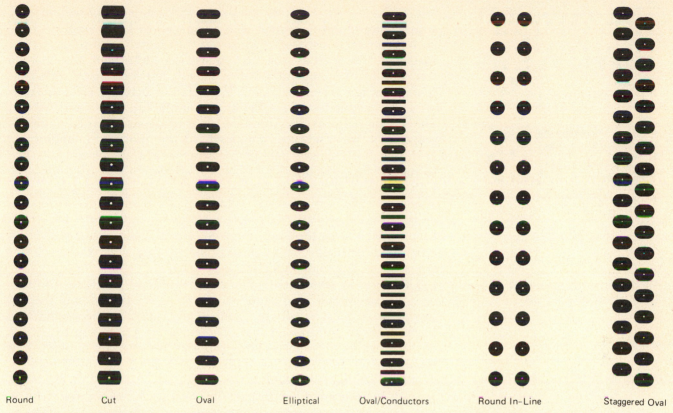

Round	Cut	Oval	Elliptical	Oval/Conductors

FIGURE 12–17 Selection of multipurpose prespaced pad styles.

Round In-Line	Staggered Oval

FIGURE 12–18 Popular continuous connector pattern styles.

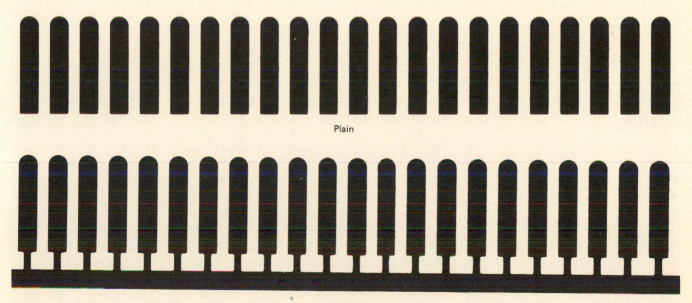

Plain

Plating Bar

FIGURE 12–19 Insertion type connector patterns without feed-through holes.

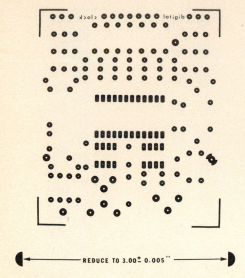

REDUCE TO 3.00 ± 0.005"

FIGURE 12–20   Pad master of a simple digital clock circuit PCB.

fingers have rounded tops (with or without feed-through holes), straight sides, and terminate with square ends. Some patterns have no interconnections, while others are terminated with finger extensions and a plating bus which is required for nickel/gold plating. Both styles of finger connector patterns have a selection of six center-to-center spacings. These are 0.050, 0.100, 0.125, 0.150, 0.156, and 0.200 in. The available lengths of the fingers are 0.250, 0.500, and 1.00 in. The finger patterns are supplied on continuous pressure-sensitive polyester strips. They are positioned as a group using the underlying grid and drawn finger pattern to achieve accurate placement. The technique for applying this preform is similar to that used for large DIP patterns. The pattern is repositioned on the liner, exposing only a small portion of the adhesive backing along one edge. When the ends of the fingers are accurately positioned, the liner is rolled out from under the pattern and removed. Finally, the entire preform is burnished. Available sizes of insertion-type edge connector preforms are given in Appendix XXXII.

The order in which individual pads, multipad preforms, continuous prespaced pads, and connector patterns are placed onto the artwork master is essentially the same as presented in this section. All single pads are positioned first. Then, working from top to bottom of the artwork, all preforms and connector patterns are applied. This technique minimizes the possibility of scratching the inked patterns as others are being applied.

With the placement of all pads, preforms, and connector strips to the pad master containing corner brackets, reduction scale, targets, and so on, completed, the result would be the artwork shown in Fig. 12–20. The artwork is now ready for the taping of conductor paths.

12-4

Conductor Path Taping

Conductor paths are taped using precision-cut tapes. Although three types of tape are available (black opaque, red transparent, and blue transparent), our discussion of artwork taping will be limited to the use of the black opaque type. This tape is flexible and has a pressure-sensitive adhesive backing. It has a thickness of 0.005 in. and comes in 20-yard rolls with a selection of widths, varying from 0.015 to 2 in. Six-inch widths are also available in 5- and 6-yard rolls. A complete listing of available tape widths is given in Appendix XXXIII.

The standard tolerance for tape widths from 0.015 to 0.312 in. is ±0.002 in. and for tapes from 0.312 to 6 in., it is ±0.005 in. The tape is photographically opaque with a nonreflective mat finish. Since it is flexible, it can be shaped and contoured to follow the conductor path pattern shown on the underlying component layout drawing. Because of this flexibility, however, the tape can be stretched as it is being shaped. Stretching must be avoided, as it results in a nar-

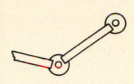

(a) Overlapping pad centers with tape

(b) Positioning knife blade for cutting tape to re-open pad centers

(c) Tape cut allowing correct overlap

FIGURE 12–21 Applying tape between pads.

rower conductor path to be formed in the copper, which reduces its current-carrying ability. In addition, stretched tape will tend to creep, even if it has been firmly pressed against the polyester surface. Creeping will cause the tape to change from its original position, thus altering spacings or air gaps, which could cause serious problems in the manufacturing phase of the board.

When applying the tape between two pads with a straight interconnection, the end of the tape is first unrolled and overlapped onto the first pad. It is helpful to place the index finger of one hand through the roll of tape, letting it act as a spindle, while the thumb guides the rotation and tape feed. The roll should be oriented so that the tape is fed from the top. Note in Fig. 12–21a that the tape is initially placed across the pad center. Once placed in this position, it is firmly pressed down with the index finger of the opposite hand. The tape is next unrolled, care being taken to avoid stretching. The underlying single-line widths representing conductor paths on the component layout drawing, together with the grid lines, are used as guides for accurate tape routing. A very slight tension is applied as the tape is run straight across to the second pad center. It is then pressed into place by running the index finger over the entire length of the run, including the second pad. The tape is cut off by placing the scalpel blade perpendicular to the edge of the tape and approximately halfway between the point of tape overlap and the center of the pad. See Fig. 12–21b. The tape is then pulled upward while pressing the blade down firmly. This results in a clean cut at the desired point. Excessive pressure or movement of the blade should be avoided since this could result in cutting through the tape and pad or scratching the inked preform. The result of a correctly cut tape is shown in Fig. 12–21c. Note that the tape overlaps the pad to a distance of at least halfway from the edge of the pad to the edge of the center hole (or to the center of the pad if it is solid).

Tape paths that are not straight must be contoured to follow those shown on the component layout. Any change of path direction should only be done with narrow tapes up to 0.040 in. wide, and then forming large radius curves. Contour taping begins with routing the tape up to the point at which it is to change direction. The tape is then gently swung into a broad curve while simultaneously pressing it downward with a finger of the other hand. The tape is next raised and formed into a curve with a smaller radius, again applying finger pressure. The tape will not lose its partial degree of curve formation during this process nor will the adhesive be affected. This procedure of picking up the tape and pressing it down, each time reducing the curve radius, is continued until the desired contour that aligns with the drawn path is reached. This procedure is shown in Fig. 12–22a.

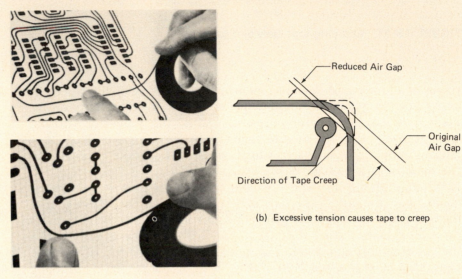

(a) Forming tape into desired path contour

(b) Excessive tension causes tape to creep

45°   60°   90°   Universal Circles

(c) Elbows and universal circles used to aid tape contouring

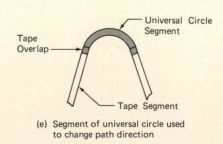

(d) Splicing an elbow onto a taped conductor path

(e) Segment of universal circle used to change path direction

FIGURE 12–22   To avoid excessive tension on tape, elbows and segments of universal circles can be used to change path direction.

One of the disadvantages of contour taping is that even after firm pressure has been applied to the curved section, the tape will have a tendency to creep. This will result in a reduced air gap as shown in Fig. 12–22b. To avoid this problem, only large-radius curves should be contoured. Where tight radii are necessary, drafting aids called *elbows* and *universal circles* should be used. These are shown in Fig. 12–22c. Elbows come on release liners in four preset angles of 30, 45, 60, and 90 degrees. They are also available in nested 90-degree configurations to facilitate the layout of sharp curves in parallel conductor traces with uniform spacings. Elbows also come in a variety of widths to accommodate all of the available tape sizes.

To use an elbow, a straight tape section is terminated slightly short of the starting point of the required bend. See Fig. 12–22d. An elbow with the correct width and angle is removed from the liner with a scalpel and one end of it made to overlap the end of the tape segment and pressed into place. The next tape section is started by overlapping it onto the other end of the elbow and continued on to its termination point.

Where a change in direction at an angle other than available with the precut elbows is required, universal circles may be used. These drafting aids are available in a wide variety of sizes and also come in cluster groupings. To use a universal circle, the appropriate circle in the cluster is first cut to the desired angle. The segment is then removed from the liner and applied to the artwork using the same technique as that just described for the application of elbows. The use of a circle segment to form a special angle is shown in Fig. 12–22e. We will describe later how a universal circle is used to isolate a smaller-diameter pad in a ground plane. Available sizes of elbows and universal circles are given in Appendix XXXIV.

Still another technique for changing conductor path direction is termed *angle taping*. It is preferred to the two methods just described because (1) it eliminates the inherent creeping problem associated with tape contouring, and (2) it requires no additional drafting aids, thereby eliminating the time-consuming process of tape alignment and splicing. Angle taping employs a simple cut-and-turn technique using only the tape to obtain the desired change in path direction. The edge of the tape is first aligned with the underlying drawn path and run beyond the point where it changes direction. See Fig. 12–23a. The blade of a scalpel is then pressed down firmly to cut the tape, leaving only a very thin segment at the opposite side to which the tape is to change direction. This tiny segment will act as a hinge from which the tape will be turned. See Fig. 12–23b. This uncut segment must be small enough so as not to cause it to buckle when the tape is turned to follow the change in direction. Any buckling will not allow that portion of the tape to stick to the artwork. When the tape is turned to the required angle and the cut edges have smoothly overlapped, it is pressed firmly in place. See Fig. 12–23c. An alternative method used for angle taping is shown in Fig. 12–23d. This is a simple angular splicing and cutting technique. Since there is no bend stress on the tape, the possibility of creeping is eliminated.

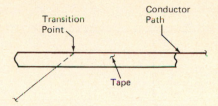

(a) Initial position of tape using underlying conductor path routing

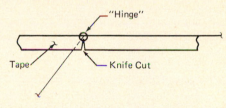

(b) Tape cut with knife to form hinge

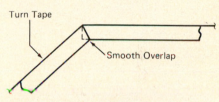

(c) Tape rotated to desired angle and pressed into position

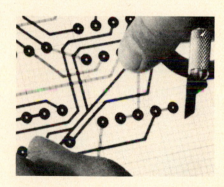

(d) Angle splicing and cutting technique

FIGURE 12–23 Angle taping techniques for changing conductor path direction. [(d) provided courtesy of Bishop Graphics, Inc., Westlake Village, California 91359]

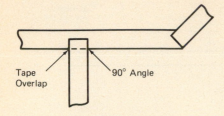

(a) Simple overlap technique

Tape Overlap 90° Angle

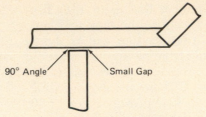

(b) Small gap to avoid tape build-up

90° Angle Small Gap

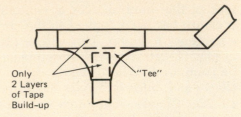

(c) Tee aligned correctly with conductor paths

Only 2 Layers of Tape Build-up "Tee"

FIGURE 12–24 Techniques for splicing conductor paths.

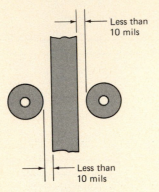

Less than 10 mils

Less than 10 mils

(a) Air gap less than specified minimum

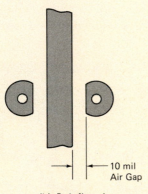

10 mil Air Gap

(b) Pads flatted

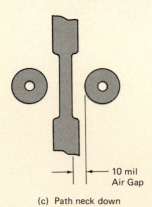

10 mil Air Gap

(c) Path neck down

FIGURE 12–25 Artwork adjustments to meet minimum air gap specifications.

Where the underlying component layout drawing requires a path to connect to another, this connection is made at a 90-degree angle. There are two methods of making these connections on the pad master. The first method is to simply overlap one tape over the other to about one-half its width before it is cut. This is shown in Fig. 12–24a. The second and often preferred method is with the use of a drafting aid called a *tee*. The tee provides a smoother filleted interface between the intersecting paths and eliminates the sharp 90-degree interior angles that result from straight overlapping. See Fig. 12–24b. Note that tape buildup (three layers or more) is avoided by not allowing the two tapes that meet at 90 degrees to overlap. The tape is cut so as to leave a small gap between them. The tee is placed over the gap with its three entry points overlapping the tapes to complete the connection. If it is found that any tape corners jut out from under the tee, this is an indication that the tapes are not at a 90-degree angle with each other or that the tee has not been properly aligned. To correct this, repositioning of the tapes or the tee is necessary. The sizes of available tees are given in Appendix XXXV.

It sometimes becomes necessary to make adjustments in the artwork so that air gaps may conform to specifications. For example, Fig. 12–25a shows that the air gaps between the pads and conductor path are violating the minimum 10-mil specification. Assume that there is no alternate route for this path, that a via hole cannot be used, and that the pads cannot be moved. One method of overcoming this problem is to cut each pad just enough so that the air gap between the tape and the flatted sides meets the minimum specifications. This is shown in Fig. 12–25b. Caution should be exercised to avoid cutting either pad to an amount that will violate the minimum annular ring specifications.

Another method that may be used to increase the air gap shown in Fig. 12–25a is termed *neck down*, where the conductor width is reduced in the area between the pads. This is shown in Fig. 12–25c. Again, care must be taken in this method of solution that the minimum conductor width in the neck area be maintained. The two methods of increasing the air gap just discussed should be avoided whenever possible and used only when all other acceptable solutions have been examined.

Another common problem, especially in the design of digital circuits, is the layout of accurate and uniform spacings between long conductor paths running in parallel. A technique for accomplishing

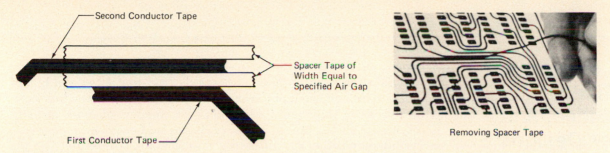

FIGURE 12–26 **Accurate spacing of closely positioned long, parallel, conductor paths.**

this quickly and efficiently is shown in Fig. 12–26. The first conductor path is taped into position. A spacer tape, having a width equal to the desired air gap, is then placed adjacent to the first path. The next conductor path tape is placed onto the pad master with its edges just contacting the spacer tape. This process is continued until all of the parallel conductors have been laid out. The spacer tapes are then removed, resulting in an accurate and uniformly spaced design.

To serve as a guide, Table 12–1 compares conventionally acceptable portions of taped layouts with those which are not recommended. Note that the oval pad style should be avoided in high-density packaging because they take up considerable space. In addition, the oval shapes are prone to heat sinking, which results in nonsymmetrical soldered connections, reducing their quality. After all the taping is complete, the artwork is carefully turned over and pressure is applied with a hard rubber roller to improve the tape adhesion.

The final step in completing a taped artwork master is the application of any additional literal and numerical information, such as component and device reference designations and positions. This is done with the use of dry transfer lettering. When applying this information, it is essential to note which side (component or circuit) of the artwork master is being labeled. Most labeling is done on the component side. When information needs to appear on the circuit side, the artwork master is removed from the work surface and reversed so that the taped side is facing down. When completed, all lettering will appear as *right-reading* on the component side and *reverse-reading* on the circuit side when viewed from the taped side. This is the result of viewing both sides of the board from the component side. With all of the reference designators added to the taped artwork, a title block decal may be applied for identification, such as board number, revision level, and so on.

When a taped artwork master is completed, it should be stored flat. Rolling it will upset the adhesive bonding between the tapes and the polyester film.

TABLE 12–1 Do's and Don'ts Guide for Taping Artwork Masters

Acceptable	Not Recommended
Avoid sharp external angles which can cause foil delamination.	
Avoid acute internal angles.	
Always use the shortest practical circuit routing.	
Maintain equal spacing where conductors pass between terminal areas.	
Avoid large multiple hole terminal areas which may cause thermal soldering problems and non-symmetrical solder fillets.	
Maintain uniform pattern around hole to produce symmetrical solder fillets.	
Avoid using conductors the same size as terminals which will cause solder to flow away from terminal.	
Plating bar should extend out from & beyond board edge to facilitate fabrication.	

12-5

Artwork Masters for Single-Sided, Double-Sided, and Multilayer Designs

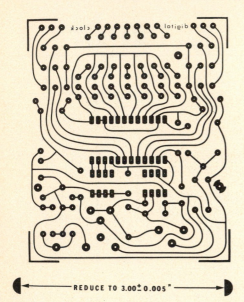

REDUCE TO 3.00 ± 0.005"

FIGURE 12–27 Completed taped artwork master of a simple digital clock PCB.

Artwork masters for single-sided designs are the simplest to lay out. The initial setup is as outlined in Sections 12–2 and 12–4. All taping is included on one sheet of polyester film, including corner brackets, reduction scale, targets (if a solder mask or component marking mask is required), datum points, pads, preforms and connector patterns, and conductor paths. Also, the circuit side may include dry transfer lettering for any literal or numerical information which is to appear in copper on the finished board. A typical taped artwork master for a single-sided board is shown in Fig. 12–27.

For double-sided designs, a minimum of two taped artworks is required. Each of these artworks include all of the pads and conductor paths as well as other required information for each side. The problem with using just two sheets of film is that it is exceedingly time consuming to achieve the required perfect registration between the two sets of pads, preforms, and connector strips so that all centers are in precise alignment. The preferred method of reducing the layout time and ensuring perfect registration is with the use of a third sheet of polyester film, called the *pad master*. The other two sheets would be designated as the component side and the circuit side. This procedure of using three sheets begins by initially using targets and registration pins to align all of the sheets with each other and with the underlying component layout drawing. Two of the polyester films are removed, leaving only one registered film on the drawing. This will become the pad master and will contain the following: (1) registration targets, (2) datum pads, (3) corner brackets, (4) reduction scale, (5) all individual terminal pads, (6) preforms, and (7) connector strips. Note that all of these items are common to *both* sides of the board. A completed pad master is shown in Fig. 12–28a. A second sheet of polyester film is then registered over the pad master and the conductor path taping for the component side is done with the use of the underlying drawing and pad master. Taped paths are overlapped onto the pads shown on the pad master just as though they appeared on the same sheet. The component side artwork contains registration targets, conductor paths that are to appear on the component side, and any literal or numerical information for that side only. This artwork is labeled *Component Side*, shown as right-reading. The completed artwork for the component side is also shown in Fig. 12–28a.

To produce the photographically reduced 1:1 scale component side phototool used in the manufacture of the board, the pad master *and* the component side artwork are registered together on a copy camera and photographed. The result is a composite of both artworks produced on one sheet, which is the complete component side artwork master. This is shown in Fig. 12–28b.

The circuit-side artwork is made in the same manner as that used for the layout of the component-side artwork. Using only the pad master and the layout drawing, the third sheet of polyester film is registered and secured. This sheet will contain all of the conductor paths that are to appear on the circuit side of the board, in addition

FIGURE 12—28 Taped artwork masters for double-sided PC boards.

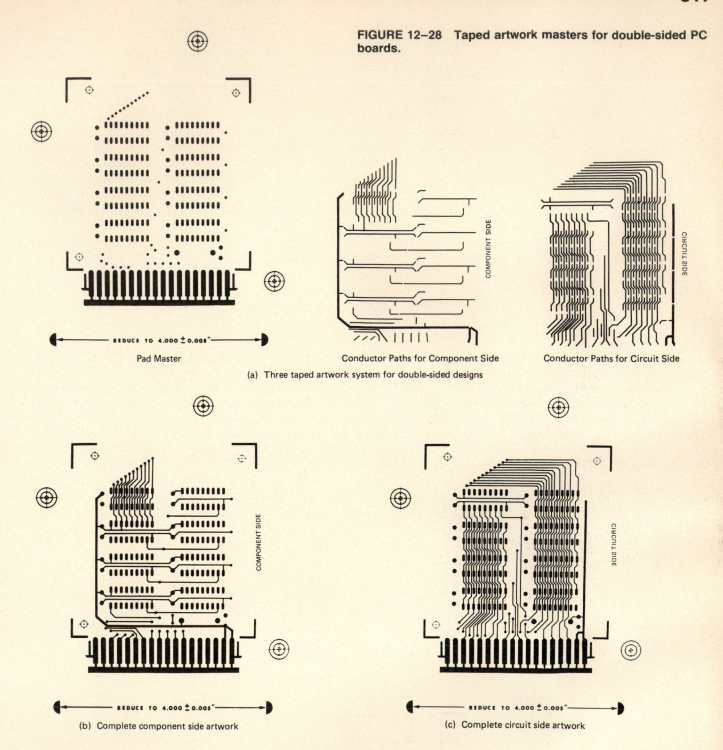

REDUCE TO 4.000 ± 0.005"

Pad Master

Conductor Paths for Component Side

Conductor Paths for Circuit Side

(a) Three taped artwork system for double-sided designs

REDUCE TO 4.000 ± 0.005"

(b) Complete component side artwork

REDUCE TO 4.000 ± 0.005"

(c) Complete circuit side artwork

to all of the literal and numerical information for this side. Finally, it is labeled *Circuit Side* on the reverse of the taped side so that these words appear reverse-reading when viewed from the taped side.

Again, this circuit-side artwork is registered with the pad master and photographically reduced. The finished composite circuit side artwork master, shown in Fig. 12–28c, results from this process.

There are two distinct advantages in the use of the pad master. First, perfect layer-to-layer pad registration is ensured. Second, since the scalpel blade does not come into direct contact with the pads during the conductor path taping phase, there is no concern as to cutting or scratching the pads.

The taping of a four-layer multilayer board requires five sheets of polyester film registered with registration pins and/or targets. Since the outside layers of a multilayer board are effectively the two sides of a double-sided board, we begin by producing a pad master in the same manner just described. The second and third sheets of film, used in conjunction with the underlying pad master and the layout drawing, will be the component side and the circuit side, respec-

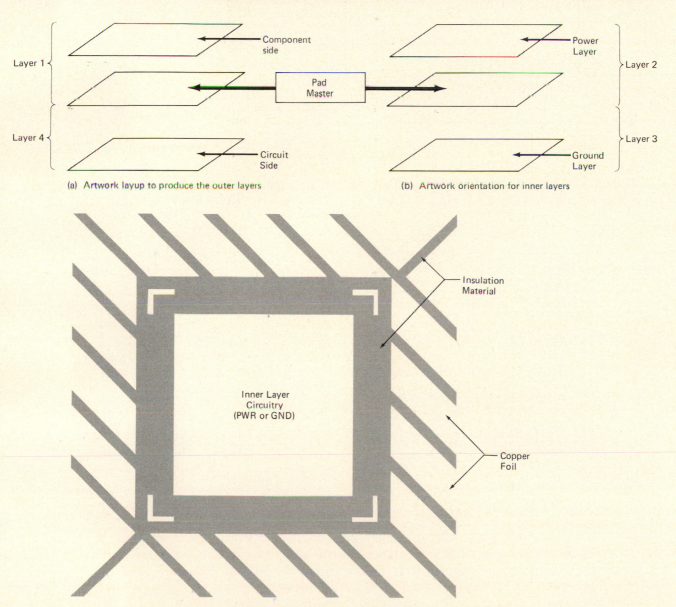

(a) Artwork layup to produce the outer layers

(b) Artwork orientation for inner layers

(c) Relieved copper border around perimeter of MLB prevents twisting during the manufacturing process

FIGURE 12–29   Layup of multilayer artworks.

tively. These are shown in Fig. 12–29a. The component side of the taped artwork will be labeled *Layer 1* and the reverse side of the circuit side artwork will be labled *Layer 4*. When these artworks are photographically processed with the pad master, the complete patterns for layers 1 and 4 will be produced.

The fourth sheet of polyester film is used for the inner power layer. It is registered over the pad master and the layout drawing. This sheet will become *Layer 2*. Before beginning the artwork of the inner layers, it is important to point out several differences when compared to the photographically *positive* artworks of layers 1 and 4. The majority of area of the inner layers of the board will be copper rather than many narrow copper paths connecting terminal pads. For this reason, they are produced as *negative* artworks. In this case, black will represent insulation instead of copper and the clear area will represent the copper. This is the opposite of the artworks made for the component and circuit sides.

For the efficient layout of the negative artworks used for layers 2 and 3, the two drafting aids shown in Fig. 12–30 are used. These are called *negative ground-plane donut pads*. The style shown in Fig. 12–30a is a *thermal relief* pad and is used when a lead is to be connected to either copper power- or ground planes (refer to Section 11–3). The four narrow paths connecting the pad to the surrounding copper serve two purposes. They not only make the electrical connection between the pad and the plane but also prevent heat from being drawn away from the pad by the large copper plane during the soldering operation.

An *antipad* is shown in Fig. 12–30b. It is used when a lead passing through the board must be electrically isolated from either power or ground planes. Available sizes of each of the pads shown in Fig. 12–30 are given in Appendix XXXVI.

Pads requiring thermal relief are placed onto the layer 2 sheet using the pad master and registration guides. The use of a comparator is recommended for this layout. All pads that do not connect to power require antipads which are positioned with the use of the pad master and a comparator. Next, using the underlying drawing as a guide, a black border is made with lengths of wide tape on the layer 2 sheet. This border is aligned with the usable board area as defined on the drawing. This area is shown as a dashed line in Fig. 11–5. The width of the border should extend beyond the outside edges of the corner brackets. Recall that black will represent the insulation and the copper is shown as clear. The black border will result in insulation surrounding the useful area of layer 2. Finally, the copper that remains outside the insulation border is relieved with strips of tape. This is shown in Fig. 12–29c. The reason for this is to prevent the board from twisting during the manufacturing process.

With the artwork for the power layer completed, it is removed and the fifth sheet of polyester, which will serve as the ground plane (layer 3), is placed in registration over the pad master. This is laid out in a similar fashion to that of layer 2. See Fig. 12–29b.

Using the negative artwork technique for the inner layers, where the majority of the area is to be copper, saves a great deal of layout time when compared to black representing copper as was done for layers 1 and 4. Any reversal (positive to negative or negative to pos-

(a) Thermal Relief   (b) Antipad

FIGURE 12–30 Negative ground-and power-plane donut pads.

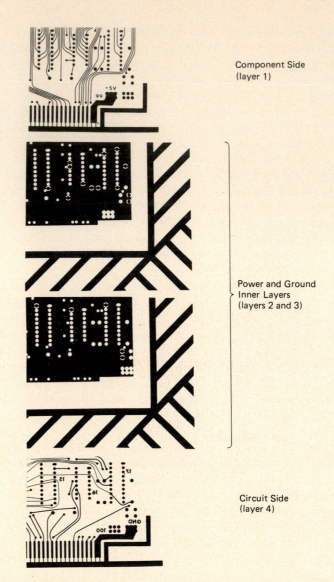

Component Side
(layer 1)

Power and Ground
Inner Layers
(layers 2 and 3)

Circuit Side
(layer 4)

FIGURE 12–31   Corner section of artworks for a 4-layer MLB. [Courtesy of Computer Consoles Incorporated.]

itive) of these scaled artworks for making phototools are quickly made with photographic equipment.

The corner sections of the artworks required to fabricate a typical 4-layer multilayer power/ground board are shown in Fig. 12–31.

12-6

Special Tapings

A common problem encountered in the layout of artworks is the need to improve the distribution of the copper areas which appear on both sides of a double-sided board or on multilayer boards. Recall that a more equal area of copper appearing on both sides of a board will greatly improve the quality and thickness of electroplated copper and

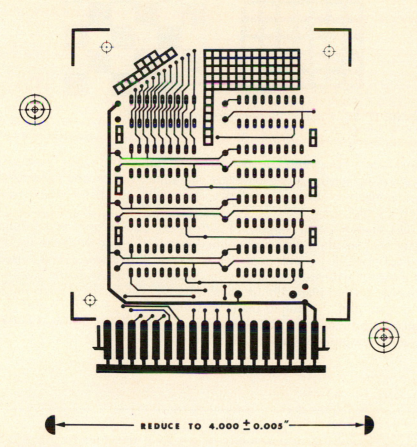

FIGURE 12–32 Layout of component side of memory board with addition of nonfunctional copper.

solder. For this reason, the completed artworks which represent the two sides of a board should be compared to see if the areas of copper are approximately equal. If not, corrective measures, which will be described here, need to be taken. An example of unequal copper areas is shown in the layout of the digital memory board of Fig. 12–28. Note that the circuitry shown on the circuit side is considerably more dense and involved than that of the component side.

Where an imbalance in the copper area of the two sides of a board exists, the PCB designer needs to estimate how much more copper needs to be added to the lesser side to improve the balance. An appropriate amount of *nonfunctional* foil is then added to the required side. This additional copper is added for plating purposes only and plays no role in the functioning of the circuit. See Fig. 12–32. Note that a crosshatched pattern has been added to the taped artwork in the areas which are void of functional circuitry. The amount of nonfunctional copper allowed should result in a more balanced amount on both sides of the board. This process of adding nonfunctional copper to the design should only be done using broken patterns, such as the ones shown in Fig. 12–33. Solid foil areas, in excess of ap-

FIGURE 12–33 Patterns for nonfunctional foil areas.

Thermal Relief   Antipad

FIGURE 12–34   Positive ground- and power-plane donut pads.

proximately $\frac{1}{2}$ in, should be avoided since they could cause the board to warp during the wave soldering and solder reflow processes used in manufacture. Large copper areas also tend to blister when used with a solder mask.

The nonfunctional copper foil pattern can also be used as a ground plane by providing electrical connections to the grid with the use of ground-plane donut pads as shown in Fig. 12–34. These aids are used as positive patterns. See Appendix XXXVI for the available sizes of positive ground-plane donut pads. In some cases, the design engineer may prefer a ground-plane grid for common reference and circuit return points over the ground-plane style shown in Fig. 12–29b to achieve optimum circuit performance.

12-7

The Documentation Package

It is generally the responsibility of the PCB designer to provide a complete documentation package to the board manufacturer. Although the contents of the package may vary from one company to another, following is a list of items normally included:

1. A minimum of one complete set of either positive or negative prints photographically reduced to a 1:1 scale. These should be on 0.007-in. high-quality polyester base and no more than second generation (i.e., contact prints from original 1:1 reductions).

Prints included for single-sided designs are:

    a. Circuit side (labeled)
    b. Component marking mask* (if required).
    c. Solder mask* (if required).

Prints included for double-sided designs are:

    a. Circuit side (labeled).
    b. Component side (labeled).
    c. Component marking mask (if required).
    d. Solder mask (if required).

* Discussed in Chapters 13 and 14.

Letter Symbol	Finished Hole Size	Plated	Quantity
A	0.033	Yes	42
B	0.042	Yes	13
	0.060	Yes	132
C	0.075	Yes	80
W	0.128	No	4

Hole Tolerance		
Finished Hole Size	Plated	Nonplated
0-to-0.031″	0.004″	0.002″
0.032″-to-0.061″	0.004″	0.002″
0.062″-to-0.188″	0.006″	0.004″
0.189″-to-0.250″	0.008″	0.005″

FIGURE 12–35   Drill sheet with typical tolerances.

Prints included for multilayer designs are:

    a. Component side (labeled Layer 1).
    b. Circuit side (labeled Layer 4).
    c. Power layer (labeled Layer 2).
    d. Ground layer (labeled Layer 3).
    e. Component marking mask (if required).
    f. Solder mask (if required).

2. At least one print showing the dimensional information for the board outline. This should be clearly marked as to which side (component or circuit) is viewed and should be complete in every respect. An example of such a print is shown in Fig. 10–11b.

3. A *drill print* and a *drill sheet*. The drill print is a copy of the 1:1 scale circuit side of the board with each hole size identified. A typical identification system is with the use of letters, as follows. The most common drill hole size is left unlabeled. The next most common hole size is labeled with the letter A, the next common with the letter B, and so on, until all holes have been identified. The plated holes use the letters A, B, C, D, and so on, while all nonplated holes make use of the last letters of the alphabet, for example, W X, Y, and Z. A drill sheet is then constructed, which is a table listing the letters, finished hole sizes, and the number of holes associated with each of the letters used in the drill print. The typical order of listing is to show the plated holes first and then the nonplated holes. The appropriate tolerances for each of the holes should be included. A typical drill sheet including the tolerances is shown in Fig. 12–35.

4. For multilayer designs, a *layup print*, as shown in Fig. 12–36, provides the manufacturer with all the information required to lam-

FIGURE 12–36 Layup print of a typical 4-layer multilayer PCB.

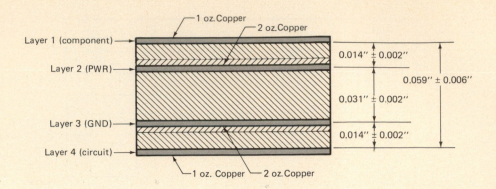

inate the board. Each layer is identified by a number (i.e., 1, 2, 3, 4, etc.) as well as their order by function (i.e., component, power, ground, and circuit). The finished copper thickness is specified in terms of oz/ft^2 for each of the layers. In addition, the required spacing between layers is specified together with the thickness of the finished board and the tolerances. The design engineer needs to work closely with the manufacturer with regard to the spacing between layers. This is because the B-stage material, used in laminating the board, differs widely in thickness from one supplier to another. For this reason, the thickness of the B-stage material be best left to the discretion of the manufacturer so as to result in the specified finished board thickness.

EXERCISES

12–1 What is the preferred work surface on which taped artworks are constructed?

12–2 What is the preferred medium upon which taped artworks are constructed?

12–3 Does the field of view of a comparator increase with increasing magnification?

12–4 Explain the purpose of Carlson pins in MLB taped artwork construction.

12–5 How are pads with center holes and solid pads aligned with those on the underlying component layout?

12–6 Explain the advantages and disadvantages of using pads with red eyes.

12–7 Explain the difference between the techniques for applying small and large preforms to the taped artwork master.

12–8 Edge connector patterns with finger extensions and a plating bus are required for what process?

12–9 List the order of application of preforms and tapes onto the artwork master.

12–10 What is the main disadvantage of contour taping?

12–11 List the two main advantages of angle taping.

12–12 Explain the term *neck down*.

12–13 What is the orientation of the nomenclature on sides 1 and 2 when the board is viewed from the component side?

12–14 What type of special pattern is used to make electrical interconnection between inner and outer layers of a MLB?

12–15 What is the purpose of nonfunctional foil on outer layers of a PCB?

13

Component Marking Artwork

Upon completion of this chapter on preparing component marking artworks, the student should be able to:

1. Accurately align the marking artwork with the component layout drawing.

2. Be familiar with component reference designators and their orientation on the artwork.

3. Properly letter the marking artwork.

4. Correctly position body outlines on the artwork.

5. Design component marking artworks.

13-0

Introduction

It is common practice to manufacture PCBs with a considerable amount of information screen-printed on the component side using nonconductive inks. This information typically takes the form of parts identification, component and device body outlines, component polarity identification, and device keying for correct assembly orientation. Component-side labeling is of significant aid to the technicians who assemble the parts to the board.

This chapter provides the necessary information for producing *component marking artworks* used to screen-print part identification onto a board. Although the use of automatic parts assembly equipment for large-volume mass production essentially eliminates the need for component marking as a guide for assembly, it still serves as a useful aid in servicing and repair. This is true even if many component bodies obscure their reference designators. As board designs become more dense, the efficient use of component marking becomes more difficult and at the same time, more necessary, especially for boards that are assembled manually.

13-1

Preparation of the Component Marking Artwork

The setup for making a component marking artwork begins by taping the completed component layout drawing to a work surface, preferably a light table. A sheet of 0.007-in. clear polyester film, somewhat larger in overall size than the intended artwork, is then positioned and taped securely to the work surface.

The following are features that are common to all component marking artworks: (1) alignment targets, (2) corner brackets, and (3) reduction scale. These are reproduced on the marking artwork using the layout drawing as a guide for their placement.

The alignment targets are placed outside the edges of the board for artwork registration. The same size and type of targets drawn on the component layout and used on the taped artworks are used on the marking artwork. They are carefully aligned and pressed onto the surface of the polyester film in accordance with their locations on the layout drawing. With accurate registration, the component marking artwork may be removed and then quickly and accurately reregistered onto the drawing as needed.

The corner brackets are next added to the polyester film. Again, the same size tape or artwork stick-ons used on the other taped artworks are added to the film. Remember that corner brackets are so positioned that their *inside* edges represent the *outside* edges of the finished board.

Because the component marking artwork is produced at the same scale as the original component layout, it also has to be photographically reduced. For this reason, a reduction scale identical to that used on the pad master artwork (discussed in Chapter 12) must be duplicated on the component marking artwork. As discussed previously, the accuracy of the reduced film artwork depends on the accuracy of placement of the reduction scale on any of the artworks. Therefore, it is essential to precisely measure the separation between the vertical sides of the half-round reduction marks as they are placed into position.

One aspect of component marking that must be considered at this point is how the alignment between a finished PCB and the printing screen, which is made from the component marking artwork, is achieved. Remember that after the board is manufactured, the peripheral registration targets are eliminated since they are placed outside the edges of the board. Using the corner brackets alone for the alignment of a screen is generally not suitable. What is typically required, in addition to the corner brackets, are internal features, such as mounting or tooling holes, which are not symmetrical with the board edges. Small targets or pads can be centered over two widely separated holes, preferably on a diagonal. They would be shown on

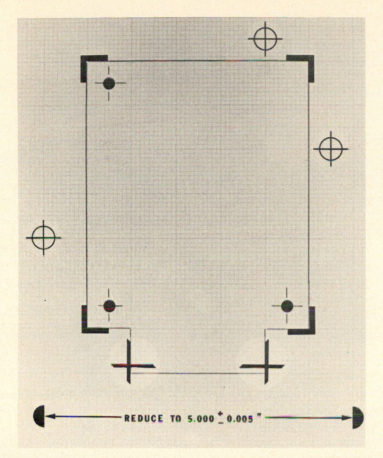

FIGURE 13–1   Initial setup of features for a typical component marking mask.

the underlying drawing to serve as a guide for the placement of the preforms onto the marking artwork. Lead access holes should *not* be used for this purpose since screen-printing ink could flow onto these holes and cause problems with the soldering of the leads. The initial setup for preparing a component marking artwork is shown in Fig. 13–1.

13-2

Component Identification

Components are identified on the marking artwork with the same reference designators as those used on the circuit schematic. Ideally, the orientation of all the reference designators should be on the same axis, reading from left to right as shown on the schematic. This is rarely possible to achieve, however, due to the position of the components and the small spaces between them which are overlaid with component bodies. For this reason, an additional reading axis perpendicular to the first is commonly used. The lettering of the second axis should be oriented in such a way as to be read more

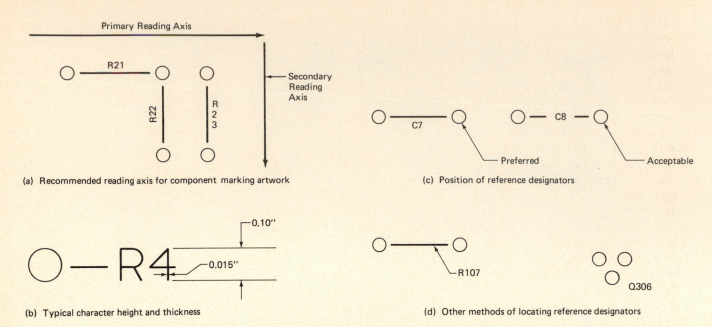

(a) Recommended reading axis for component marking artwork

(c) Position of reference designators

(b) Typical character height and thickness

(d) Other methods of locating reference designators

FIGURE 13–2 Component identification using reference designators.

easily by either placing the letters in the same direction as the primary reading plane or at a 90-degree angle with this plane. Both of these orientation schemes are shown in Fig. 13–2a.

For suitable size and clarity, the thickness and width of the characters used with component identification must be considered. The information produced on an enlarged scale marking artwork will be screen-printed onto the component side of the board at a 1:1 scale. Some of this printing may, in many cases, appear over a conductor path. For this reason, the size and width of the characters used may be critical.

Lettering heights range from 0.059 to 0.2 in., with proportional thicknesses from 0.012 to 0.031 in., all at a 1:1 scale. A typical size used in the industry is a height of 0.10 in. with a thickness of 0.015 in. for the screen printing of the smooth component side of single-sided boards. This character size is shown in Fig. 13–2b. Screen printing over the "hills" and "valleys" (conductor paths and air gaps) on double-sided boards could result in blurred lettering if the thickness of the characters is not appropriate. In general, character thickness increases proportionately with character height. Because screen printing is an art form, the manufacturer should be consulted as to the minimum-size lettering that can be printed given the nature of the board surface and the character spacing.

Whenever possible, the reference designators should be placed slightly above or below the components' body positions after they are mounted so that they will be visible after assembly. This is not always possible, however, due to high-density packaging. It then becomes acceptable to place the designator so that it becomes concealed by the component body after assembly. See Fig. 13–2c. In these cases, it is necessary that any identifying nomenclature on the component body itself remain visible after assembly.

For two-lead axial or radial components, adding lines to identify the pads associated with the component along with the reference designator aids in the proper placement of the parts. For other types of components, it becomes necessary to modify these methods of labeling because of restricted space. In many instances, these modifications will result in a reduction of clarity in the part's identification. For example, in Fig. 13–2d, it is shown that R107 is to be placed between the two pads as denoted by an arrow drawn from its reference designator. This identification method is used when there is not sufficient space to fit the designator between the pads. Again, in Fig. 13–2d, the designator Q306 is positioned to the side of the transistor pad grouping. Here it is implied that the three closest pads to the designator are those to which that designator refers. Although it is not always possible to locate a designator that is totally free of possible misrepresentations, the general rule is that it should be included rather than leaving the part void of identification.

Reference designators or pad-locating lines should be positioned so that they will not extend onto a terminal pad or a plated hole. A minimum clearance for these characters is 0.020 in. from the hole rim. The ink used for screen printing contains an epoxy-based material and will result in soldering problems if any is allowed to flow into a plated hole. In addition, epoxy inks act as electrical insulators when cured.

Two common methods of applying reference designators to the marking polyester sheet are employed. The first is with the use of adhesive-backed precut transparent stick-ons that are applied in a fashion similar to that used for the device patterns discussed in Chapter 12. These designators are printed with opaque Gothic-style numerical and alphabetical characters in sequential number and letter groups. They are also available in common designators, such as R1, C1, Q1, U1, VR5, and so on. The letter heights of these designators are 0.080, 0.100, 0.125, 0.200, and 0.375 in. Numbers and individual letters are also available in 0.400-in. heights. See Fig. 13–3. The designators are applied by simply peeling the stick-ons from the backing sheet with a scalpel, placing over the desired position, and pressing firmly onto the polyester sheet.

The other common method of applying reference designators is with the use of dry transfer lettering. Refer to Section 2–5 and Fig. 2–12

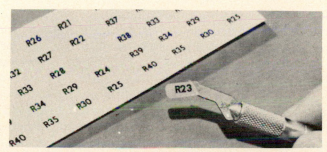

FIGURE 13–3 Removing stick-on designators from backing sheet with knife. [Provided courtesy of Bishop Graphics, Inc. Westlake Village, California 91359]

for a discussion on the application of this lettering and numbering technique. Dry transfer characters are not as durable as stick-on designators. They are especially susceptible if handled improperly. Like all PCB artworks, the marking artwork should never be rolled since dry transfers tend to crack resulting in poor line definition.

13-3

Component Body Outlines

It is often the case with high-density PCB layouts that it becomes impossible to show all the body outlines on the component marking artwork. However, where space permits, it is good practice to show these outlines, especially for parts that could easily be improperly positioned during assembly, such as polarized capacitors, diodes, transistors, and integrated circuits. These outlines should be accompanied by any critical keying shown on the component layout drawing.

Body outlines are available as inked symbols on a special 0.0015-in. transparent pressure-sensitive acetate material. Some are provided with keying designations as part of the body outline. They are applied in a fashion similar to the drafting aids discussed in Chapter 12. Like precut patterns or stick-on lettering, a body outline is first removed from its protective backing sheet with a scalpel. It is then positioned on the marking artwork using the underlying component layout drawing as a guide. When aligned with its respective component, it is firmly pressed into place.

The addition of body outlines and reference designators to a PCB is extremely helpful in correctly placing the part into its designated holes on the board. Thus confusion is eliminated as to which holes are associated with specific components, not only by their designators but also by the body shape.

Similar to lettering and numbering, the minimum line width used for the body outlines on the marking artwork must be considered. One of the major factors used in the determination of line width is the nature of the surface to which the marking is to be screen-printed. The manufacturer should be consulted as to the optimum line thickness required for obtaining thin, crisp lines on a given board surface.

A number of body outlines representing a variety of components and devices is shown in Fig. 13–4. The pads are *not* a part of the outlines but are shown as they appear on the layout drawing for purposes of orientation. Cylindrical and axial-lead-type components, such as resistors and tubular nonpolarized capacitors, can be represented by either of the body outlines shown in Fig. 13–4a. A rectangle alone or a smaller rectangle with extenders to identify its pads both serve to represent these component body outlines. As mentioned previously, the reference designators may be included inside (R1) or outside (R2) the outline. Positioning the designator outside the outline is preferred where space permits because it is more visible after the components have been assembled to the board.

The body outlines of Fig. 13–4a can also be used to represent axial-style electrolytic capacitors. For this application, however, the

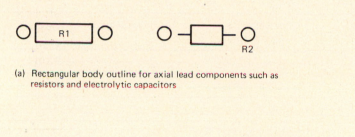

(a) Rectangular body outline for axial lead components such as resistors and electrolytic capacitors

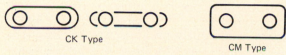

CK Type   CM Type

(b) Body outlines for radial lead ceramic and mica capacitors

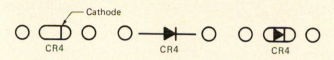

Cathode

CR4   CR4   CR4

(c) Body outline for DO–16 or DO–17 diodes

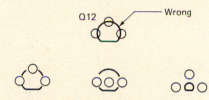

Q5

Emitter Key

(d) Body outline for the TO–5 package

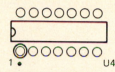

Q12   Wrong

(e) Various body outlines for the TO–92 package

1   U4

(f) Body outline for the TO–116 style package (14 pin DIP)

FIGURE 13–4 Various component body outlines.

appropriate reference designator would have to be included and the positive terminal identified with a + sign. This keying designation may be placed inside or outside the body outline, but close to the appropriate end which is nearest the hole that is to accept the positive lead of the capacitor.

A variety of body outline sizes used to represent cylindrical axial lead components on marking artworks is shown in Table 13–1. These outlines are used for representing the body shapes of resistors, capacitors, and diodes. The prefixes CK and CS refer to ceramic and tantalum capacitors, while RL and RN, RC, and RW represent metal glaze or metal film, composition, and wire-wound resistors, respectively. The body outlines shown are also available for use on 2:1 and 4:1 scaled artworks.

The body outlines with 1:1 scale dimensions for different-style potentiometers are shown in Table 13–2. Notice again that, unlike preform patterns used for taped artworks, no pads are shown. Pads are not used with body outlines because they could allow ink to run into holes and are really unnecessary since the underlying layout drawing serves as a guide for outline placement.

Body outlines for radial lead nonpolarized capacitors of the ceramic (CK) or mica (CM) types are shown in Fig. 13–4b. These outlines are also available in a wide range of sizes, as shown in Table 13–3, which are needed to satisfy a variety of capacitors. Because the CK-type capacitor body outlines have narrow widths, care must be exercised in their application. It is strongly recommended that the outline be relieved in the area of the pads as shown in Fig. 13–4b. If the solid outline is used, any misregistration or undue stretching of the screen while printing could result in marking ink entering a plated hole.

TABLE 13–1 Body Outlines for Axial Lead Components

Body Outlines	For MIL-STD Components	1:1 Scale Width x Length* (in.)
	RC05 AB Type (⅛ W)	0.07 x 0.13
	DO-16, DO-18, DO-34	0.08 x 0.15
	RC05, RC06, RN50	0.08 x 0.16
	RC07, RL07, RW81 (¼ W) CSR09 (A1), CKR12, DO7	0.10 x 0.28
	DO-7	0.11 x 0.29
	RC08, CK11	0.12 x 0.23
	RC20, RL20, RN60 (½ W) CK14, CSR09 (B1)	0.16 x 0.42
	RC32 (1 W)	0.24 x 0.59
	RL32, RW89	0.21 x 0.59
	RC42, RL42 (2 W) CK16	0.34 x 0.73

*Body outlines shown are available 2X and 4X scale.

TABLE 13–2 Body Outlines for Variable Resistors

Body Outlines	For MIL-STD Components	1:1 Scale Width x Length* (in.)
	¾-in. rect.	0.19 x 0.75
	RT10, type P	0.31 x 1.00
	RT11, type P	0.28 x 1.25
	RT24, type P	0.38 x 0.38
	RT24, type X	0.19 x 0.38

*Most body outlines shown are available 2X and 4X scale.

TABLE 13–3   Body Outlines for Radial Lead Components

Body Outline	For MIL-STD Components	1:1 Scale Width x Length* (in.)
	CK08	0.06 x 0.15
	CKR05, CK05, CKR06, CK06	0.10 x 0.30
	CK60, CKR07, CMR04, CK61	0.15 x 0.40
	CK62	0.15 x 0.60
	CK63, CKR64	0.15 x 0.80
	CM05, CMR05	0.22 x 0.47
	CM06, CMR06	0.30 x 0.70

*Body outlines shown are available 2X and 4X scale.

Three methods of representing the body outline of a diode together with the appropriate keying to identify the cathode and anode leads are shown in Fig. 13–4c. The outline for the DO-16 and DO-7 diodes shown in Table 13–1 is used with a solid bar added close to one end to identify the cathode terminal. This is usually satisfactory for keying a diode on the marking artwork. Using the diode symbol inside the body outline, as shown in Fig. 13–4c, is another method, which not only provides an indication of the size and shape of CR4 but also is a more positive keying identification of both diode leads. The flat side of the triangular portion of the symbol represents the anode end of the diode. If no body outlines are desired, the diode symbol alone may be used with extender lines for pad location and for keying. This is also shown in Fig. 13–4c.

The body outlines used for transistors and integrated circuits are shown in Table 13–4. These devices are prone to being installed incorrectly into the board if careful attention is not given to their keying. Due to the density of device holes, there is even more concern that the outline is formed with sufficient hole clearance. The body outline of a transistor packaged in a TO-5-style case, shown in Fig. 13–4d, will serve as one example of correct keying. The keying tab must be oriented close to the emitter pad on the underlying layout drawing before the outline is pressed into position on the marking artwork. This type of screened marking information on the board helps to eliminate improper device mounting.

The TO-92-style plastic case, shown in Fig. 13–4e, is another common transistor package which is prone to misassembly. The straight section of the semicircular body outline must be aligned with its same orientation on the layout drawing to ensure correct lead location and device installation.

TABLE 13–4   Body Outlines for Transistors and Integrated Circuits

Body Outlines	For MIL-STD Components	1:1 Scale Package Style*
	8-pin DIP	0.15 x 0.43
	14-pin DIP	0.15 x 0.78
	16-pin DIP	0.15 x 0.88
	18-pin DIP	0.15 x 0.98
	24-pin DIP	0.45 x 1.30
	40-pin DIP	0.45 x 2.05
	14-pin flat pack	0.26 x 0.26
	16-pin flat pack	0.26 x 0.30
	TO-92	0.2 dia./ 0.15 flat
	TO-18	0.2 dia.
	TO-5	0.35 dia.
	TO-3	1.0 x 1.55 (max.) inner circle dia. 7/8
	TO-66	0.75 x 1.25 (max.) inner circle dia. 1/2

*Body outlines shown are available 2X scale and many 4X.

Depending on the diameter of the transistor body outline and that of its lead circle, inking problems can result on the marking artwork. Note in Fig. 13–4e that if a solid body outline would not allow sufficient clearance to the pads, it must be relieved in the terminal areas. An alternative method is to use smaller body outlines to provide the necessary clearance yet still show the proper keying for the transistor's correct installation.

The final body outline that we will consider is that of the dual-in-line package (DIP), shown in Fig. 13–4f and Table 13–4. This configuration is used extensively in the packaging of both analog and digital integrated circuits. Note that the body outlines are made sufficiently narrow to allow proper pad clearance. As discussed in Chapter 8, the numbering order of pins for this package begins with pin 1 in the upper left corner when viewed from the component side. If it is preferable not to use a body outline, pin 1 must be keyed on the marking artwork with a *#1* close to the appropriate pin, a dot or a circle having sufficient clearance (0.020 in. minimum) around pin 1. Any of these methods of keying are acceptable. If the body outline is used, note in Fig. 13–4f that a *case key*, which is a small half-circle indentation, is at one end of the outline. This indentation is the same as that which appears on the actual device case and indicates that the pin at the left of the key is pin 1, thereby eliminating the need for any other keying method.

13-4

Completing the Component Marking Artwork

Before the marking artwork is removed from the light table, the following should be checked to ensure completeness:

1. No marking should extend to within 0.1 in. of any board edge.
2. Omit any information that, due to board density, could result in confusion.
3. Use the reference designators as shown on the circuit schematic wherever possible or at least use the component value.
4. Label or show the polarity of all electrolytic capacitors.
5. Label or show the cathode terminal of all diodes.
6. Label or show the emitter of all transistors.
7. Label or show pin 1 of all ICs.
8. Label connectors with their reference designators and in their numerical positions.
9. No marking should be positioned closer than 0.020 in. from any plated hole edge.
10. Maintain the minimum widths of feature detail and separation between features recommended by the manufacturer.
11. Omit any body outlines that interfere with clarity.

After final inspection of the completed component marking artwork using the list above as a guide, it may be removed from the work table and stored flat. A typical completed component marking artwork is shown in Fig. 13–5.

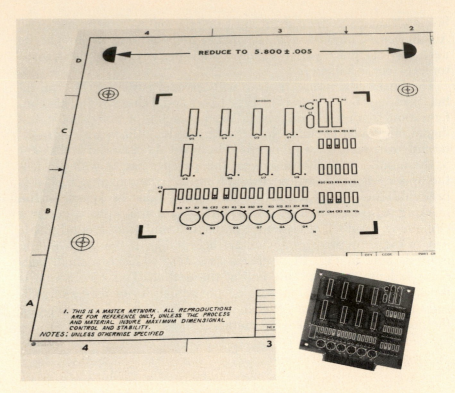

FIGURE 13–5 Completed marking mask artwork. [Provided courtesy of Bishop Graphics, Inc., Westlake Village, California 91359]

EXERCISES

13–1 List three features that are common to all marking artworks.

13–2 What is the convention when reference designators cannot be placed along one axis?

13–3 Explain why the letter width and height are critical.

13–4 What is the acceptable positions of reference designators in relationship to component body positions?

13–5 Name the two common methods of applying reference designators to the component marking mask.

13–6 Show three methods of representing a diode together with its appropriate keying to identify its cathode.

13–7 Briefly explain how component marking inks are applied to printed circuit boards.

13–8 Explain the difficulties encountered if marking ink enters holes in the printed circuit board.

13–9 What is the typical marking mask edge limit with respect to the outside edge of the board?

13–10 How close can the marking ink come to the edge of a plated-through hole?

14

The Solder Mask Artwork

LEARNING OBJECTIVES

Upon completion of this chapter on preparing solder mask artworks, the student should be able to:

1. Have an appreciation of the need for a solder mask.

2. Be familiar with the application of a solder mask.

3. Be familiar with the two types of solder masks.

4. Be familiar with the factors to be considered in solder mask artwork construction.

5. Design solder mask artworks.

14-0

Introduction

Printed circuit boards require soldering after the components have been assembled. In volume production, automatic *wave soldering* is the method most commonly used for this purpose. In wave soldering, molten solder is applied simultaneously to all component leads, exposed terminal pads, and conductor paths on the circuit side of the board. This method of soldering has the distinct advantage that all the required electrical interconnections between the components and the conductor pattern can be formed in a matter of seconds, dramatically reducing the time that it would take to hand-solder each connection individually. This has contributed largely to the sharp reduction in the cost of PCB assembly.

There is one disadvantage in the use of wave soldering methods, however, which is the tendency to form *solder bridges*. These are unwanted solidified solder which is formed between closely spaced conductor paths and pads on the circuit side of the board. For this reason, solder bridges are commonly referred to as electrical *shorts*, since they result in short circuits wherever they appear. The formation of solder bridges is difficult to eliminate, even with the most careful monitoring of the wave soldering operation. This is largely due to the fact that modern high-density boards have spacings of 10 mils or less between conducting features. These tight air gaps make it difficult for molten solder, with its high surface tension, to "pull back" and form good definition with the circuit pattern as the board exits from the surface of the molten solder wave.

Solder bridges are often difficult to detect and repair. Considerable time can be spent in isolating electrical faults resulting from the shorts formed. For this reason, every effort is made to minimize the possibility of the formation of solder bridges. Recall that conductor paths on the circuit side are oriented so that they are parallel to the direction of board motion over the solder wave. This orientation aids in minimizing the formation of bridges. Another design feature is often added to further reduce the bridging problem. This is the use of a *solder mask,* often specified for high-density designs.

The solder mask is a high-temperature-resistant material applied to the processed board by either a screening or a laminating process. A solder mask may be applied to just the circuit side or to both sides of a double-sided PCB. Solder masks are designed to cover all areas of the board except fingers, mounting holes, terminal pad areas, and holes in which component leads are to be electrically interconnected by soldering. Thus closely spaced paths, protected by the solder mask material, are not exposed to the molten solder during the wave soldering operation, greatly reducing the possibility of the formation of bridges. Because of extremely small air gaps which may exist in some areas of high-density boards, it is rarely possible to achieve complete solder mask coverage between all conductor features. For this reason, critical areas will require careful inspection even when a solder mask is employed.

This chapter provides a detailed discussion of the requirements and techniques for producing solder mask artworks to be used on high-density boards intended for wave soldering operations.

14-1

The Solder Mask

In order to be effective in minimizing the formation of solder bridges during the wave soldering process, the solder mask material must be applied to the processed board in such a way as to cover all conductor paths and other areas not intended to be exposed to the molten solder wave and to expose all other features to the solder. A portion of a processed board to which a solder mask has been applied is shown in Fig. 14–1. Note that all conductor paths and the insulating surface area of the board are protected by the solder mask material. Terminal pads and their plated-through lead access holes remain exposed for soldering. Further note that the diameter of the hole in the solder mask provides a space (distance A) around the terminal pad. Also, the solder mask does not extend to the finished edge of

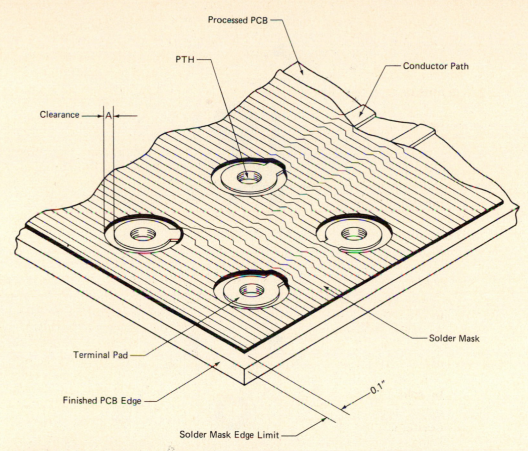

Clearance

A

Processed PCB

PTH

Conductor Path

Solder Mask

0.1"

Terminal Pad

Finished PCB Edge

Solder Mask Edge Limit

FIGURE 14–1 Processed PCB with solder mask.

the board but is terminated at a specified distance from the edge. In Fig. 14–1, this distance is shown as 0.1 in. It can thus be seen that when this board is processed through the wave soldering machine, only the terminal pads and their associated component leads would directly contact the solder wave. No solder bridges would form between a terminal pad and an unrelated conductor path since the masking material covers all of these potential problem areas.

Refer again to the space between the terminal pad and the edge of the hole in the solder mask material (distance A) shown in Fig. 14–1. This is also referred to as *air gap* or *clearance*. It is apparent that the larger the air gap around the pad, the greater the exposure to molten solder. There are minimum and maximum limits to the size of this clearance, however. If it were made too large, the edge of an adjacent conductor path could be exposed, inviting the formation of a solder bridge. See Fig. 14–2a. If, on the other hand, the diameter of the hole in the solder mask artwork is equal to or less than the pad diameter (see Fig. 14–2b), some of the screened solder mask material would be introduced onto the pad, or worse, into the hole. This could result in unreliable electrical connections. It can thus be seen that the clearance distance around the pads is a critical factor in the design of solder mask artworks. Providing proper clearance around mounting holes and board edges is also of concern to

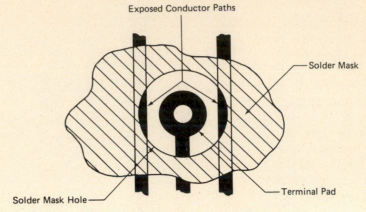

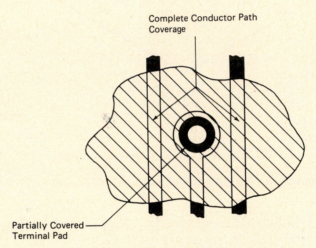

FIGURE 14–2 **Incorrect solder mask hole sizes.**

avoid chipping the mask in these areas during parts assembly and when cutting the board to its finished size.

To aid in determining the optimum size of air gaps in the design of a solder mask artwork, an understanding of the methods of applying the masking material to the board is essential. These are discussed in the following section.

14-2

Types of Solder Masks

There are two types of solder mask materials common to the PCB industry. These are heat-resistant *dry film* and *screen ink*. The dry film material is laminated onto the board and the mask is then photographically processed. The screen ink is a high-temperature resistant epoxy, normally dark green in color, which is applied to the board by screen printing to form the mask. For both methods, solder mask artwork is required.

The dry film solder mask process is relatively new and requires equipment that is extremely expensive. For this reason, a vast majority of the industry continues to use the screen printing process. Costs aside, however, the dry film solder masks are superior to those made with screening inks in terms of accuracy of masking. The dry film solder mask process begins by laminating the boards with a high-temperature-resistant photosensitive material using a special laminator that applies the film with heat and pressure. The laminated boards are then imaged with a 1:1 scale positive of the solder mask artwork. The boards are then developed and cured to result in the solder mask. The close tolerance of registering the solder mask artwork into the processed board cannot be duplicated by the screening method. In fact, the accuracy of registration could allow for an air gap of 0 in. (i.e., the solder mask hole diameter will exactly equal the pad diameter). One technical disadvantage of the dry film method is that some film residue can seep into the plated holes while it is being developed. Unless it is thoroughly removed, this residue can result in unreliable electrical connections.

The screen printing solder mask method requires a fine-mesh (200 to 330 filaments per inch) stainless steel or polyester screen which is tightly stretched onto a metal frame. The solder mask artwork is photoreduced and contact-printed onto a photosensitive screen film. After processing, it will leave all of the screen's mesh openings completely clear within the solder mask edge limit border *except* those areas directly over terminal pads, fingers, and mounting holes. The mesh above these areas is filled and will not permit the screening ink to be transferred to the board, thus allowing them to be exposed to the wave of solder. The round filled holes on the screen must have larger diameters than the features they are covering, for reasons discussed previously. After the screen has been processed, it is positioned over the circuit side of the board and a hard rubber or plastic squeegee is drawn along the screen to force the epoxy ink through the open mesh and onto the board surface. The result is that the ink will cover all areas of the board with the exception of terminal pads, fingers, holes, and the border. After the screen printing has been completed, the solder mask is cured at an elevated temperature in a convection oven.

Although the screen printing of solder masks is quick and economical, it has the major disadvantage of resulting in masking inaccuracy. A generally accepted figure for accuracy of positioning is ± 0.0075 in. In addition, the screen is unavoidably stretched when the squeegee is drawn across to apply the ink. This stretching causes the screen to become distorted, resulting in misregistration of the solder masking pattern on the terminal pads. To compensate for this distortion and the inherent 0.0075-in. placement tolerance, the proper air gap dimensions must be provided to prevent the screening ink from leaving exposed conductor paths or covering portions of terminal pads or holes as shown in Fig. 14–2. A commonly acceptable minimum air gap for solder masks is 0.012 in., as shown in Fig. 14–3.

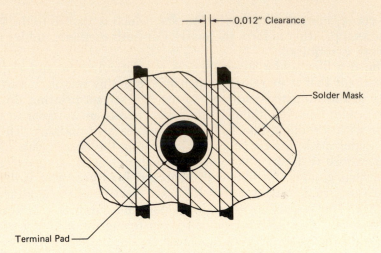

FIGURE 14–3   Typical solder mask hole to pad clearance.

14-3

Requirements of Solder Masks

Even though the primary purpose of a solder mask is to cover all conductor paths to prevent the formation of solder bridges during the wave soldering process, there are other factors to be considered before artwork construction may begin. Some of these are cosmetic in nature.

If the screened ink process of solder masking is used, the artwork must be prepared in such a way as to prevent masking ink from running onto the board edges. This is achieved by defining a distance between the edge limits of the mask and the board edges. This distance is not critical and varies from one manufacturer to another. An acceptable range of this clearance is 0.100 to 0.200 in. Of course, the defined distance would have to be adjusted to cover any conductor paths that may be closer to the board edge than this distance.

Another factor that requires attention in artwork construction is the unplated holes into which hardware or assemblies are to be mounted. An area around these holes must be sufficiently relieved of masking material. If ink is allowed to enter these holes, it may reduce the inside diameter and impede parts assembly but more important, it would have an adverse cosmetic effect. There are no strict specifications for this clearance distance, but an accepted guideline is to allow a minimum space around each mounting hole of twice the finished hole diameter. If the hole is to accept a machine screw, the clearance area should be at least equal to the maximum diameter of the screw head or the maximum dimensions of the nut. When mounting holes are relieved in this way, solder mask chipping is minimized during assembly, thus enhancing the appearance of the final product.

A critical consideration in the preparation of the solder mask artwork is to prevent the screened ink from making contact with any of the gold-plated fingers. Since masking ink is an electrical insulator, any ink allowed to spread onto the fingers will result in severe electrical problems. The artwork must provide a minimum distance above the crown of all fingers so as to prevent masking ink from being screened beyond this distance. A clearance distance of 0.100 in. above the finger crown points is adequate in most applications.

The most important concern in the design of solder mask artwork is the optimum clearance distance to be allowed around all terminal pads. As mentioned previously, a dry film solder mask material can be made with the mask openings exactly equal to the diameter of the pads (i.e., zero air gap). Excellent registration and repeatability is another advantage of this method. With the use of screen printing of the mask, clearance must be provided around all terminal pads to compensate for the registration accuracy tolerance and the inherent stretching of the screen during its use. Other factors that affect the quality of the screened mask are (1) type of screen mesh and its filaments per inch, (2) the screen tension on the frame, (3) the hardness of the squeegee and the condition of its working edge, (4) the viscosity of the ink, (5) temperature, (6) humidity, (7) off-contact distance (i.e., height of the screen above the work surface), and (8) the operator's pressure and smoothness in screening the ink onto the board. It is apparent when considering all the variables that screen printing is a relatively sensitive operation which requires careful control to achieve satisfactory results. To reduce the difficulty of the task and to produce a good-quality product, the PCB manufacturer would like as large a mask clearance as possible. On the other hand, the design engineer strives for the absolute minimum clearance (0.010 in. or less) to ensure complete coverage of all conductor paths so as to result in the minimum possible areas of solder bridging. It is apparent that a trade-off of specifications is established between the two parties. To demonstrate the effects of those trade-offs, we have shown in Fig. 14–4 the reduction of the air gaps due to mask hole size selection and manufacturing alignment tolerances. This figure uses 0.012 in. as an example of worst-case misalignment between the solder mask and the terminal pad. Noncritical areas (e.g., where there are widely spaced conductors) should be provided with as much clearance as possible, equal to or more than 0.020 in. See Fig. 14–4a. Note that in both columns (Perfect Registration and Worst-Case Misalignment) the conductor path is completely covered by the solder mask, using a 0.020-in air gap.

In more critical areas, such as that shown in Fig. 14–4b, the air gap must be reduced to a minimum of 0.012 in. to ensure mask coverage of the conductor path. With this reduced air gap, the worst-case misalignment results in marginal clearances between the solder mask and the pad and conductor path. Note the tangency between the edge of the mask and the terminal pad. Also observe that the left side of the conductor path is completely covered to within 0.001 in. of the mask. Complete coverage of the center conductor path would not be possible using a 0.020-in. clearance under worst-case misalignment conditions if this option were considered for correcting the tangency. See Fig. 14–4c. Again, a 0.010-in. air gap would improve the path coverage to within 0.003 in. of the mask but would

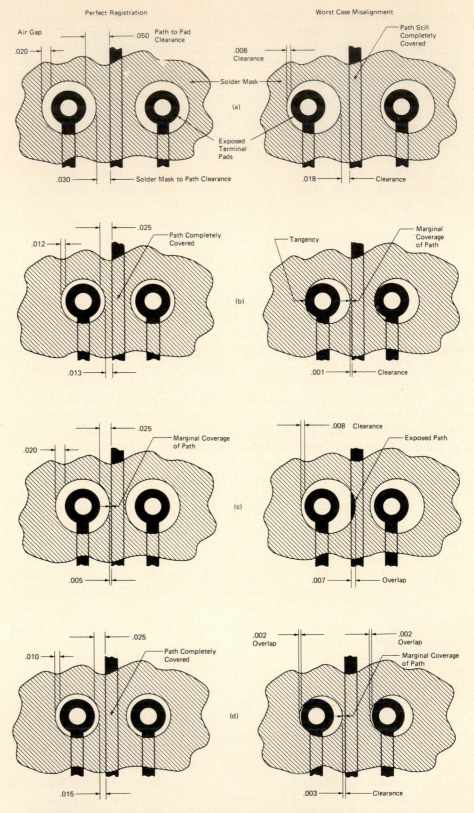

FIGURE 14–4 Reduction of air gaps and solder mask coverage due to mask hole size selection and manufacturing tolerances.

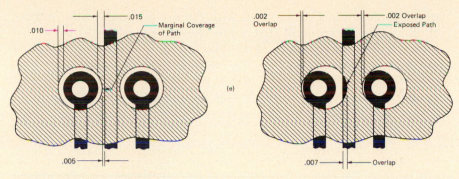

FIGURE 14—4 (continued)

result in the terminal pad edges being overlapped by the mask by 0.002 in. This is shown in Fig. 14—4d. It is thus seen that the 0.012-in. air gap shown in Fig. 14—4b is an optimum compromise for this specific case.

An extremely critical area, having 0.015-in. air gaps between the center path and pads, is shown in Fig. 14—4e. It is impossible to provide complete coverage of the center conductor even with a minimum masking clearance of 0.010 in. If the masking air gap were reduced to overcome this problem, applying ink with a screen would very likely result in it being printed onto the pad and even into the hole, causing unreliable connections. There are two techniques sometimes employed in an attempt to achieve complete coverage and avoid screening ink from being printed onto the pads or holes. One method, called *eccentric positioning*, is to offset the centers of the terminal pads. This is shown in Fig. 14—5a. The second method, shown in Fig. 14—5b, is to trim and flatten the edges of the solder mask holes adjacent to the conductor path. Eccentric positioning sacrifices air gap width for better conductor path coverage, while trimming the solder mask hole marginally covers the path but results in a larger air gap. Experience and company policy dictate if either of these alternative methods should be employed. If they are deemed unfeasible, the only satisfactory solution to the masking

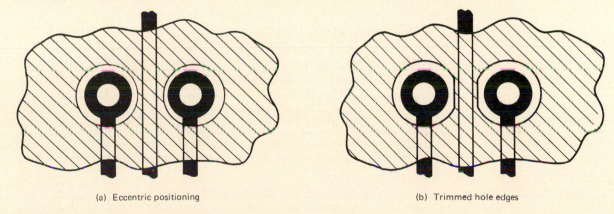

(a) Eccentric positioning   (b) Trimmed hole edges

FIGURE 14—5 Solder masking alternatives in critical areas.

problem shown in Fig. 14–4e is with the use of the dry film process. It should be understood that some conductors may still be unprotected by the solder mask, and solder bridging is possible in critical areas of high-density boards. Nevertheless, solder masks are used extensively since they result in vast savings on rework even though complete coverage of all conductor path edges is not always possible.

14-4

Preparation of the Solder Mask Artwork

Laying out the solder mask artwork begins by taping the appropriate grid system onto a light table or work surface. The pad master and taped conductor path artwork for the circuit side of the board are aligned with each other and with the grid. They are then taped into place at all four corners. A sheet of 0.007-in. clear polyester film larger than the intended layout is finally positioned and taped over the setup.

Since the solder mask requires no preliminary design, its layout may begin immediately. Alignment targets are first applied to the solder mask artwork. The same type and size targets as those used on the underlying artworks are carefully applied using the pad master targets and the grid as positioning guides. It is essential that the best possible registration of targets be achieved.

The same reduction scale that was used on the pad master is reproduced onto the solder mask artwork. Because, like all artworks, the solder mask will be photographically reduced, the vertical section of the half-circles used in the reduction scale must be spaced as accurately as possible.

From this point in the layout and on, we must reverse the concept used in the preparation of all other artworks. That is, all areas onto which solder-resistant ink is to appear on the board will remain *clear* in this artwork. These clear areas represent the open portions of the screen through which the ink will pass and be deposited onto the board to mask conductor paths. Those areas where no solder-resistant ink is to appear (terminal pads, fingers, etc.) will be made *opaque* (masked off) on this artwork. These opaque areas represent portions of the screen in which the mesh openings are filled and will prevent the ink from being deposited onto the board.

Either black or red tapes may be used for the layout of a solder mask artwork. The red tape has a thickness of 0.002 in. and is photographically opaque (i.e., it is reproduced as black). The advantage to using the red tape is that it is transparent, which allows the area being masked to be clearly viewed while the tape is being applied. Both the black crepe tape and the red tape are available in a variety of widths, ranging from 0.015 to 6 in. The red tape is slightly more expensive and more difficult to handle than the black tape. However, its transparency is of significant value, especially when taping board edges or fingers.

The next portion of the solder mask artwork to be considered is the edges of the board as defined by the corner brackets on the underlying artworks. Recall that no masking ink should be deposited closer than approximately 0.1 in. of any board edge. To do this on the solder mask artwork, the specified distance (times the reduction scale) is measured in from each inside edge of the corner brackets. For example, if the solder mask is to terminate 0.1 in. from all board edges and the reduction scale is 2X, the underlying grid is used to measure 0.2 in. from the edges. Tape is then applied from this masking edge limit to beyond the outside edges of the corner brackets. One of the final steps in preparing a screen for masking is to apply a filler (block-out) over all unused areas outside the corner brackets of the board. This is to contain the ink over the image while the squeegee makes its pass. For this reason, the greater the distance already taped beyond the corner brackets, the easier it is to apply the block-out onto the screen. The tape is aligned 0.2 in. in from each edge and allowed to extend beyond the corner brackets. This is shown in Fig. 14–6.

The gold-plated finger area of the board must also be masked to prevent solder-resistant ink from being deposited. As mentioned previously, the solder mask should not extend closer than 0.1 in. above the finger crowns. This distance of 0.1 in. times the reduction scale is first determined. A strip of tape, wide enough to extend from this line to beyond the outer edges of the finger delineation markers, is then positioned onto the artwork. A portion of a solder mask artwork overlaying the finger area of the underlying component drawing is shown in Fig. 14–6. Corner brackets, registration markers, and other board delineation marks have been properly positioned using the underlying drawing as a guide. Note that a wide section of tape is positioned so that it extends from the clearance distance above the finger crowns to the outside edges of the tongue delineation marker, which defines the finished length and width of the tongue. Also, for the sake of appearance, the upper corner of the tape has been cut at

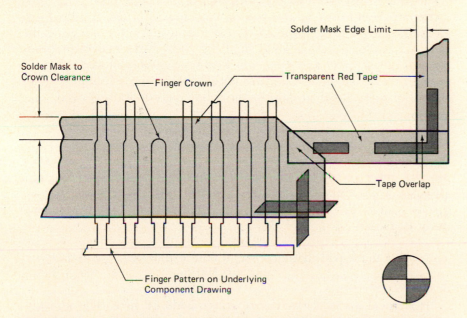

FIGURE 14–6 Transparent red tape used to mask finger area and solder mask edge limit on solder mask artwork.

a 45-degree angle. To define the solder mask edge limit, strips of narrower tape are positioned so that they extend beyond the outside edges of the corner brackets and board delineation markers. There is no proper order in which these tapes are applied. However, it is important that the ends of all of the individual strips of tape overlap to ensure that a solid, unbroken opaque pattern appears on the photographically reduced solder mask.

To allow solder mask clearance for the mounting holes, their finished hole size diameter is first multiplied by the reduction scale. A pad whose diameter is equal to or greater than this scaled value is then selected. For example, if the finished mounting hole diameter is 0.125 in. and the reduction scale is 2X, the pad selected should be 0.125×2 or 0.250 in. in diameter if a dry film solder mask is to be used. This value will need to be increased (0.270 to 0.290) if the mask is to be screened.

After the proper pad size has been selected, it is carefully positioned over each mounting hole location using the underlying grid intercepts and hole as guides. Although solid black and transparent pads are available, it is extremely difficult to position them accurately by hand, especially when 0.010- and 0.020-in. air gaps are specified. Instead, black pads with the smallest possible center holes *or* black pads with red transparent film centers (in a limited size selection) are preferred. These will aid significantly in accurate pad positioning, due to the underlying grid intercepts being visible. Recall that the red-centered pads will photograph as opaque. Where the appropriate size is not available and the black center-holed pad is used, the center hole must be made opaque with a small piece of tape. The entire pad area must be opaque so as to prevent screening ink from flowing onto any portions of the pad. A faster and more effective photographic method of masking the center holes of solder mask pads is discussed later in this section.

Attention will now be focused on masking all of the terminal pad areas. To aid in this, Table 14–1 provides a guide for selecting the proper sizes for masking some standard pads. Note that the table is divided into *critical* and *noncritical* categories as well as 2X and 4X scale artworks. To obtain pad sizes for other than those listed for either critical or noncritical situations, the following equation is used:

$$\text{solder mask pad diameter}$$
$$= \text{scale factor } [(1\text{:}1 \text{ terminal pad diameter}) + (2) \text{ (air gap)}] \quad \textbf{(14–1)}$$

For example, on 2X scale artwork, the pad size on the solder mask artwork for a 0.050-in. terminal pad on the underlying layouts where the air gap is to be 0.020 in. (noncritical area) is found as follows using Equation 14–1:

$$\text{solder mask pad diameter} = 2[(0.050) + (2)(0.020)] = 2[0.090]$$
$$= 0.180 \text{ in.}$$

Noncritical areas of the layout are considered first. The underlying circuit-side conductor pattern and pad master are inspected to

TABLE 14–1   Guide for Selecting Donut Pad Sizes for Solder Mask Artwork*

(a) Donut Pad Selection for Noncritical Conductor Pattern Areas

1:1 Scale Terminal Pad Dia.	1:1 Scale Solder Mask Pad Dia.	2:1 Scale Solder Mask Pad Dia.	Donut Pad Dia.	Resulting 1:1 Scale Air Gap†	4:1 Scale Solder Mask Pad Dia.	Donut Pad Dia.	Resulting 1:1 Scale Air Gap†
0.050	0.090	0.180	0.180	0.020	0.360	0.360	0.020
0.062	0.102	0.204	0.208	0.021	0.408	0.416	0.021
0.070	0.110	0.220	0.220	0.020	0.440	0.437	0.020
0.075	0.115	0.230	0.230	0.020	0.460	0.468	0.021
0.080	0.120	0.240	0.240	0.020	0.480	0.500	0.023
0.085	0.125	0.250	0.250	0.020	0.500	0.500	0.020
0.093	0.133	0.266	0.270	0.021	0.532	0.531	0.020
0.100	0.140	0.280	0.280	0.020	0.560	0.562	0.020
0.125	0.165	0.330	0.340	0.023	0.660	0.687	0.023

(b) Donut Pad Selection for Critical Conductor Pattern Areas

1:1 Scale Terminal Pad Dia.	1:1 Scale Solder Mask Pad Dia.	2:1 Scale Solder Mask Pad Dia.	Donut Pad Dia.	Resulting 1:1 Scale Air Gap†	4:1 Scale Solder Mask Pad Dia.	Donut Pad Dia.	Resulting 1:1 Scale Air Gap†
0.050	0.074	0.148	0.150	0.013	0.296	0.300	0.013
0.062	0.086	0.172	0.170	0.012	0.344	0.343	0.012
0.070	0.094	0.188	0.187	0.012	0.376	0.375	0.012
0.075	0.099	0.198	0.200	0.013	0.396	0.400	0.013
0.080	0.104	0.208	0.208	0.012	0.416	0.416	0.012
0.085	0.109	0.218	0.218	0.012	0.436	0.437	0.012
0.093	0.117	0.234	0.230	0.011	0.468	0.468	0.012
0.100	0.124	0.248	0.250	0.013	0.496	0.500	0.013
0.125	0.145	0.290	0.300	0.013	0.596	0.600	0.013

*All dimensions are in inches.

†Values rounded off to third place.

identify those pads which have no conductor paths or other pads in close proximity (i.e., equal to or less than 0.010 in.). The appropriate-size pad is selected from the *noncritical* section of Table 14–1, with or without red centers, as available. The pad is placed directly over the underlying terminal pad, paying particular attention to its precise alignment using the grid intercepts, as viewed from the center hole, as guides. When all pads have been properly positioned for the noncritical areas, the critical pad areas of the layout are masked in the same manner, using the *critical* section of Table 14–1. Recall that in extremely dense areas, it may improve the coverage of close conductor paths by slightly misaligning the masking pad, or trimming it. Both of these techniques are shown in Fig. 14–5. Of course, the preferred method of masking these highly critical areas is with a dry film solder mask.

A list of other available pads that will provide clearances closer to 0.010 in. is given in Appendix XXII. It may be found that no pad diameter is available for the exact value calculated. In this case, obtain the closest diameter above and below that calculated and then determine the resulting 1:1 scale air gap for each one. The air gap is calculated as follows:

$$\text{air gap} = \frac{\text{solder mask pad diameter} - (\text{scale factor}) \times (1{:}1 \text{ terminal pad diameter})}{(2)(\text{scale factor})} \quad (14\text{--}2)$$

For example, for 4X scale artwork in a critical area (0.012 air gap), the pad size of the solder mask artwork for a 0.050-in. terminal pad is calculated as follows using Equation 14–1:

$$\text{solder mask pad diameter} = 4[0.050 + (2)(0.012)] = 0.296 \text{ in.}$$

The listing in Appendix XXII shows that the closest available pad diameters are 0.281 and 0.300 in. Using Equation 14–2, we calculate the 1:1 scale air gap for a 0.281- and a 0.300-in. pad:

$$\text{air gap } (0.281) = \frac{0.281 - (4)(0.050)}{(2)(4)} = 0.0101 \text{ in.}$$

$$\text{air gap } (0.300) = \frac{0.300 - (4)(0.050)}{(2)(4)} = 0.0125 \text{ in.}$$

If the manufacturing tolerance of 0.010 in. can be met, the 0.281-in. pad should be used. However, if the 0.012-in. tolerance is the best obtainable, the 0.300-in. pad must be used.

Finally, the most difficult area to mask is the DIP pattern. If the board density is high, then typically the IC pattern used on the pad master would be individual pad sizes of either 0.050×0.125 in. in the oval style or 0.043×0.125 in. in the narrow-cut style. Either one of these pad sizes and styles can be masked quite easily by using the standard 0.070×0.125 in. pattern. This is shown in Fig. 14–7. This would provide a critical clearance of 0.010 in. for the underlying 0.050-in. oval terminal pad or, as shown, a 0.0135-in. clearance for the 0.043-in. narrow-cut pad. Note also that the lengths of both pads are a standard 0.125 in. This results in tangency at the ends of the solder mask pads and the underlying IC pads with perfect registration. There will thus be some solder-resistant ink deposited at the extreme ends of these pads due to screen stretch. This is unavoidable and marginally acceptable as long as ink is not deposited in the terminal pad holes.

Black precut patterns with red transparent centers should be used on the solder mask for all IC patterns. Since IC areas are normally critical with closely spaced conductor paths, great care must be exercised in aligning these patterns with those on the underlying artworks.

Recall that all masked terminal pad areas must be completely opaque to avoid ink from contacting any pads or holes. Where red-centered pads are used, nothing more need be done since red photographs as opaque. However, where black center-hole pads are used for grid intercept alignment, the transparent holes must be masked. Although covering them with a piece of tape will accomplish this, it

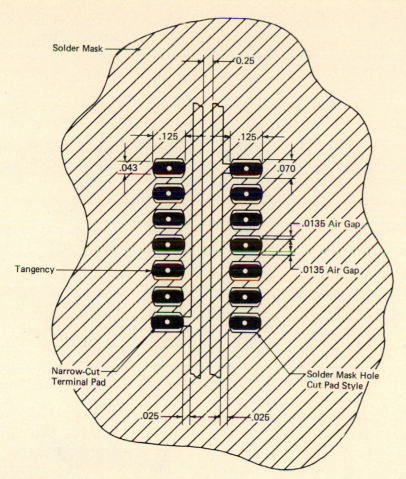

FIGURE 14-7 Solder mask for IC pattern.

is a slow and tedious process. A more effective and efficient method of making the transparent center holes opaque is through photographic processes. It is much less time consuming than manually covering each pad center with tape. The procedure for this photographic method is shown in Fig. 14–8. The solder mask artwork is first photoreduced and the resulting 1:1 scale negative is contact printed. See steps I through III in Fig. 14–8. A piece of blank PCB stock is then drilled with all holes accurately positioned. Only one drill hole size is used, the diameter of which is larger than that of the largest 1:1 scale transparent hole diameter but not equal to or greater than the smallest pad diameter. After the holes are drilled, the board is contact-printed. See steps IV and V. The positives obtained in steps III and V are then accurately overlaid, registered, and contact printed for a second time. The result of this process is shown in step VI. This is the 1:1 scale negative of the original scaled solder mask with all pads having the specified size and completely transparent. When this negative is contact-printed with screen film and processed, it results in the formation of the screen shown in step VII. Solder-resistant ink squeegeed through this screen results in masking the circuit side of the board.

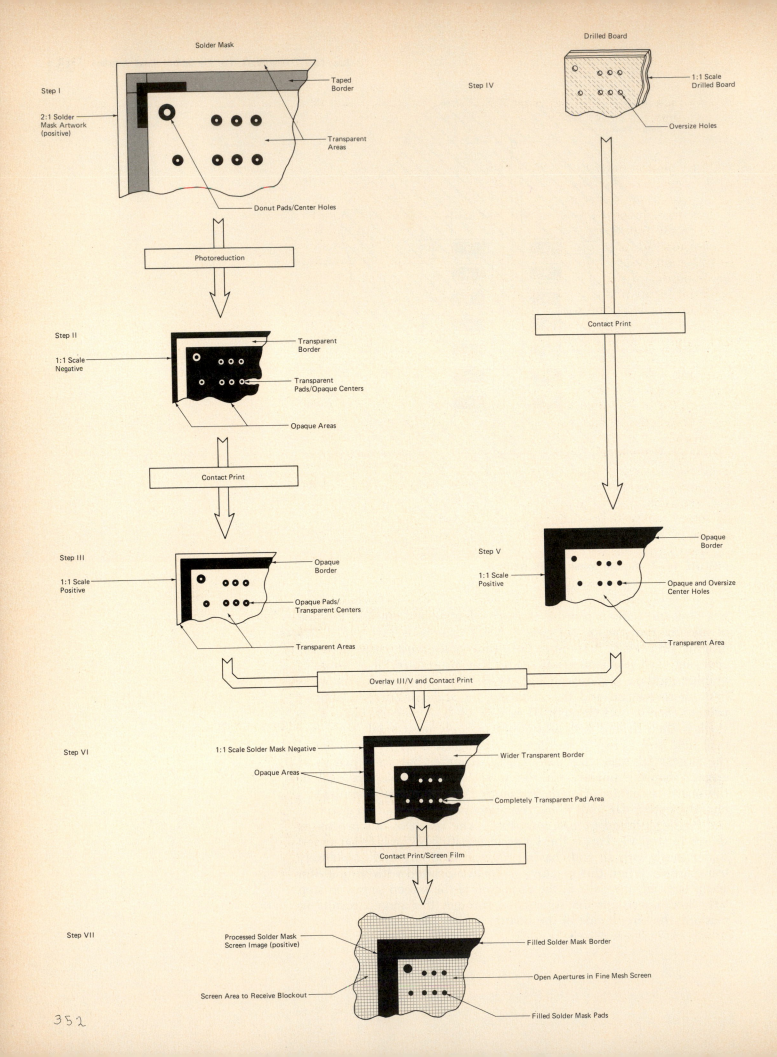

Solder Mask

Step I

2:1 Solder Mask Artwork (positive)

Taped Border

Transparent Areas

Donut Pads/Center Holes

Photoreduction

Step II

1:1 Scale Negative

Transparent Border

Transparent Pads/Opaque Centers

Opaque Areas

Contact Print

Step III

1:1 Scale Positive

Opaque Border

Opaque Pads/ Transparent Centers

Transparent Areas

Drilled Board

Step IV

1:1 Scale Drilled Board

Oversize Holes

Contact Print

Step V

1:1 Scale Positive

Opaque Border

Opaque and Oversize Center Holes

Transparent Area

Overlay III/V and Contact Print

Step VI

1:1 Scale Solder Mask Negative

Opaque Areas

Wider Transparent Border

Completely Transparent Pad Area

Contact Print/Screen Film

Step VII

Processed Solder Mask Screen Image (positive)

Filled Solder Mask Border

Open Apertures in Fine Mesh Screen

Screen Area to Receive Blockout

Filled Solder Mask Pads

352

FIGURE 14–8 (facing page) Photographic process for opaquing transparent pad centers in solder masks.

When photoreducing or contact printing, the film size used is always larger than the overall size of the image. As a result, there is an increase in the solder mask border width beyond the outside edges of the corner bracket positions. This is shown in step VI of Fig. 14–8. Since the unused area of the screen between the image and the frame must be blocked out, there is concern that no block-out material be allowed to contact the image as it is spread. The resulting wider border of step VI allows a safer distance since the block-out need not be applied close to the image.

Before removing the solder mask artwork from the work surface, it should be carefully checked to see (1) if it is in the best possible registration with the underlying component layout drawing and pad master; (2) if a solder mask pad exists for each terminal pad and mounting hole; (3) if a masking pad with a 0.020-in. minimum air gap has been provided for all noncritical areas; (4) if a masking pad with a 0.010-in. minimum air gap has been provided for all critical areas; (5) whether a slight shifting of the masking pad in critical areas, as shown in Fig. 14–5a, would aid in more complete coverage of closely spaced conductor paths; and (6) if all border tape ends are of proper width and overlap appropriately.

A completed solder mask artwork is shown in Fig. 14–9a and its negative is shown in Fig. 14–9b. The reduced negative will be the phototool that will be used to process the masking image on the screen. Note on the negative that the dark areas inside and outside the transparent border represent those on the finished board where

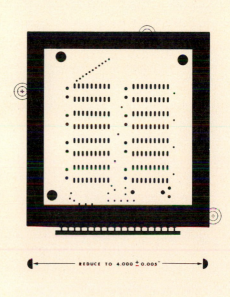

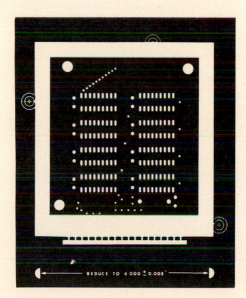

(a) Positive   (b) Negative

FIGURE 14–9 Solder mask artworks.

solder-resistant ink will be deposited. The transparent areas are those that will be protected from the ink while the board is being screened.

Now that you have a basic understanding of a solder mask and its preparation, there is one point worth considering. When a solder mask is specified as part of the design, it may be highly advantageous to alter the order of artworks from that described in this book. That is, if the solder mask were constructed immediately after the pad master, it can be used as an underlying guide, together with the pad master, for optimizing path routing on the taped artwork. Using the solder mask in this manner could result in avoiding unnecessary generation of critical conductor pattern areas.

EXERCISES

14–1 What is the typical material used for a solder mask on printed circuit boards?

14–2 How is the solder mask applied to the board?

14–3 What is the purpose of a solder mask?

14–4 How close to the edge of the board should the solder mask extend?

14–5 Explain the term *air gap* as it applies to solder masks.

14–6 What is the typical mesh count for screens used to apply solder masks?

14–7 List the factors that effect the positional accuracy of solder masks applied with a screen.

14–8 List five factors that effect the quality of a screened solder mask.

14–9 How close should the solder mask extend to the crown of the plated fingers?

14–10 Explain the procedure for masking an IC pattern with pads that measure 0.050×0.125 in.

15

Computer-Aided Drafting and Design of Printed Circuit Boards

Upon completion of this chapter on computer-aided drafting and design of printed circuit boards, the student should be able to:

1. Understand the basic component parts of a digitizing system.

2. Understand the basic component parts of a CAD system used for designing printed circuit boards.

3. Understand the difference between semiautomated and totally automated design systems.

15-0

Introduction

One of the more difficult demands continually placed on the printed circuit board industry is to increase component density. This requires a corresponding increase in the ability to produce even more precise artworks with dimensional accuracies that are beyond the capability to perform manually. In order for the manufacturer to successfully fabricate new generations of higher-density and higher-quality PCBs, such factors as more precise pad locations and front-to-back (or layer-to-layer) registration on both double-sided and multilayer boards must be incorporated initially onto the artwork masters. These, in addition to more accurately spaced conductor paths with uniform minimum widths, are required on modern artworks on a regular basis. It is becoming more difficult and time consuming to attain these levels of design when they are manually prepared and for some layouts, impossible to produce.

To obtain the more demanding-quality artworks of the types shown in Chapters 7 through 14, the electronics industry has been shifting away from manual preparation and moving toward artworks that are produced with the aid of computer systems. These artwork generation systems range from relatively inexpensive and easy-to-operate units *(digitizers)* to very expensive and sophisticated interactive computer-aided design and drafting (CADD, or more commonly, *CAD*) systems. Both of these modern drafting methods have a digital minicomputer and memory in which the PCB design data are stored in digital form. The outputs of these units are then used to input *photoplotters* which photographically generate very precise and high-quality 1:1 scale artwork films required for high-density boards.

Artworks produced by photoplotting can routinely achieve typical accuracies of 0.001 in. Even though several types of photoplotters are available, they are all driven by the same kind of x and y coordinate information supplied to them in digital form from either the digitizer or the CAD system.

This chapter is devoted to a discussion of computer-aided drafting (digitizing) and computer-aided design (CAD) used to produce high-quality PCB artworks.

15-1

The Basic Digitizer System

The digitizer is essentially a very accurate x-y coordinate measuring system. There are basically two types, both of which have a digitizing control (cursor) which is used with a large drafting-board-type work table. One system has the cursor attached to a vertically positioned carriage (y dimension) which can be moved along a horizontal track (x dimension). The second type is designed with a hand-held cursor.

The use of the digitizer begins with a manually prepared component layout drawing, similar to those described in Chapters 7 through 11. All of the creative aspects of PCB design, such as parts placement and orientation as well as conductor path routing are prepared manually by the designer on the drawing. The digitizer eliminates the need for the less accurate manual preparation of the taped artworks.

A typical digitizer is shown in Fig. 15–1. It consists of (1) a track-type digitizer/table, (2) a computer-controlled interactive display console with floppy disk memory, and (3) a flat bed photoplotter. Digitizing systems are divided into four functional parts:

1. *Input equipment:* digitizer table and console with function and alphanumeric keyboards.
2. *Processor equipment:* CRT and floppy disk memory.
3. *Software:* set of programmed instructions on floppy disk memory which converts the hardware into a functional PCB digitizing tool.
4. *Output equipment:* photoplotter or line plotter.

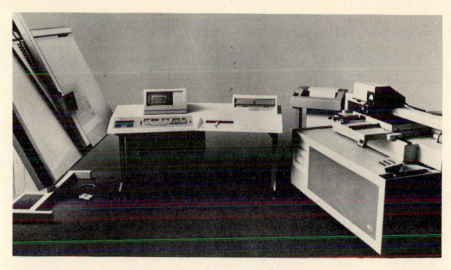

FIGURE 15-1   Complete system used to digitize printed circuit board layouts. [Courtesy of Gerber Scientific Instrument Company.]

To begin the actual digitizing procedure, the operator first registers and aligns the component layout drawing onto the digitizer table. These tables are provided with grids and all the scales common to PCB design. With the drawings accurately registered and aligned, the datum point is located with the digitizer cursor. Its cross-hairs are aligned with the desired location for the datum point and the register is cleared. Next, using the cursor and its pushbutton switches, the computer is instructed to record the x and y coordinates of each terminal pad center as well as their diameters. The digitizer shown in Fig. 15–1 is a *track-type* model. This type determines positions through the use of x and y incremental photoelectric encoders which are connected to up/down counters and are driven by a precision gear system. The computer converts the encoded signals that it receives into x and y coordinates.

The hand-held cursor is not mechanically connected to a carriage, which eliminates interference of movement and allows complete freedom to follow complex contours accurately. The hand-held cursor is coupled to the system through an electromagnetic field.

The operator can use either type of cursor to record, in digital form, the start, directional change points, and end-of-line coordinates of all conductor paths. This information is stored in the floppy disk memory. Other information that is stored in memory is the board outline, target shapes, corner brackets, and other special shapes, as well as letter and number symbols which can be entered directly from the alphanumeric keyboard.

The digitizing process described can transform a manually designed component and conductor layout drawing into a digital data base of information sufficiently detailed and capable of driving a photoplotter to produce high-quality 1:1 scale artwork films. Thus the need for the time-consuming and less accurate manual preparation of artworks is eliminated.

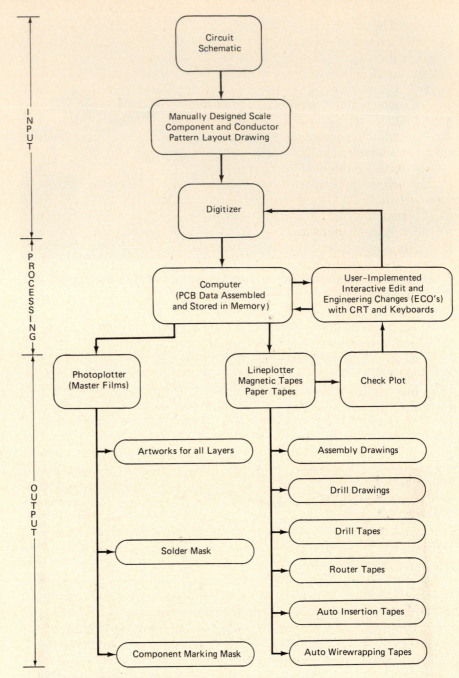

FIGURE 15–2   Flow chart for computer-aided drafting.

Digitizing is also referred to as *computer-assisted drafting* and the new generation of machines and systems have features which allow the operator to do real-time editing as well as design modifications as the digitizing progresses. This capability is termed *user-implemented interactive*, which means that the system has no decision-making ability but does provide the means for a designer to work easily at a CRT display console to design a PCB. This allows immediate monitoring of the effects of layout decisions during the design process. The interactivity between the designer and the system can greatly increase the versatility of the digitizer. Editing is achieved through the use of a set of programmed software instructions available to the designer in *menu* form through a function keyboard. This keyboard is an array of microswitches aligned in a matrix. The function of each switch is predefined on the menu layout of the keyboard and identified with a symbol or abbreviated command notation. When the designer depresses a microswitch key, a specific system command is activated. In the *edit* mode of operation, the designer can view selected individual areas of the layout on the CRT and perform such functions as *find* and *change* a pad size or location, *change* or *add* text material, *break* a conductor path line segment, *move* and reconstruct the line, and *insert* additional features or *move* existing features. Any changes are added instantly to the data base, which is stored in memory and can be verified on the CRT display at once.

After the digitizing process is completed, the design is stored on a floppy disk for future use. A flowchart describing the sequential steps in using a computer-aided drafting system is shown in Fig. 15–2.

Typically, the digitizing designer will run a *check plot* of the design for the engineer, who will review and verify the design as well as make any changes. Check plots are often run on *line plotters*, which are available in three types: *flat bed*, *belt bed*, and *drum*. A drum-style line plotter is shown in Fig. 15–3. The drawing media used on the drum plotter may be vellum, polyester, or paper. As the drum

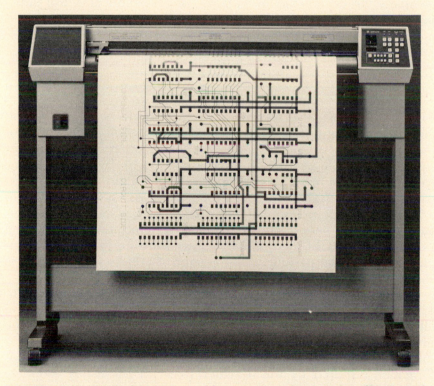

FIGURE 15–3   Typical plotter used in computer-aided design. [Photograph courtesy of Hewlett-Packard.]

rotates, pens move left and right along the length of the drum. Information is fed to the plotter from the digitized data base. The plotter converts the data from the computer into a finished drawing. After the drawings have been verified, artwork films for all layers as well as solder masks and component marking masks can be produced with the photoplotter.

Photoplotting (see Fig. 15–1 for equipment) is accomplished by first placing a sheet of photographic film on the bed of the plotter under safe-light conditions. A light source and aperture wheel with up to 24 different size holes or stations are located above the bed on a movable track system which has x and y directional capabilities. The digitized data base directs the plotter to set the aperture wheel to the appropriate size to construct the proper pad diameter as well as establishing how long to flash the light to expose the film at each recorded set of coordinates. Conductor path widths and routes are constructed by selecting the aperture station with the appropriate setting. As the head moves over the film, it is exposed and the path is formed as directed by the digitizer. Special shapes can also be photoplotted in this manner.

As seen from the flowchart of Fig. 15–2, additional documentation, often referred to as *deliverables*, is available as an automatic output of the digitized data base. Tapes (paper or magnetic) can be generated to command numerically controlled (NC) driller/router units as well as automatic insertion and wire-wrapping equipment. This type of capability is called *computer-aided manufacturing* (CAM) and is welcomed by the board manufacturer since established data consistency is assured. Also available from the digitized data base, to complete the documentation package, are assembly and drill drawings.

15-2

The Basic CAD System

A more advanced level of complexity and sophistication over the digitizing system described in the preceding section is the ability of the computer to complement the creative thinking of the PCB designer in both the placement of parts and the routing of conductor paths. A system with this capability is generally referred to as *computer-aided design* or *CAD*. Typically, a CAD system consists of a CRT color graphics terminal, input keyboards, system table menu, and light pen as peripheral equipment to a minicomputer or central processing unit (CPU) with the necessary software to form the complete system. The peripheral equipment of a basic CAD system is shown in Fig. 15–4.

Most of the design of a PCB can be done at the CAD workstation. Prior to the actual design, however, the physical shape and size of all component parts and hardware to be used in the design must be entered and stored in the computer's memory. This library of information must include all pin identifications and other electronic information required by the designer. Next, a *net* list is added to memory. This is a detailed summary of all interconnections associ-

FIGURE 15-4   Typical computer-aided design workstation. [Courtesy of Applicon, Inc.]

ated with a signal path (obtained from the circuit schematic), all electrical restrictions, board outline, and all other mechanical features.

With the net list entered, *parts positioning* software can be run on the CAD system, which will allow the computer to attempt optimum positioning within the board outline of all components appearing on the circuit schematic. At any time, the CAD operator can enter the loop to stop the parts positioning program and alter the position or orientation of one or more components to simplify the path routing. Access to the computer is done through the keyboard or the light pen and table menu.

Upon the completion of the parts placement, which requires interaction between the computer and the operator, the next step is to run the *automatic routing program*, where the computer will attempt to route all the conductor paths guided by programmed instructions. In this phase, the computer is programmed to duplicate the creative thinking of the designer. For a typical design, approximately 90% of the routing connections can be made by the computer, which performs a *maze search*. It takes each lead, one at a time, and searches for the most direct way to route it. When it finds that all available paths are blocked, it will list the lead as *unroute*. The computer will then go on to another lead. This type of program is termed *semiautomated*. Unlike a user-implemented interactive system, a semiautomatic system will make decisions. Some of the routing, however, is not always optimal. But with 90% of the routing completed, it provides the designer with an advanced starting point from which to work. Again, at any time, the operator can enter the loop and manually insert those paths which have not been connected on the screen. This may require that some parts be repositioned.

The most sophisticated and most expensive CAD systems have *totally automated* capability, which allows the system, when provided with all the required information, to make all decisions necessary to complete the design in optimal fashion without manual intervention. This is the result of the programs having strategies which enable the computer to have "second thoughts" about any initial decisions when a block occurs in the maze search. Thus the computer responds to the problem in much the same way as would the manual designer. Programs with advanced strategies are available that are capable of better than 99% automatic completion of many PCB designs.

After all conductor paths have been routed, the CAD system is able to run checks on the design, such as verification of design rules,

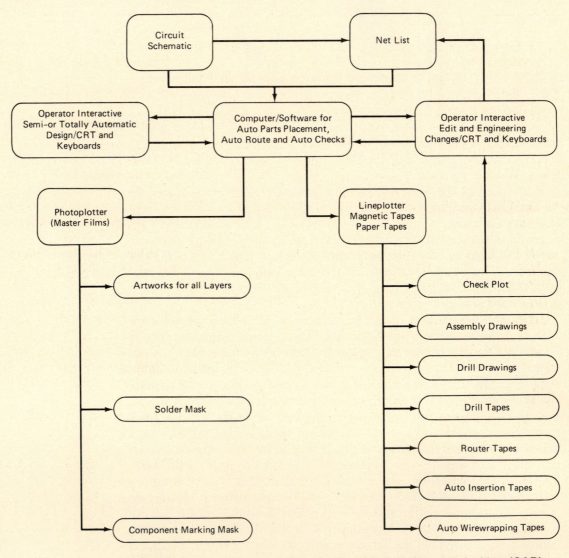

FIGURE 15–5 Flow chart of computer-aided design and drafting (CAD) process.

spacings, interconnections, and so on. With the checks completed, all the data required to produce the entire documentation package are stored in memory and can be processed into artwork films and other documentation, such as shown in Fig. 15–2.

A flowchart summarizing the design process using a CAD system is shown in Fig. 15–5. It should now be apparent that the use of computers in the design of PCBs is of significant benefit to both the designer and the manufacturer. The vast amount of information that is assembled and stored during the design process can automatically output film artworks and programmed tapes to save enormous amounts of time by eliminating the tedious and error-prone manual procedures.

EXERCISES

15–1 Give the main reasons for the need of computer-aided design and drafting.

15–2 Artworks produced by photoplotting can routinely achieve what kind of accuracy?

15–3 Explain the purpose of a digitizer.

15–4 What is a cursor?

15–5 What are the four functional parts of a digitizer system?

15–6 Explain the difference between *hardware* and *software*.

15–7 What major step in the design of a PCB does a digitizer system eliminate?

15–8 Define the term *user-implemented interactive*.

15–9 Explain the difference between a line plotter and a photoplotter.

15–10 What are deliverables?

15–11 What important characteristic does the automatic output documentation have that is of great importance to the designer and manufacturer?

15–12 To what does the acronym *CAM* refer?

15–13 What is a net list?

15–14 What hardware does the CAD operator use to enter commands into the computer?

15–15 Give some of the types of programs available to the designer for interactive editing.

15–16 What is the basic difference between semi-automated and totally automated design systems?

15–17 What is the basic difference between computer-aided drafting and CAD?

15–18 Explain how the vast amount of information that a computer has assembled and stored during the design phase is used.

15–19 List the major benefits of CAD to the designer.

15–20 List the benefits of CAD to the manufacturer.

BIBLIOGRAPHY

ANSI Standard Y32.2, *Graphic Symbols for Electrical and Electronics Diagrams*, 1975.

American Society of Mechanical Engineers, *Drawing Sheet Size and Format*, ANSI Y14.1, 1975.

Baer, Charles J., and John R. Ottoway, *Electrical and Electronics Drawing*, 4th ed., McGraw-Hill Book Company, New York, 1980.

Bethune, James D., *Essentials of Drafting*, Prentice-Hall, Inc., Englewood Cliffs, N.J., 1977.

Bishop Graphics Printed Circuit Technical Manual 107, Bishop Graphics, Inc., Westlake Village, Calif., 1981.

Circuits Manufacturing, published monthly by Benwill Publishing Corp., Boston.

Coombs, Clyde F., Jr., et al., *Printed Circuits Handbook*, 2nd ed., McGraw-Hill Book Company, New York, 1979.

Coughlin, Robert F., and Frederick F. Driscoll, *Operational Amplifiers and Linear Integrated Circuits*, 2nd ed., Prentice-Hall, Inc., Englewood Cliffs, N.J., 1982.

CSA Standard 799, *Graphic Symbols for Electrical and Electronics Diagrams*, 1975.

Driscoll, Frederick F., *Microprocessor–Microcomputer Technology*, Breton Publishers, North Scituate, Mass., 1983.

Giesecke, Frederick E., et al., *Technical Drawing*, 7th ed., Macmillan Publishing Company, New York, 1980.

IEEE Standard 315, *Graphic Symbols for Electrical and Electronics Diagrams*, 1975.

IPC Standards: The Institute for Interconnecting and Packaging Electronic Circuits, Evanston, Ill.

    IPC-A-600B   *Acceptability of Printed Wiring Boards.*

    IPC-CM-770A   *Printed Board Component Mounting.*

    IPC-D-300F   *Standard Specification Printed Wiring Board Dimensions and Tolerances Single and Two-Sided Rigid Boards.*

    IPC-D-310A   *Artwork Generation and Measurement Techniques.*

    IPC-D-320   *Standard Specification, Printed Board, Rigid, Single-and Double-Sided, End Product Specification.*

    IPC-D-350A   *Standard Specification, End Product Description in Numeric Form.*

IPC-D-390 *Guidelines for Design Layout and Artwork Generation on Computer Automated Equipment for Printed Wiring.*

IPC-ML-910A *Design and End Production Specification for Rigid Multilayer Printed Circuit Boards.*

IPC-ML-950A *Performance Specifications for Multilayer Printed Wiring Boards.*

IPC-ML-975 *End Product Documentation Specification for Multilayer Printed Wiring Boards.*

IPC-SM-840 *Qualification and Performance of Permanent Polymer Coating (Solder Mask) for Printed Boards.*

IPC-T-50A *Standard Specification Industry Standard Terms and Definitions.*

IPC-TC-500A *Specification for Copper Plated Through Hole Connection for Double-Sided Boards, Rigid.*

IPC-TP-219 *CADMON: Improving the CAD System Human Interface.*

IPC-TP-221 *Additional Data on Realistic Values for Printed Board Conductor Spacings.*

IPC-TP-226 *Fine Line Reproduction in Dry Film Photo Resists.*

IPC-TP-230 *Innerlayer Production Using Dry Film Photoresists.*

IPC-TP-238 *Swaging of Terminals and Interconnecting Pins on Printed Wiring Boards.*

Lindsey, Darryl, *The Design and Drafting of Printed Circuits*, 2nd ed., Bishop Graphics, Inc., Westlake Village, Calif., 1982.

Lund, Preben, *Generation of Precision Artwork for Printed Circuit Boards*, John Wiley & Sons, Inc., New York, 1978.

Markus, John, *Modern Electronic Circuits Reference Manual*, McGraw-Hill Book Company, New York, 1980.

Military Standard MIL-STD-275, *Printed Wiring for Electronic Equipment.*

Printed Circuit Fabrication, published monthly by PMS Industries, Roswell, Ga.

Raskhodoff, Nicholas M., *Electronic Drafting and Design*, 4th ed., Prentice-Hall, Inc., Englewood Cliffs, N.J., 1981.

Spence, William P., *Drafting Technology and Practice*, Charles A. Bennett Co., Inc., Peoria, Ill., 1973.

Villanucci, Robert S., Alexander W. Avtgis, and William F. Megow, *Electronic Techniques: Shop Practices and Construction*, 2nd ed., Prentice-Hall, Inc., Englewood Cliffs, N.J., 1981.

Villanucci, Robert S., Alexander W. Avtgis, and William F. Megow, *Electronic Shop Fabrication*, Prentice-Hall, Inc., Englewood Cliffs, N.J., 1982.

Wallach, Paul, *Metric Drafting*, Glencoe Publishing Co., Inc., Encino, Calif., 1979.

APPENDIXES

APPENDIX I   Comparison of Flat Size Drafting Paper: Metric to English

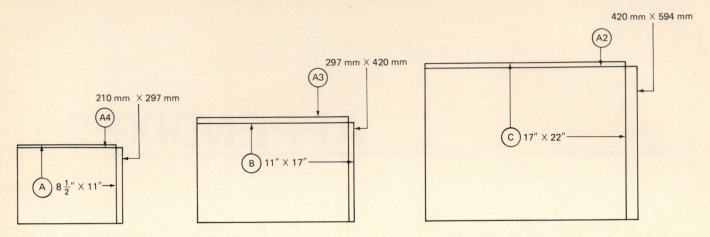

210 mm × 297 mm

297 mm × 420 mm

420 mm × 594 mm

A4

A3

A2

(A) 8½″ × 11″

(B) 11″ × 17″

(C) 17″ × 22″

APPENDIX II   Comparison of Metric and English Drafting Paper Sizes

Metric Designation	Length–Width (mm)	Length–Width (in.)
2A	1189 x 1682	46.81 x 66.22
A0	841 x 1189	33.11 x 46.81
A1	594 x 841	23.39 x 33.11
A2	420 x 594	16.54 x 23.39
A3	297 x 420	11.69 x 16.54
A4	210 x 297	8.27 x 11.69
A5	148 x 210	5.83 x 8.27
A6	105 x 148	4.13 x 5.83
A7	74 x 105	2.91 x 4.13
A8	52 x 74	2.05 x 2.91
A9	37 x 52	1.46 x 2.05
A10	26 x 37	1.02 x 1.46

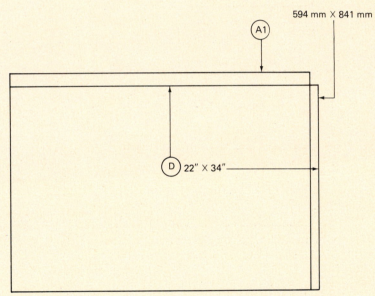

594 mm × 841 mm

A1

(D) 22″ × 34″

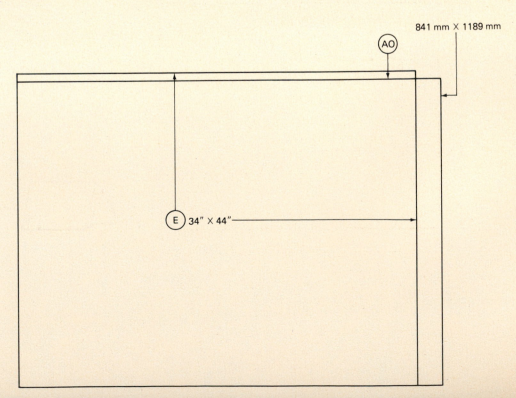

841 mm × 1189 mm

AO

(E) 34″ × 44″

APPENDIX III Procedure for Folding Flat-Size Drafting Papers for Storage: Sizes B through E

B
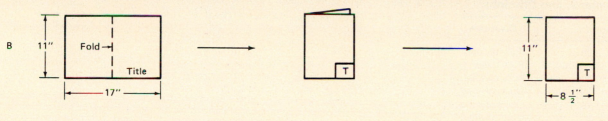

C
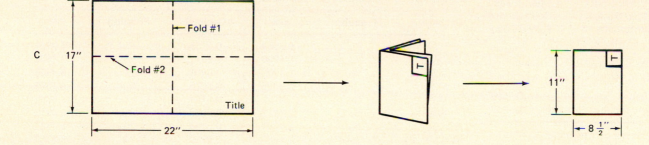

D
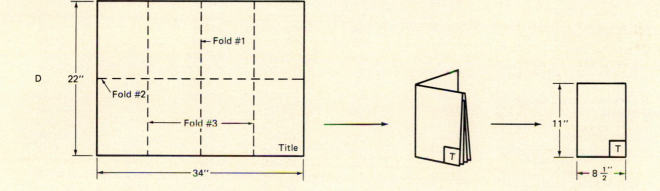

E
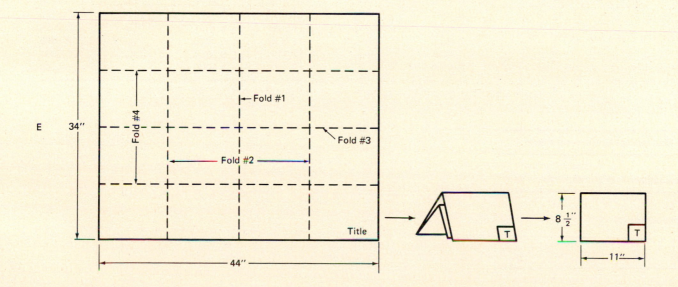

APPENDIX IV Drafting Machine Scales, Graduations, and Divisions

Graduations	Divisions
Full and half size inch	Fully divided
Half and quarter size inch	Fully divided
Quarter and eighth size inch	Fully divided
1½" and 3" to the foot	Open divided
⅜" and ¾" to the foot	Open divided
½" and 1" to the foot	Open divided
¼" and ⅛" to the foot	Open divided
⅛", ¼", ½", 1" to the foot	Open divided
⅜", ¾", 1½", 3" to the foot	Open divided
32nds and 50ths to an inch	Fully divided
10ths and 50th to an inch	Fully divided
20ths and 40ths to an inch	chain
30ths and 60th to an inch	scales
1 mm and ½ mm	Fully divided
Metric 1:250 and 1:500	Fully divided

APPENDIX VI Measurement Guide for Fully Divided Scales

(a) Architect's Scale

Scale Designator	From 0 Index Each Minor Division* Represents:
3	⅛"
1½	¼"
1	¼"
¾	½"
½	½"
⅜	1"
¼	1"
3/16	1"
3/32	2"
⅛	2"

(b) Mechanical Engineer's Scale

Scale Designator	From 0 Index Each Mark* Represents:
⅛	¼"
¼	⅛"
⅜	1/16"
½	1/16"
¾	1/32"

*Fully divided unit.

APPENDIX V Decimal to Fractional to Metric Numerical Values

Decimal	Fraction	mm	Decimal	Fraction	mm
0.0156	1/64	0.3969	0.5156	33/64	13.0969
0.0312	1/32	0.7938	0.5312	17/32	13.4938
0.0469	3/64	1.1906	0.5469	35/64	13.8906
0.0625	1/16	1.5875	0.5625	9/16	14.2875
0.0781	5/64	1.9844	0.5781	37/64	14.6844
0.0938	3/32	2.3812	0.5938	19/32	15.0812
0.1094	7/64	2.7781	0.6094	39/64	15.4781
0.1250	⅛	3.1750	0.6250	⅝	15.8750
0.1406	9/64	3.5719	0.6406	41/64	16.2719
0.1562	5/32	3.9688	0.6562	21/32	16.6688
0.1719	11/64	4.3656	0.6719	43/64	17.0656
0.1875	3/16	4.7625	0.6875	11/16	17.4625
0.2031	13/64	5.1594	0.7031	45/64	17.8594
0.2188	7/32	5.5562	0.7188	23/32	18.2562
0.2344	15/64	5.9531	0.7344	47/64	18.6531
0.2500	¼	6.3500	0.7500	¾	19.0500
0.2656	17/64	6.7469	0.7656	49/64	19.4469
0.2812	9/32	7.1438	0.7812	25/32	19.8438
0.2969	19/64	7.5406	0.7969	51/64	20.2406
0.3125	5/16	7.9375	0.8125	13/16	20.6375
0.3281	21/64	8.3344	0.8281	53/64	21.0344
0.3438	11/32	8.7312	0.8438	27/32	21.4312
0.3594	23/64	9.1281	0.8594	55/64	21.8281
0.3750	⅜	9.5250	0.8750	⅞	22.2250
0.3906	25/64	9.9219	0.8906	57/64	22.6219
0.4062	13/32	10.3188	0.9062	29/32	23.0188
0.4219	27/64	10.7156	0.9219	59/64	23.4156
0.4375	7/16	11.1125	0.9375	15/16	23.8125
0.4531	29/64	11.5094	0.9531	61/64	24.2094
0.4688	15/32	11.9062	0.9688	31/32	24.6062
0.4844	31/64	12.3031	0.9844	63/64	25.0031
0.5000	½	12.7000	1.0000	1	25.4000

APPENDIX VII Scale Selector Guide for Scale Drawing

Divisions	Type	Scale Designator	Scale Reduction: Percent of Full Scale	Scale Reduction: Scale Size (Fractional)	Actual Measurement Between Major Graduations	Measurement Represented
Fully divided	Engineer's	10	100	Full size	1″	10′ or 100′ or 1000′ or 1 mi. etc.
		20	50	½ size	1″	20′ or 200′ or 2000′ or 2 mi. etc.
		30	33.3	⅓ size	1″	30′ or 300′ or 3000′ or 3 mi. etc.
		40	25	¼ size	1″	40′ or 400′ or 4000′ or 4 mi. etc.
		50	20	⅕ size	1″	50′ or 500′ or 5000′ or 5 mi. etc.
		60	16.7	⅙ size	1″	60′ or 600′ or 6000′ or 6 mi. etc.
Open divided	Mechanical	⅛	12.5	⅛ size	⅛″	1″
		¼	25	¼ size	¼″	1″
		⅜	37.5	⅜ size	⅜″	1″
		½	50	½ size	½″	1″
		¾	75	¾ size	¾″	1″
Open divided	Architect's	16	100	Full size	1″	1′ (or 12″ = 1′)
		3	25	¼ size	3″	1′
		1½	12.5	⅛ size	1½″	1′
		1	8.3	1/12 size	1″	1′
		¾	6.25	1/16 size	¾″	1′
		½	4.2	1/24 size	½″	1′
		⅜	3.1	1/32 size	⅜″	1′
		¼	2	1/48 size	¼″	1′
		9/16	1.5	1/64 size	3/16″	1′
		⅛	1	1/96 size	⅛″	1′
		3/32	0.78	1/128 size	3/32″	1′
Fully divided	Metric*	1:1	100	Full size	10 mm	10 mm
		1:2	50	½ size	10 mm	20 mm
		1:2.5	40	⅖ size	10 mm	25 mm
		1:5	20	⅕ size	10 mm	50 mm
		1:10	10	1/10 size	10 mm	100 mm
		1:20	5	1/20 size	10 mm	200 mm
		1:25	4	1/25 size	10 mm	250 mm
		1:33.3	3	1/33 size	10 mm	333 mm

*Metric ratios may be enlarged or reduced as desired by multiplying or dividing by 10.

APPENDIX VIII   Selector Guide for Circle Template

Circle Template Style	Number of Holes	Available Hole Diameters
Perfect circle	42	Range: 1/32″ to 2″ Increments: 1/64″, 1/32″, 1/16″, 1/8″, 1/4″, 1/2″
Senior circle	42	Range: 1/32″ to 2″ Increments: 1/64″, 1/32″, 1/16″, 1/8″, 1/4″, 1/2″
Large circle	45	Range: 1/16″ to 2¼″ Increments: 1/64″, 1/32″, 1/16″, 1/8″, 1/4″, 1/2″
Circle master	44	Range: 1/16″ to 3″ Increments: 1/64″, 1/32″, 1/16″, 1/8″, 1/4″, 1/2″
Extra-large circle	13	Range: 1¼″ to 3½″ Increments: 1/8″, 1/4″
Decimal perfect*	66	0.100″ to 0.500″ in increments of 0.010″ 0.500″ to 1.000″ in increments of 0.020″
Small increment	82	Range: 1/32″ to 1¼″ to 1/4″ in increments of 1/128″, to 1″ in increments of 1/64″
Metric	37	Range: 2 mm to 30 mm to 10 mm in increments of 0.5 mm, to 30 mm in increments of 1 mm
Large metric	44	Range: 2 mm to 50 mm Increments: 0.5 mm, 1 mm, 2 mm, 3 mm, 5 mm
Combinational	39	Range: (3/64″)(1 mm)(0.04″) to (1⅜″)(35 mm)(1.38″) Increments: increasing
Circle guide	46	Range: 1/16″ to 2¾″ Increments: 1/64″, 1/32″, 1/16″, 1/8″
Circle indicator	40	Range: 3/64″ to 2″ to 1/4″ by 1/64″, 9/32″ to 5/8″ by 1/32″, 11/16″ to 1″ by 1/16″ 1⅛″ to 2″ by 1/8″

*All increments of 0.010″ and 0.020″ are in horizontal rows and all increments of 0.100″ are in vertical rows to facilitate selection of size.

APPENDIX IX   Graphic Symbols

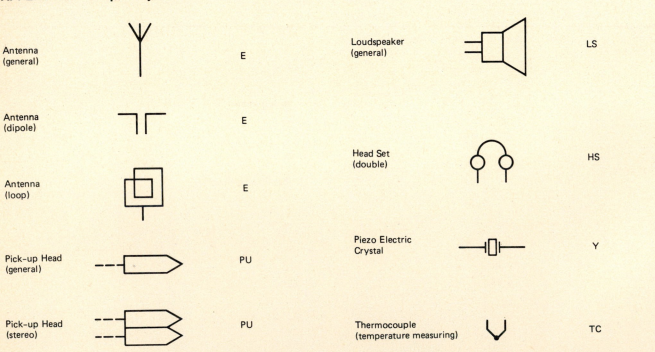

Antenna (general)		E
Antenna (dipole)		E
Antenna (loop)		E
Pick-up Head (general)		PU
Pick-up Head (stereo)		PU
Loudspeaker (general)		LS
Head Set (double)		HS
Piezo Electric Crystal		Y
Thermocouple (temperature measuring)		TC

Selector Switch (break-before-make, nonshorting)		S	
	OR		
Selector Switch (make-before-break, shorting)		S	
	OR		
Selector Switch (rotary)		S	
	OR		
Operational Amplifier (general)		AR	
Summing Amplifier (general)		AR	
	OR	AR	
Integrating Amplifier		A	
	IC		
	OR		
	IC	A	

Multiplier — A

Divider — A

Amplifier (dual input, single ended output) — A

Amplifier (single ended input, dual output) — A

Amplifier (dual input and output) — A

Amplifier (with adjustable gain) — A

Amplifier (with attenuator) — A

Amplifier (with power supply) — A
PS

Amplifier (with external feedback path) — A
NET

Transistors

Application: NPN Transistor with Multiple Emitters (with 4 emitters shown) — Q
(E)(E)(E)(E) (C)
(B)

NPN Transistor with Transverse-Biased Base — Q
(E) (C)
(B1) (B2)

373

APPENDIX IX (continued)

8.6.4 PNIP Transistor with Ohmic Connection to the Intrinsic Region Q

PNIN Transistor with Ohmic Connection to the Intrinsic Region Q

NPIN Transistor with Ohmic Connection to the Intrinsic Region Q

NPIP Transistor with Ohmic Connection to the Intrinsic Region Q

Semiconductor Diodes

Capacitive Diode (varactor) D, CR

Photosensitive Type: 2-segment, with Common Cathode Lead D, CR

Temperature-Dependent Diode D, CR

Photosensitive Type: 4-quadrant with Common Cathode Lead D, CR

NPN-type

Bidirectional Photodiode; Photo- Duo-Diode (photosensitive type) 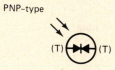 D, CR

Storage Diode D, CR

PNP-type

Field-Effect Transistors

N-channel Junction Gate

OR

P-channel Junction Gate

OR

 Q

 Q

N-channel Insulated-Gate, Depletion-Type, Single-Gate, Passive-Bulk (substrate), Three-Terminal Device

P-channel Insulated-Gate, Depletion-Type, Single-Gate, Passive-Bulk (substrate), Three-Terminal Device Q

N-channel Insulated-Gate, Depletion-Type,
Single-Gate, Active-Bulk (substrate)
Internally Terminated to Source, Three-
Terminal Device

(G) (D) (S)

P-channel Insulated-Gate, Depletion-Type,
Single-Gate, Active-Bulk (substrate) Internally
Terminated to Source, Three-Terminal Device

(G) (D) (S) Q

N-channel Insulated-Gate, Depletion-Type,
Single-Gate, Active-Bulk (substrate)
Externally Terminated, Four-Terminal
Device

(G) (D) (U) (S)

P-channel Insulated-Gate, Depletion-Type,
Single-Gate, Active-Bulk (substrate) Externally
Terminated, Four-Terminal Device

(G) (D) (U) (S) Q

N-channel Insulated-Gate Depletion-Type,
Two-Gate, Five-Terminal Device

(G2) (G1) (D) (U) (S)

P-channel Insulated-Gate, Depletion-Type,
Two-Gate, Five-Terminal Device

(G2) (G1) (D) (U) (S) Q

N-channel Insulated-Gate, Enhancement-
Type, Single-Gate, Active-Bulk (substrate)
Externally Terminated, Four-Terminal
Device

(G) (D) (U) (S)

P-channel Insulated-Gate, Enhancement Type,
Single-Gate, Active-Bulk (substrate) Externally
Terminated, Four-Terminal Device

(G) (D) (U) (S) Q

N-channel Insulated-Gate Enhancement-
Type, Two-Gate, Five-Terminal Device

(G2) (G1) (D) (U) (S)

P-channel Insulated-Gate, Enhancement-
Type, Two-Gate, Five-Terminal Device

(G2) (G1) (D) (U) (S)

APPENDIX X   Standard Resistance Values (Ohms) for Fixed Resistors*

Tens	Hundreds	Thousands	Ten Thousands	Hundred Thousands	Millions
10	**100**	**1000**	**10K**	**100K**	**1M**
11	110	1100	11K	110K	1.1M
12	**120**	**1200**	**12K**	**120K**	**1.2M**
13	130	1300	13K	130K	1.3M
15	**150**	**1500**	**15K**	**150K**	**1.5M**
16	160	1600	16K	160K	1.6M
18	**180**	**1800**	**18K**	**180K**	**1.8M**
20	200	200	20K	200K	2.0M
22	**220**	**220**	**22K**	**220K**	**2.2M**
24	240	2400	24K	240K	2.4M
27	**270**	**2700**	**27K**	**270K**	**2.7M**
30	300	3000	30K	300K	3.3M
33	**330**	**3300**	**33K**	**330K**	**3.3M**
36	360	3600	36K	360K	3.6M
39	**390**	**3900**	**39K**	**390K**	**3.9M**
43	430	4300	43K	430K	4.3M
47	**470**	**4700**	**47K**	**470K**	**4.7M**
51	510	5100	51K	510K	5.1M
56	**560**	**5600**	**56K**	**560K**	**5.6M**
62	620	6200	62K	620K	6.2M
68	**680**	**6800**	**68K**	**680K**	**6.8M**
75	750	7500	75K	750K	7.5M
82	**820**	**8200**	**82K**	**820K**	**8.2M**
91	910	9100	91K	910K	9.1M

*K = 1000, M = 1,000,000; ±10% values in bold type, ±5% values in standard type.

APPENDIX XI   Resistor Color Code

Color	1st Band (number)	2nd Band (number)	3rd Band (number of zero's added)	4th Band (tolerance)
Black	0	0	No Zero's	
Brown	1	1	0	
Red	2	2	00	
Orange	3	3	000	
Yellow	4	4	0,000	
Green	5	5	00,000	
Blue	6	6	000,000	
Violet	7	7		
Gray	8	8		
White	9	9		
Gold				5%
Silver				10%
No Color				20%

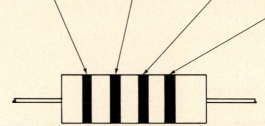

Example: Position group of color bands to left.
Value of resistor determined as shown.

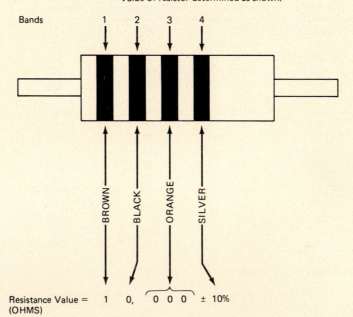

Bands   1   2   3   4

BROWN   BLACK   ORANGE   SILVER

Resistance Value =   1   0,   0 0 0   ± 10%
(OHMS)

APPENDIX XII   Standard Capacitor Values with ±10% Tolerance*

Ceramic Disc						Electrolytics (μF)		
pF*			μF†					
1.0	10	100	0.001	0.01	0.1	1.0	10	100
1.2	12	120	0.0012	0.012	0.12	1.2	12	120
1.5	15	150	0.0015	0.015	0.15	1.5	15	150
1.8	18	180	0.0018	0.018	0.18	1.8	18	180
2.2	22	220	0.0022	0.022	0.22	2.2	22	220
2.7	27	270	0.0027	0.027	0.27	2.7	27	270
3.3	33	330	0.0033	0.033	0.33	3.3	33	330
3.9	39	390	0.0039	0.039	0.39	3.9	39	390
4.7	47	470	0.0047	0.047	0.47	4.7	47	470
5.6	56	560	0.0056	0.056	0.56	5.6	56	560
6.8	68	680	0.0068	0.068	0.68	6.8	68	680
8.2	82	820	0.0082	0.082	0.82	8.2	82	820

*pF $= 10^{-12} = \mu\mu$F or MMF

†μF $= 10^{-6} =$ MF

Note: $\mu\mu$F, MMF, MF and MFD are old notation symbols being replaced by pF and μF.

APPENDIX XIII   Selection Guide for Styles of Cross-Sectioning

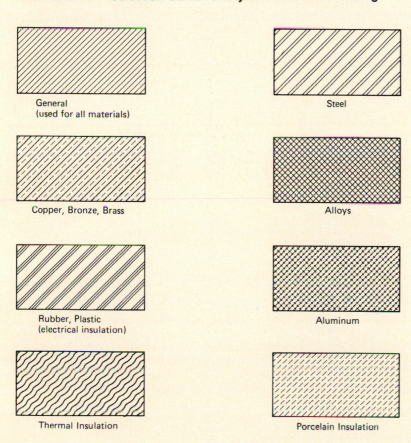

General
(used for all materials)

Steel

Copper, Bronze, Brass

Alloys

Rubber, Plastic
(electrical insulation)

Aluminum

Thermal Insulation

Porcelain Insulation

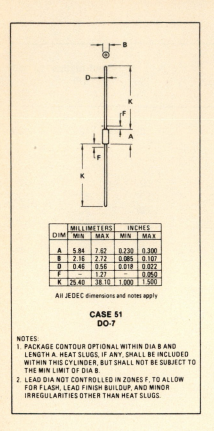

DIM	MILLIMETERS MIN	MAX	INCHES MIN	MAX
A	5.84	7.62	0.230	0.300
B	2.16	2.72	0.085	0.107
D	0.46	0.56	0.018	0.022
F	–	1.27	–	0.050
K	25.40	38.10	1.000	1.500

All JEDEC dimensions and notes apply

CASE 51
DO-7

NOTES:
1. PACKAGE CONTOUR OPTIONAL WITHIN DIA B AND LENGTH A. HEAT SLUGS, IF ANY, SHALL BE INCLUDED WITHIN THIS CYLINDER, BUT SHALL NOT BE SUBJECT TO THE MIN LIMIT OF DIA B.
2. LEAD DIA NOT CONTROLLED IN ZONES F, TO ALLOW FOR FLASH, LEAD FINISH BUILDUP, AND MINOR IRREGULARITIES OTHER THAN HEAT SLUGS.

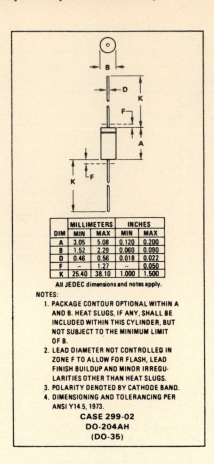

DIM	MILLIMETERS MIN	MAX	INCHES MIN	MAX
A	3.05	5.08	0.120	0.200
B	1.52	2.29	0.060	0.090
D	0.46	0.56	0.018	0.022
F	–	1.27	–	0.050
K	25.40	38.10	1.000	1.500

All JEDEC dimensions and notes apply.

NOTES:
1. PACKAGE CONTOUR OPTIONAL WITHIN A AND B. HEAT SLUGS, IF ANY, SHALL BE INCLUDED WITHIN THIS CYLINDER, BUT NOT SUBJECT TO THE MINIMUM LIMIT OF B.
2. LEAD DIAMETER NOT CONTROLLED IN ZONE F TO ALLOW FOR FLASH, LEAD FINISH BUILDUP AND MINOR IRREGULARITIES OTHER THAN HEAT SLUGS.
3. POLARITY DENOTED BY CATHODE BAND.
4. DIMENSIONING AND TOLERANCING PER ANSI Y14.5, 1973.

CASE 299-02
DO-204AH
(DO-35)

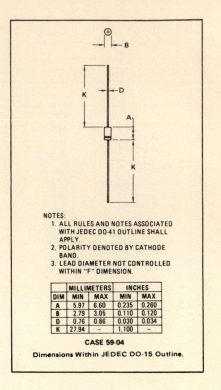

NOTES:
1. ALL RULES AND NOTES ASSOCIATED WITH JEDEC DO-41 OUTLINE SHALL APPLY.
2. POLARITY DENOTED BY CATHODE BAND.
3. LEAD DIAMETER NOT CONTROLLED WITHIN "F" DIMENSION.

DIM	MILLIMETERS MIN	MAX	INCHES MIN	MAX
A	5.97	6.60	0.235	0.260
B	2.79	3.05	0.110	0.120
D	0.76	0.86	0.030	0.034
K	27.94	–	1.100	–

CASE 59-04
Dimensions Within JEDEC DO-15 Outline.

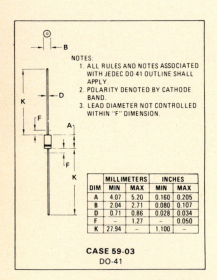

NOTES:
1. ALL RULES AND NOTES ASSOCIATED WITH JEDEC DO 41 OUTLINE SHALL APPLY.
2. POLARITY DENOTED BY CATHODE BAND.
3. LEAD DIAMETER NOT CONTROLLED WITHIN "F" DIMENSION.

DIM	MILLIMETERS MIN	MAX	INCHES MIN	MAX
A	4.07	5.20	0.160	0.205
B	2.04	2.71	0.080	0.107
D	0.71	0.86	0.028	0.034
F	–	1.27	–	0.050
K	27.94	–	1.100	–

CASE 59-03
DO-41

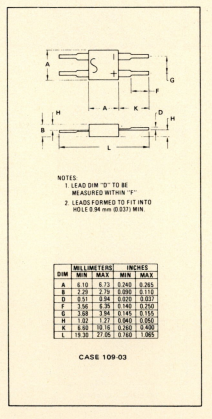

NOTES:
1. LEAD DIM "D" TO BE MEASURED WITHIN "F"
2. LEADS FORMED TO FIT INTO HOLE 0.94 mm (0.037) MIN.

DIM	MILLIMETERS MIN	MAX	INCHES MIN	MAX
A	6.10	6.73	0.240	0.265
B	2.29	2.79	0.090	0.110
D	0.51	0.94	0.020	0.037
F	3.56	6.35	0.140	0.250
G	3.68	3.94	0.145	0.155
H	1.02	1.27	0.040	0.050
K	6.60	10.16	0.260	0.400
L	19.30	27.05	0.760	1.065

CASE 109-03

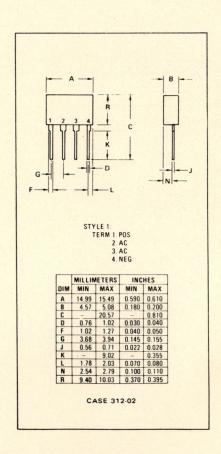

STYLE 1:
TERM 1. POS
2. AC
3. AC
4. NEG

DIM	MILLIMETERS MIN	MAX	INCHES MIN	MAX
A	14.99	15.49	0.590	0.610
B	4.57	5.08	0.180	0.200
C	–	20.57	–	0.810
D	0.76	1.02	0.030	0.040
F	1.02	1.27	0.040	0.050
G	3.68	3.94	0.145	0.155
J	0.56	0.71	0.022	0.028
K	–	9.02	–	0.355
L	1.78	2.03	0.070	0.080
N	2.54	2.79	0.100	0.110
R	9.40	10.03	0.370	0.395

CASE 312-02

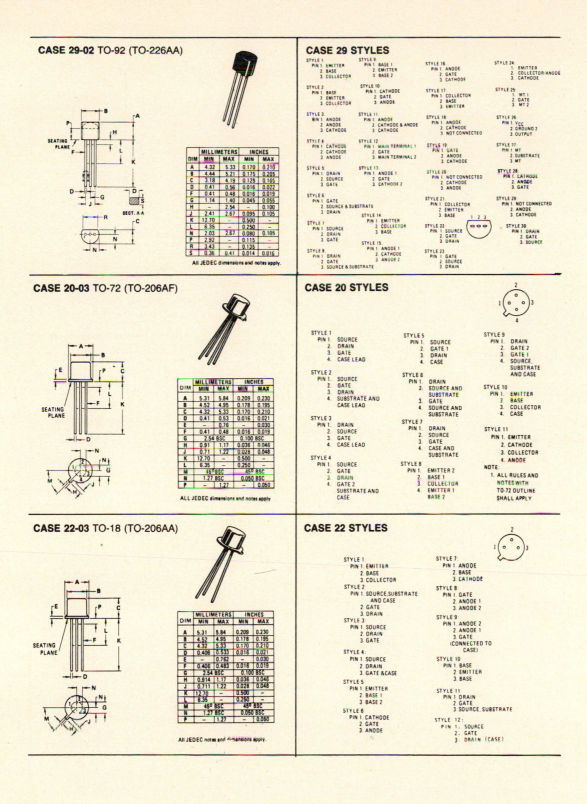

CASE 29-02 TO-92 (TO-226AA)

DIM	MILLIMETERS		INCHES	
	MIN	MAX	MIN	MAX
A	4.32	5.33	0.170	0.210
B	4.44	5.21	0.175	0.205
C	3.18	4.19	0.125	0.165
D	0.41	0.56	0.016	0.022
F	0.41	0.48	0.016	0.019
G	1.14	1.40	0.045	0.055
H	—	2.54	—	0.100
J	2.41	2.67	0.095	0.105
K	12.70	—	0.500	—
L	6.35	—	0.250	—
N	2.03	2.67	0.080	0.105
P	2.92	—	0.115	—
R	3.43	—	0.135	—
S	0.36	0.41	0.014	0.016

All JEDEC dimensions and notes apply.

CASE 29 STYLES

STYLE 1:
PIN 1. EMITTER
2. BASE
3. COLLECTOR

STYLE 2:
PIN 1. BASE
2. EMITTER
3. COLLECTOR

STYLE 3:
PIN 1. ANODE
2. ANODE
3. CATHODE

STYLE 4:
PIN 1. CATHODE
2. CATHODE
3. ANODE

STYLE 5:
PIN 1. DRAIN
2. SOURCE
3. GATE

STYLE 6:
PIN 1. GATE
2. SOURCE & SUBSTRATE
3. DRAIN

STYLE 7:
PIN 1. SOURCE
2. DRAIN
3. GATE

STYLE 8:
PIN 1. DRAIN
2. GATE
3. SOURCE & SUBSTRATE

STYLE 9:
PIN 1. BASE 1
2. EMITTER
3. BASE 2

STYLE 10:
PIN 1. CATHODE
2. GATE
3. ANODE

STYLE 11:
PIN 1. ANODE
2. CATHODE & ANODE
3. CATHODE

STYLE 12:
PIN 1. MAIN TERMINAL 1
2. GATE
3. MAIN TERMINAL 2

STYLE 13:
PIN 1. ANODE 1
2. GATE
3. CATHODE 2

STYLE 14:
PIN 1. EMITTER
2. COLLECTOR
3. BASE

STYLE 15:
PIN 1. ANODE 1
2. CATHODE
3. ANODE 2

STYLE 16:
PIN 1. ANODE
2. GATE
3. CATHODE

STYLE 17:
PIN 1. COLLECTOR
2. BASE
3. EMITTER

STYLE 18:
PIN 1. ANODE
2. CATHODE
3. NOT CONNECTED

STYLE 19:
PIN 1. GATE
2. ANODE
3. CATHODE

STYLE 20:
PIN 1. NOT CONNECTED
2. CATHODE
3. ANODE

STYLE 21:
PIN 1. COLLECTOR
2. EMITTER
3. BASE

STYLE 22:
PIN 1. SOURCE
2. GATE
3. DRAIN

STYLE 23:
PIN 1. GATE
2. SOURCE
3. DRAIN

STYLE 24:
PIN 1. EMITTER
2. COLLECTOR/ANODE
3. CATHODE

STYLE 25:
1. MT 1
2. GATE
3. MT 2

STYLE 26:
PIN 1. VCC
2. GROUND 2
3. OUTPUT

STYLE 27:
PIN 1. MT
2. SUBSTRATE
3. MT

STYLE 28:
PIN 1. CATHODE
2. ANODE
3. GATE

STYLE 29:
PIN 1. NOT CONNECTED
2. ANODE
3. CATHODE

STYLE 30:
PIN 1. DRAIN
2. GATE
3. SOURCE

CASE 20-03 TO-72 (TO-206AF)

DIM	MILLIMETERS		INCHES	
	MIN	MAX	MIN	MAX
A	5.31	5.84	0.209	0.230
B	4.52	4.95	0.178	0.195
C	4.32	5.33	0.170	0.210
D	0.41	0.53	0.016	0.021
E	—	0.76	—	0.030
F	0.41	0.48	0.016	0.019
G	2.54 BSC		0.100 BSC	
H	0.91	1.17	0.036	0.046
J	0.71	1.22	0.028	0.048
K	12.70	—	0.500	—
L	6.35	—	0.250	—
M	45° BSC		45° BSC	
N	1.27 BSC		0.050 BSC	
P	—	1.27	—	0.050

ALL JEDEC dimensions and notes apply

CASE 20 STYLES

STYLE 1:
PIN 1. SOURCE
2. DRAIN
3. GATE
4. CASE LEAD

STYLE 2:
PIN 1. SOURCE
2. GATE
3. DRAIN
4. SUBSTRATE AND CASE LEAD

STYLE 3:
PIN 1. DRAIN
2. SOURCE
3. GATE
4. CASE LEAD

STYLE 4:
PIN 1. SOURCE
2. GATE
3. DRAIN
4. GATE 2 SUBSTRATE AND CASE

STYLE 5:
PIN 1. SOURCE
2. GATE 1
3. DRAIN
4. CASE

STYLE 6:
PIN 1. DRAIN
2. SOURCE AND SUBSTRATE
3. GATE
4. SOURCE AND SUBSTRATE

STYLE 7:
PIN 1. DRAIN
2. SOURCE
3. GATE
4. CASE AND SUBSTRATE

STYLE 8:
PIN 1. EMITTER 2
2. BASE 1
3. COLLECTOR
4. EMITTER 1 BASE 2

STYLE 9:
PIN 1. DRAIN
2. GATE 2
3. GATE 1
4. SOURCE, SUBSTRATE AND CASE

STYLE 10:
PIN 1. EMITTER
2. BASE
3. COLLECTOR
4. CASE

STYLE 11:
PIN 1. EMITTER
2. CATHODE
3. COLLECTOR
4. ANODE

NOTE:
1. ALL RULES AND NOTES WITH TO-72 OUTLINE SHALL APPLY

CASE 22-03 TO-18 (TO-206AA)

DIM	MILLIMETERS		INCHES	
	MIN	MAX	MIN	MAX
A	5.31	5.84	0.209	0.230
B	4.52	4.95	0.178	0.195
C	4.32	5.33	0.170	0.210
D	0.406	0.533	0.016	0.021
E	—	0.762	—	0.030
F	0.406	0.483	0.016	0.019
G	2.54 BSC		0.100 BSC	
H	0.914	1.17	0.036	0.046
J	0.711	1.22	0.028	0.048
K	12.70	—	0.500	—
L	6.35	—	0.250	—
M	45° BSC		45° BSC	
N	1.27 BSC		0.050 BSC	
P	—	1.27	—	0.050

All JEDEC notes and dimensions apply.

CASE 22 STYLES

STYLE 1:
PIN 1. EMITTER
2. BASE
3. COLLECTOR

STYLE 2:
PIN 1. SOURCE, SUBSTRATE AND CASE
2. GATE
3. DRAIN

STYLE 3:
PIN 1. SOURCE
2. DRAIN
3. GATE

STYLE 4:
PIN 1. SOURCE
2. DRAIN
3. GATE & CASE

STYLE 5:
PIN 1. EMITTER
2. BASE 1
3. BASE 2

STYLE 6:
PIN 1. CATHODE
2. GATE
3. ANODE

STYLE 7:
PIN 1. ANODE
2. BASE
3. CATHODE

STYLE 8:
PIN 1. GATE
2. ANODE 1
3. ANODE 2

STYLE 9:
PIN 1. ANODE 2
2. ANODE 1
3. GATE (CONNECTED TO CASE)

STYLE 10:
PIN 1. BASE
2. EMITTER
3. BASE

STYLE 11:
PIN 1. DRAIN
2. GATE
3. SOURCE, SUBSTRATE

STYLE 12:
PIN 1. SOURCE
2. GATE
3. DRAIN (CASE)

APPENDIX XIV (continued)

CASE 1-03
TO-204AA
(TO-3)
Metal Package
$R_{\theta JA}$ = 45° C/W(Typ)
$R_{\theta JC}$ = See Data Sheet

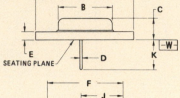
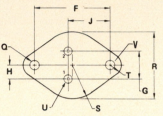

DIM	MILLIMETERS		INCHES	
	MIN	MAX	MIN	MAX
B	–	22.23	–	0.875
C	6.35	11.43	0.250	0.450
D	0.97	1.09	0.038	0.043
E	–	3.43	–	0.135
F	30.15 BSC		1.187 BSC	
G	10.92 BSC		0.430 BSC	
H	5.46 BSC		0.215 BSC	
J	16.89 BSC		0.665 BSC	
K	7.92	–	0.312	–
Q	3.84	4.09	0.151	0.161
S	–	13.34	–	0.525
T	–	4.78	–	0.188
V	3.84	4.09	0.151	0.161

CASE 29-02
TO-226AA
(TO-92)
Plastic Package
$R_{\theta JA}$ = 200° C/W

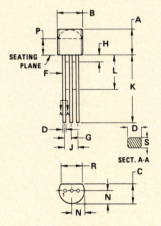

DIM	MILLIMETERS		INCHES	
	MIN	MAX	MIN	MAX
A	4.32	5.33	0.170	0.210
B	4.44	5.21	0.175	0.205
C	3.18	4.19	0.125	0.165
D	0.41	0.56	0.016	0.022
F	0.41	0.48	0.016	0.019
G	1.14	1.40	0.045	0.055
H	–	2.54	–	0.100
J	2.41	2.67	0.095	0.105
K	12.70	–	0.500	–
L	6.35	–	0.250	–
N	2.03	2.67	0.080	0.105
P	2.92	–	0.115	–
R	3.43	–	0.135	–
S	0.36	0.41	0.014	0.016

CASE 79-02
TO-205AD
(TO-39)
Metal Package
$R_{\theta JA}$ = 185° C/W(Typ)

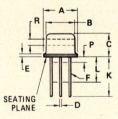

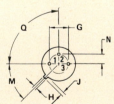

DIM	MILLIMETERS		INCHES	
	MIN	MAX	MIN	MAX
A	8.89	9.40	0.350	0.370
B	8.00	8.51	0.315	0.335
C	6.10	6.60	0.240	0.260
D	0.406	0.533	0.016	0.021
E	0.229	3.18	0.009	0.125
F	0.406	0.483	0.016	0.019
G	4.83	5.33	0.190	0.210
H	0.711	0.864	0.028	0.034
J	0.737	1.02	0.029	0.040
K	12.70	–	0.500	–
L	6.35	–	0.250	–
M	45° NOM		45° NOM	
P	–	1.27	–	0.050
Q	90° NOM		90° NOM	
R	2.54	–	0.100	–

CASE 80-02
TO-213AA
(TO-66)
Metal Package
$R_{\theta JA} = 45°C/W(Typ)$
$R_{\theta JC} =$ See Data Sheet

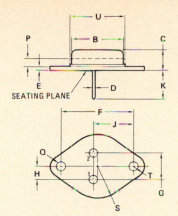

DIM	MILLIMETERS		INCHES	
	MIN	MAX	MIN	MAX
B	11.94	12.70	0.470	0.500
C	6.35	8.64	0.250	0.340
D	0.71	0.86	0.028	0.034
E	1.27	1.91	0.050	0.075
F	24.33	24.43	0.958	0.962
G	4.83	5.33	0.190	0.210
H	2.41	2.67	0.095	0.105
J	14.48	14.99	0.570	0.590
K	9.14	–	0.360	–
P	–	1.27	–	0.050
Q	3.61	3.86	0.142	0.152
S		8.89		0.350
T	–	3.68°	–	0.145
U	–	15.75	–	0.620

CASE 221A-02
TO-220AB
Plastic Power
$R_{\theta JA} = 65° C/W(Typ)$
$R_{\theta JC} =$ See Data Sheet

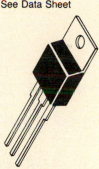

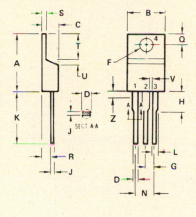

DIM	MILLIMETERS		INCHES	
	MIN	MAX	MIN	MAX
A	15.11	15.75	0.595	0.620
B	9.65	10.29	0.380	0.405
C	4.06	4.82	0.160	0.190
D	0.64	0.89	0.025	0.035
F	3.61	3.73	0.142	0.147
G	2.41	2.67	0.095	0.105
H	2.79	3.30	0.110	0.130
J	0.36	0.56	0.014	0.022
K	12.70	14.27	0.500	0.562
L	1.14	1.27	0.045	0.050
N	4.83	5.33	0.190	0.210
Q	2.54	3.04	0.100	0.120
R	2.04	2.79	0.080	0.110
S	1.14	1.39	0.045	0.055
T	5.97	6.48	0.235	0.255
U	0.76	1.27	0.030	0.050
V	1.14	.	0.045	.
Z	–	2.03	–	0.080

CASE 601-04
Metal Package
$R_{\theta JA} = 160° C/W(Typ)$

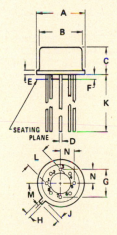

DIM	MILLIMETERS		INCHES	
	MIN	MAX	MIN	MAX
A	8.51	9.40	0.335	0.370
B	7.75	8.51	0.305	0.335
C	4.19	4.70	0.165	0.185
D	0.41	0.48	0.016	0.019
E	0.25	1.02	0.010	0.040
F	0.25	1.02	0.010	0.040
G	5.08 BSC		0.200 BSC	
H	0.71	0.86	0.028	0.034
J	0.74	1.14	0.029	0.045
K	12.70	–	0.500	–
L	3.05	4.06	0.120	0.160
M	45° BSC		45° BSC	
N	2.41	2.67	0.095	0.105

CASE 603-04
TO-100
Metal Can
$R_{\theta JA} = 160° C/W$

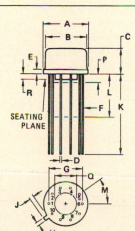

DIM	MILLIMETERS		INCHES	
	MIN	MAX	MIN	MAX
A	8.51	9.39	0.335	0.370
B	7.75	8.51	0.305	0.335
C	4.19	4.70	0.165	0.185
D	0.407	0.533	0.016	0.021
E	–	1.02	–	0.040
F	0.406	0.483	0.016	0.019
G	5.84 BSC		0.230 BSC	
H	0.712	0.864	0.028	0.034
J	0.737	1.14	0.029	0.045
K	12.70	–	0.500	–
L	6.35	12.70	0.250	0.500
M	36° BSC		36° BSC	
P	–	1.27	–	0.050
Q	3.56	4.06	0.140	0.160
R	0.254	1.02	0.010	0.040

APPENDIX XIV (continued)

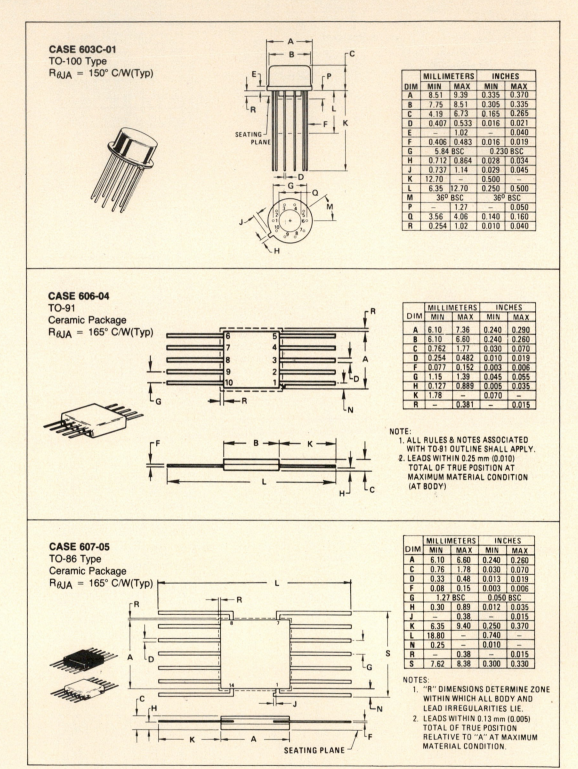

CASE 603C-01
TO-100 Type
$R_{\theta JA} = 150°$ C/W(Typ)

	MILLIMETERS		INCHES	
DIM	MIN	MAX	MIN	MAX
A	8.51	9.39	0.335	0.370
B	7.75	8.51	0.305	0.335
C	4.19	6.73	0.165	0.265
D	0.407	0.533	0.016	0.021
E	–	1.02	–	0.040
F	0.406	0.483	0.016	0.019
G	5.84 BSC		0.230 BSC	
H	0.712	0.864	0.028	0.034
J	0.737	1.14	0.029	0.045
K	12.70	–	0.500	–
L	6.35	12.70	0.250	0.500
M	36° BSC		36° BSC	
P	–	1.27	–	0.050
Q	3.56	4.06	0.140	0.160
R	0.254	1.02	0.010	0.040

CASE 606-04
TO-91
Ceramic Package
$R_{\theta JA} = 165°$ C/W(Typ)

	MILLIMETERS		INCHES	
DIM	MIN	MAX	MIN	MAX
A	6.10	7.36	0.240	0.290
B	6.10	6.60	0.240	0.260
C	0.762	1.77	0.030	0.070
D	0.254	0.482	0.010	0.019
F	0.077	0.152	0.003	0.006
G	1.15	1.39	0.045	0.055
H	0.127	0.889	0.005	0.035
K	1.78	–	0.070	–
R	–	0.381	–	0.015

NOTE:
1. ALL RULES & NOTES ASSOCIATED WITH TO-91 OUTLINE SHALL APPLY.
2. LEADS WITHIN 0.25 mm (0.010) TOTAL OF TRUE POSITION AT MAXIMUM MATERIAL CONDITION (AT BODY)

CASE 607-05
TO-86 Type
Ceramic Package
$R_{\theta JA} = 165°$ C/W(Typ)

	MILLIMETERS		INCHES	
DIM	MIN	MAX	MIN	MAX
A	6.10	6.60	0.240	0.260
C	0.76	1.78	0.030	0.070
D	0.33	0.48	0.013	0.019
F	0.08	0.15	0.003	0.006
G	1.27 BSC		0.050 BSC	
H	0.30	0.89	0.012	0.035
J	–	0.38	–	0.015
K	6.35	9.40	0.250	0.370
L	18.80	–	0.740	–
N	0.25	–	0.010	–
R	–	0.38	–	0.015
S	7.62	8.38	0.300	0.330

NOTES:
1. "R" DIMENSIONS DETERMINE ZONE WITHIN WHICH ALL BODY AND LEAD IRREGULARITIES LIE.
2. LEADS WITHIN 0.13 mm (0.005) TOTAL OF TRUE POSITION RELATIVE TO "A" AT MAXIMUM MATERIAL CONDITION.

CASE 614-02
(TO-66 Type)
Metal Package
$R_{\theta JA} = 35°$ C/W(Typ)
$R_{\theta JC}$ = See Data Sheet

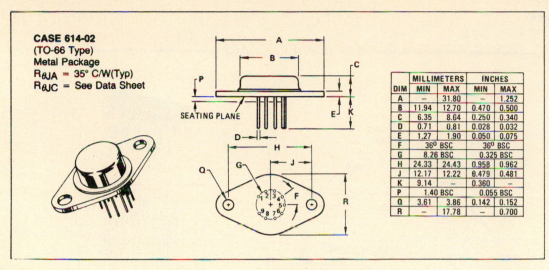

SEATING PLANE

DIM	MILLIMETERS		INCHES	
	MIN	MAX	MIN	MAX
A	–	31.80	–	1.252
B	11.94	12.70	0.470	0.500
C	6.35	8.64	0.250	0.340
D	0.71	0.81	0.028	0.032
E	1.27	1.90	0.050	0.075
F	36° BSC		36° BSC	
G	8.26 BSC		0.325 BSC	
H	24.33	24.43	0.958	0.962
J	12.17	12.22	0.479	0.481
K	9.14	–	0.360	–
P	1.40 BSC		0.055 BSC	
Q	3.61	3.86	0.142	0.152
R	–	17.78	–	0.700

CASE 607-04 CERAMIC

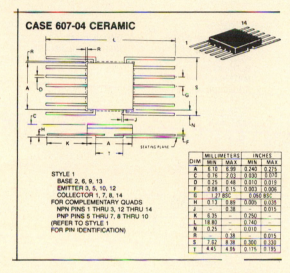

SEATING PLANE

STYLE 1
BASE 2, 6, 9, 13
EMITTER 3, 5, 10, 12
COLLECTOR 1, 7, 8, 14
FOR COMPLEMENTARY QUADS
NPN PINS 1 THRU 3, 12 THRU 14
PNP PINS 5 THRU 7, 8 THRU 10
(REFER TO STYLE 1
FOR PIN IDENTIFICATION)

DIM	MILLIMETERS		INCHES	
	MIN	MAX	MIN	MAX
A	6.10	6.99	0.240	0.275
C	0.76	2.03	0.030	0.070
D	0.25	0.48	0.010	0.019
F	0.08	0.15	0.003	0.006
G	1.27 BSC		0.050 BSC	
H	0.13	0.89	0.005	0.035
J	–	0.38	–	0.015
K	6.35	–	0.250	–
L	18.80	–	0.740	–
N	0.25	–	0.010	–
R	–	0.38	–	0.015
S	7.62	8.38	0.300	0.330
T	4.45	4.95	0.175	0.195

■ 8-PIN PACKAGE ■

PLASTIC PACKAGE
CASE 626-04

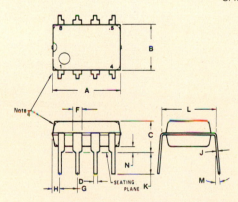

SEATING PLANE

NOTES:
1. LEAD POSITIONAL TOLERANCE:
   ⊕ ∅ 0.13 (0.005) Ⓜ T A Ⓜ B Ⓜ
2. DIMENSION "L" TO CENTER OF
   LEADS WHEN FORMED PARALLEL.
3. PACKAGE CONTOUR OPTIONAL
   (ROUND OR SQUARE CORNERS).
4. DIMENSIONS A AND B ARE DATUMS.
5. DIMENSIONING AND TOLERANCING
   PER ANSI Y14.5, 1973.

DIM	MILLIMETERS		INCHES	
	MIN	MAX	MIN	MAX
A	9.40	10.16	0.370	0.400
B	6.10	6.60	0.240	0.260
C	3.94	4.45	0.155	0.175
D	0.38	0.51	0.015	0.020
F	1.02	1.52	0.040	0.060
G	2.54 BSC		0.100 BSC	
H	0.76	1.27	0.030	0.050
J	0.20	0.30	0.008	0.012
K	2.92	3.43	0.115	0.135
L	7.62 BSC		0.300 BSC	
M	–	10°	–	10°
N	0.51	0.76	0.020	0.030

APPENDIX XIV (continued)

14-PIN PACKAGES

PLASTIC PACKAGE
CASE 646

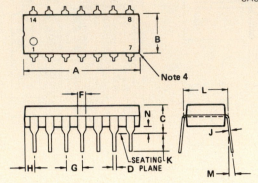

NOTES:
1. LEADS WITHIN 0.13 mm (0.005) RADIUS OF TRUE POSITION AT SEATING PLANE AT MAXIMUM MATERIAL CONDITION.
2. DIMENSION "L" TO CENTER OF LEADS WHEN FORMED PARALLEL.
3. DIMENSION "B" DOES NOT INCLUDE MOLD FLASH.
4. ROUNDED CORNERS OPTIONAL.

DIM	MILLIMETERS		INCHES	
	MIN	MAX	MIN	MAX
A	18.16	19.56	0.715	0.770
B	6.10	6.60	0.240	0.260
C	4.06	5.08	0.160	0.200
D	0.38	0.53	0.015	0.021
F	1.02	1.78	0.040	0.070
G	2.54 BSC		0.100 BSC	
H	1.32	2.41	0.052	0.095
J	0.20	0.38	0.008	0.015
K	2.92	3.43	0.115	0.135
L	7.62 BSC		0.300 BSC	
M	0°	10°	0°	10°
N	0.51	1.02	0.020	0.040

CERDIP PACKAGE
CASE 632

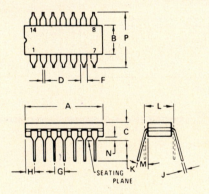

NOTES:
1. ALL RULES AND NOTES ASSOCIATED WITH MO-001 AA OUTLINE SHALL APPLY.
2. DIMENSION "L" TO CENTER OF LEADS WHEN FORMED PARALLEL.
3. DIMENSION "A" AND "B" (632-07) DO NOT INCLUDE GLASS RUN OUT.
4. LEADS WITHIN 0.25 mm (0.010) DIA OF TRUE POSITION AT SEATING PLANE AND MAXIMUM MATERIAL CONDITION.

DIM	MILLIMETERS		INCHES	
	MIN	MAX	MIN	MAX
A	19.05	19.94	0.750	0.785
B	6.10	7.49	0.240	0.295
C	–	5.08	–	0.200
D	0.38	0.58	0.015	0.023
F	1.40	1.77	0.055	0.070
G	2.54 BSC		0.100 BSC	
H	1.91	2.29	0.075	0.090
J	0.20	0.38	0.008	0.015
K	3.18	4.32	0.125	0.170
L	7.62 BSC		0.300 BSC	
M	–	15°	–	15°
N	0.51	1.02	0.020	0.040

16-PIN PACKAGES

PLASTIC PACKAGE
CASE 648

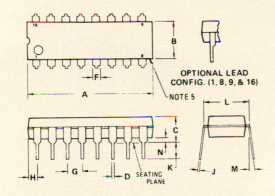

OPTIONAL LEAD
CONFIG. (1, 8, 9, & 16)

NOTE 5

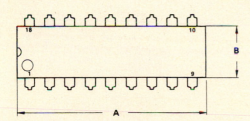

SEATING
PLANE

	MILLIMETERS		INCHES	
DIM	MIN	MAX	MIN	MAX
A	18.80	21.34	0.740	0.840
B	6.10	6.60	0.240	0.260
C	4.06	5.08	0.160	0.200
D	0.38	0.53	0.015	0.021
F	1.02	1.78	0.040	0.070
G	2.54 BSC		0.100 BSC	
H	0.38	2.41	0.015	0.095
J	0.20	0.38	0.008	0.015
K	2.92	3.43	0.115	0.135
L	7.62 BSC		0.300 BSC	
M	0°	10°	0°	10°
N	0.51	1.02	0.020	0.040

NOTES:
1. LEADS WITHIN 0.13 mm (0.005) RADIUS OF TRUE POSITION AT SEATING PLANE AT MAXIMUM MATERIAL CONDITION.
2. DIMENSION "L" TO CENTER OF LEADS WHEN FORMED PARALLEL.
3. DIMENSION "B" DOES NOT INCLUDE MOLD FLASH.
4. "F" DIMENSION IS FOR FULL LEADS. "HALF" LEADS ARE OPTIONAL AT LEAD POSITIONS 1, 8, 9, and 16).
5. ROUNDED CORNERS OPTIONAL.

18-PIN PACKAGES

PLASTIC PACKAGE
CASE 707

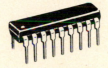

SEATING PLANE

	MILLIMETERS		INCHES	
DIM	MIN	MAX	MIN	MAX
A	22.22	23.24	0.875	0.915
B	6.10	6.60	0.240	0.260
C	3.56	4.57	0.140	0.180
D	0.36	0.56	0.014	0.022
F	1.27	1.78	0.050	0.070
G	2.54 BSC		0.100 BSC	
H	1.02	1.52	0.040	0.060
J	0.20	0.30	0.008	0.012
K	2.92	3.43	0.115	0.135
L	7.62 BSC		0.300 BSC	
M	0°	15°	0°	15°
N	0.51	1.02	0.020	0.040

NOTES:
1. POSITIONAL TOLERANCE OF LEADS (D), SHALL BE WITHIN 0.25mm(0.010) AT MAXIMUM MATERIAL CONDITION, IN RELATION TO SEATING PLANE AND EACH OTHER.
2. DIMENSION L TO CENTER OF LEADS WHEN FORMED PARALLEL.
3. DIMENSION B DOES NOT INCLUDE MOLD FLASH.

APPENDIX XIV   (continued)

20-PIN PACKAGES

PLASTIC PACKAGE
CASE 738

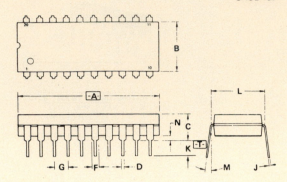

NOTES:
1. DIM -A- IS DATUM.
2. POSITIONAL TOL FOR LEADS;
   ⊕ Ø 0.25 (0.010) Ⓜ T AⓂ
3. -T- IS SEATING PLANE.
4. DIM "B" DOES NOT INCLUDE MOLD FLASH.
5. DIM -L- TO CENTER OF LEADS WHEN
   FORMED PARALLEL.
6. DIMENSIONING AND TOLERANCING
   PER ANSI Y14.5, 1973.

DIM	MILLIMETERS		INCHES	
	MIN	MAX	MIN	MAX
A	25.65	27.18	1.010	1.070
B	6.10	6.60	0.240	0.260
C	3.94	4.57	0.155	0.180
D	0.38	0.56	0.015	0.022
F	1.27	1.78	0.050	0.070
G	2.54 BSC		0.100 BSC	
J	0.20	0.38	0.008	0.015
K	2.79	3.56	0.110	0.140
L	7.62 BSC		0.300 BSC	
M	0°	15°	0°	15°
N	0.51	1.02	0.020	0.040

PLASTIC PACKAGE
CASE 708

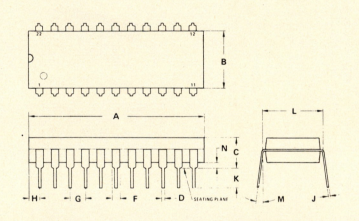

NOTES:
1. POSITIONAL TOLERANCE OF LEADS (D),
   SHALL BE WITHIN 0.25mm(0.010) AT
   MAXIMUM MATERIAL CONDITION, IN
   RELATION TO SEATING PLANE AND
   EACH OTHER.
2. DIMENSION L TO CENTER OF LEADS
   WHEN FORMED PARALLEL.
3. DIMENSION B DOES NOT INCLUDE
   MOLD FLASH.

DIM	MILLIMETERS		INCHES	
	MIN	MAX	MIN	MAX
A	27.56	28.32	1.085	1.115
B	8.64	9.14	0.340	0.360
C	3.94	5.08	0.155	0.200
D	0.36	0.56	0.014	0.022
F	1.27	1.78	0.050	0.070
G	2.54 BSC		0.100 BSC	
H	1.02	1.52	0.040	0.060
J	0.20	0.38	0.008	0.015
K	2.92	3.43	0.115	0.135
L	10.16 BSC		0.400 BSC	
M	0°	15°	0°	15°
N	0.51	1.02	0.020	0.040

CASE 708-04

PLASTIC PACKAGE
CASE 724

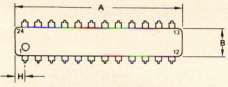

NOTE:
1. LEADS, TRUE POSITIONED WITHIN 0.25 mm (0.010) DIA AT SEATING PLANE AT MAXIMUM MATERIAL CONDITION (DIM D).

DIM	MILLIMETERS		INCHES	
	MIN	MAX	MIN	MAX
A	31.24	32.13	1.230	1.265
B	6.35	6.86	0.250	0.270
C	4.06	4.57	0.160	0.180
D	0.38	0.51	0.015	0.020
F	1.02	1.52	0.040	0.060
G	2.54 BSC		0.100 BSC	
H	1.60	2.11	0.063	0.083
J	0.18	0.30	0.007	0.012
K	2.92	3.43	0.115	0.135
L	7.37	7.87	0.290	0.310
M	–	10°	–	10°
N	0.51	1.02	0.020	0.040

CERDIP PACKAGE
CASE 758

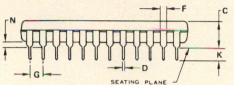

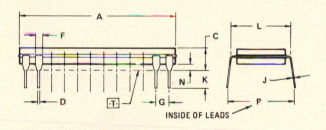

INSIDE OF LEADS

NOTES:
1. DIMENSION A IS DATUM.
2. POSITIONAL TOLERANCE FOR LEADS: 24 PLACES
   [⊕ |0.25 (0.010) Ⓜ| T | A Ⓜ|]
3. -T- IS SEATING PLANE.
4. DIMENSION L TO CENTER OF LEADS WHEN FORMED PARALLEL.
5. DIMENSIONING AND TOLERANCING PER ANSI Y14.5, 1973.

DIM	MILLIMETERS		INCHES	
	MIN	MAX	MIN	MAX
A	31.50	32.64	1.240	1.285
B	7.24	7.75	0.285	0.305
C	3.68	4.44	0.145	0.175
D	0.38	0.53	0.015	0.021
F	1.14	1.57	0.045	0.062
G	2.54 BSC		0.100 BSC	
J	0.20	0.33	0.008	0.013
K	2.54	4.19	0.100	0.165
L	7.62	7.87	0.300	0.310
N	0.51	1.27	0.020	0.050
P	9.14	10.16	0.360	0.400

Case 623-05
24-Pin Ceramic Dual In-Line

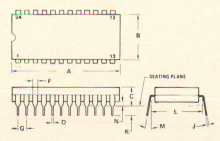

SEATING PLANE

NOTES:
1. DIM "L" TO CENTER OF LEADS WHEN FORMED PARALLEL.
2. LEADS WITHIN 0.13 mm (0.005) RADIUS OF TRUE POSITION AT SEATING PLANE AT MAXIMUM MATERIAL CONDITION. (WHEN FORMED PARALLEL).

DIM	MILLIMETERS		INCHES	
	MIN	MAX	MIN	MAX
A	31.24	32.77	1.230	1.290
B	12.70	15.49	0.500	0.610
C	4.06	5.59	0.160	0.220
D	0.41	0.51	0.016	0.020
F	1.27	1.52	0.050	0.060
G	2.54 BSC		0.100 BSC	
J	0.20	0.30	0.008	0.012
K	3.18	4.06	0.125	0.160
L	15.24 BSC		0.600 BSC	
M	0°	15°	0°	15°
N	0.51	1.27	0.020	0.050

CASE 623-05

387

APPENDIX XIV   (continued)

28-PIN PACKAGES

CERAMIC PACKAGE
CASE 719-03

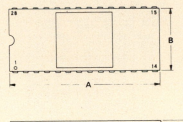

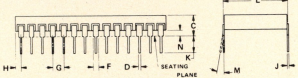

NOTES:
1. LEADS, TRUE POSITIONED WITHIN 0.25 mm (0.010) DIAMETER (AT SEATING PLANE) AT MAXIMUM MATERIAL CONDITION.
2. DIMENSION "L" TO CENTER OF LEADS WHEN FORMED PARALLEL.

DIM	MILLIMETERS		INCHES	
	MIN	MAX	MIN	MAX
A	35.20	35.92	1.386	1.414
B	14.73	15.34	0.580	0.604
C	3.05	4.19	0.120	0.165
D	0.38	0.53	0.015	0.021
F	0.76	1.40	0.030	0.055
G	2.54 BSC		0.100 BSC	
H	0.76	1.78	0.030	0.070
J	0.20	0.30	0.008	0.012
K	2.54	4.19	0.100	0.165
L	14.99	15.49	0.590	0.610
M	–	10°	–	10°
N	0.51	1.52	0.020	0.060

CERAMIC PACKAGE
CASE 719-04

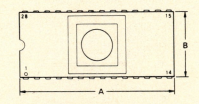

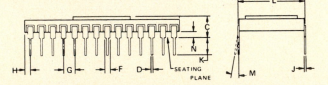

NOTES:
1. LEADS, TRUE POSITIONED WITHIN 0.25 mm (0.010) DIAMETER (AT SEATING PLANE) AT MAXIMUM MATERIAL CONDITION.
2. DIMENSION "L" TO CENTER OF LEADS WHEN FORMED PARALLEL.

DIM	MILLIMETERS		INCHES	
	MIN	MAX	MIN	MAX
A	35.20	35.92	1.386	1.414
B	14.73	15.34	0.580	0.604
C	3.18	5.08	0.125	0.200
D	0.38	0.53	0.015	0.021
F	0.76	1.40	0.030	0.055
G	2.54 BSC		0.100 BSC	
H	0.76	1.78	0.030	0.070
J	0.20	0.30	0.008	0.012
K	2.54	4.57	0.100	0.180
L	14.99	15.49	0.590	0.610
M	–	10°	–	10°
N	0.51	1.52	0.020	0.060

28-PIN PACKAGES (Continued)

PLASTIC PACKAGE
CASE 710-02

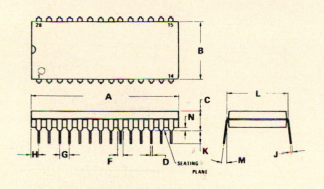

NOTES:
1. POSITIONAL TOLERANCE OF LEADS (D), SHALL BE WITHIN 0.25mm(0.010) AT MAXIMUM MATERIAL CONDITION, IN RELATION TO SEATING PLANE AND EACH OTHER.
2. DIMENSION L TO CENTER OF LEADS WHEN FORMED PARALLEL.
3. DIMENSION B DOES NOT INCLUDE MOLD FLASH.

DIM	MILLIMETERS		INCHES	
	MIN	MAX	MIN	MAX
A	36.45	37.21	1.435	1.465
B	13.72	14.22	0.540	0.560
C	3.94	5.08	0.155	0.200
D	0.36	0.56	0.014	0.022
F	1.02	1.52	0.040	0.060
G	2.54 BSC		0.100 BSC	
H	1.65	2.16	0.065	0.085
J	0.20	0.38	0.008	0.015
K	2.92	3.43	0.115	0.135
L	15.24 BSC		0.600 BSC	
M	0°	15°	0°	15°
N	0.51	1.02	0.020	0.040

CERPID PACKAGE
CASE 733-03

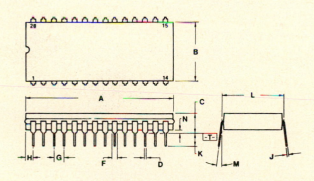

NOTES:
1. DIM -A- IS DATUM.
2. POSITIONAL TOL FOR LEADS:
   $\oplus$ | $\emptyset$ 0.25 (0.010) Ⓜ | T | A Ⓜ |
3. -T- IS SEATING PLANE.
4. DIM A AND B INCLUDES MENISCUS.
5. DIM -L- TO CENTER OF LEADS WHEN FORMED PARALLEL.
6. DIMENSIONING AND TOLERANCING PER ANSI Y14.5, 1973.

DIM	MILLIMETERS		INCHES	
	MIN	MAX	MIN	MAX
A	36.45	37.85	1.435	1.490
B	12.70	15.37	0.500	0.605
C	4.06	5.84	0.160	0.230
D	0.38	0.56	0.015	0.022
F	1.27	1.65	0.050	0.065
G	2.54 BSC		0.100 BSC	
J	0.20	0.30	0.008	0.012
K	3.18	4.06	0.125	0.160
L	15.24 BSC		0.600 BSC	
M	5°	15°	5°	15°
N	0.51	1.27	0.020	0.050

PLASTIC PACKAGE
CASE 711-03

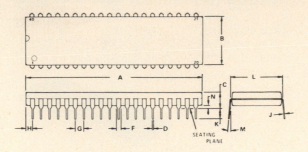

NOTES:
1. POSITIONAL TOLERANCE OF LEADS (D), SHALL BE WITHIN 0.25 mm (0.010) AT MAXIMUM MATERIAL CONDITION, IN RELATION TO SEATING PLANE AND EACH OTHER.
2. DIMENSION L TO CENTER OF LEADS WHEN FORMED PARALLEL.
3. DIMENSION B DOES NOT INCLUDE MOLD FLASH.

DIM	MILLIMETERS		INCHES	
	MIN	MAX	MIN	MAX
A	51.69	52.45	2.035	2.065
B	13.72	14.22	0.540	0.560
C	3.94	5.08	0.155	0.200
D	0.36	0.56	0.014	0.022
F	1.02	1.52	0.040	0.060
G	2.54 BSC		0.100 BSC	
H	1.65	2.16	0.065	0.085
J	0.20	0.38	0.008	0.015
K	2.92	3.43	0.115	0.135
L	15.24 BSC		0.600 BSC	
M	0⁰	15⁰	0⁰	15⁰
N	0.51	1.02	0.020	0.040

CHIP CARRIER
CASE 761-01

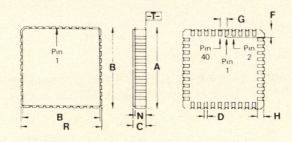

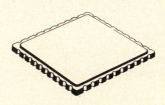

NOTES:
1. DIMENSIONS A & R ARE DATUMS.
2. -T- IS GAUGE PLANE.
3. POSITIONAL TOLERANCE FOR TERMINALS (D): 40 PLACES:
$\oplus$ 0.25 (0.010) Ⓜ T AⓈ RⓈ
4. DIMENSIONING AND TOLERANCING PER ANSI Y14.5, 1973.

DIM	MILLIMETERS		INCHES	
	MIN	MAX	MIN	MAX
A	11.94	12.57	0.470	0.495
B	11.05	11.30	0.435	0.445
C	1.60	2.08	0.063	0.082
D	0.33	0.69	0.013	0.027
F	1.07	1.47	0.042	0.058
G	1.02 BSC		0.040 BSC	
H	0.84	1.19	0.033	0.047
N	1.27	1.79	0.050	0.070
R	11.94	12.57	0.470	0.495

CHIP CARRIER
CASE 753

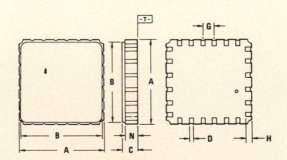

NOTES:
1. DIMENSION A IS DATUM. (2 PLACES)
2. -T- IS GUAGE PLANE.
3. POSITIONAL TOLERANCE FOR TERMINALS (D): 24 PLACES
$\oplus$ 0.25 (0.010) Ⓜ T AⓈ
4. DIMENSIONING AND TOLERANCING PER ANSI Y14.5, 1973.

DIM	MILLIMETERS		INCHES	
	MIN	MAX	MIN	MAX
A	9.91	10.41	0.390	0.410
B	9.78	9.90	0.385	0.390
C	1.63	1.93	0.064	0.076
D	0.39	0.63	0.015	0.025
G	1.27 BSC		0.050 BSC	
H	0.77	1.01	0.030	0.040
N	1.40	1.65	0.055	0.065

CERAMIC PACKAGE
CASE 740

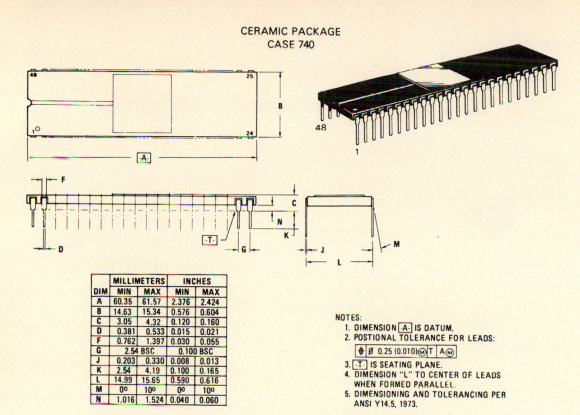

DIM	MILLIMETERS		INCHES	
	MIN	MAX	MIN	MAX
A	60.35	61.57	2.376	2.424
B	14.63	15.34	0.576	0.604
C	3.05	4.32	0.120	0.160
D	0.381	0.533	0.015	0.021
F	0.762	1.397	0.030	0.055
G	2.54 BSC		0.100 BSC	
J	0.203	0.330	0.008	0.013
K	2.54	4.19	0.100	0.165
L	14.99	15.65	0.590	0.616
M	0°	10°	0°	10°
N	1.016	1.524	0.040	0.060

NOTES:
1. DIMENSION -A- IS DATUM.
2. POSITIONAL TOLERANCE FOR LEADS:
   ⊕ Ø 0.25 (0.010) Ⓜ T A Ⓜ
3. -T- IS SEATING PLANE.
4. DIMENSION "L" TO CENTER OF LEADS WHEN FORMED PARALLEL.
5. DIMENSIONING AND TOLERANCING PER ANSI Y14.5, 1973.

CERAMIC PACKAGE
CASE 746

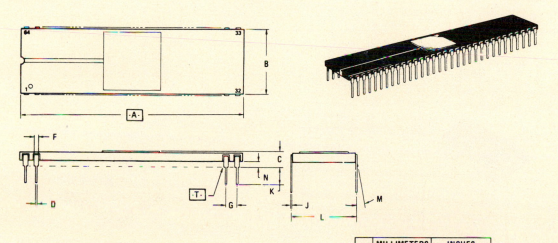

NOTES:
1. DIMENSION -A- IS DATUM.
2. POSITIONAL TOLERANCE FOR LEADS:
   ⊕ 0.25 (0.010) Ⓜ T A Ⓜ
3. -T- IS SEATING PLANE.
4. DIMENSION "L" TO CENTER OF LEADS WHEN FORMED PARALLEL.
5. DIMENSIONING AND TOLERANCING PER ANSI Y14.5, 1973.

DIM	MILLIMETERS		INCHES	
	MIN	MAX	MIN	MAX
A	80.52	82.04	3.170	3.230
B	22.25	22.96	0.876	0.904
C	3.05	4.32	0.120	0.170
D	0.38	0.53	0.015	0.021
F	0.76	1.40	0.030	0.055
G	2.54 BSC		0.100 BSC	
J	0.20	0.33	0.008	0.013
K	2.54	4.19	0.100	0.165
L	22.61	23.11	0.890	0.910
M	—	10°	—	10°
N	1.02	1.52	0.040	0.060

APPENDIX XIV (continued)

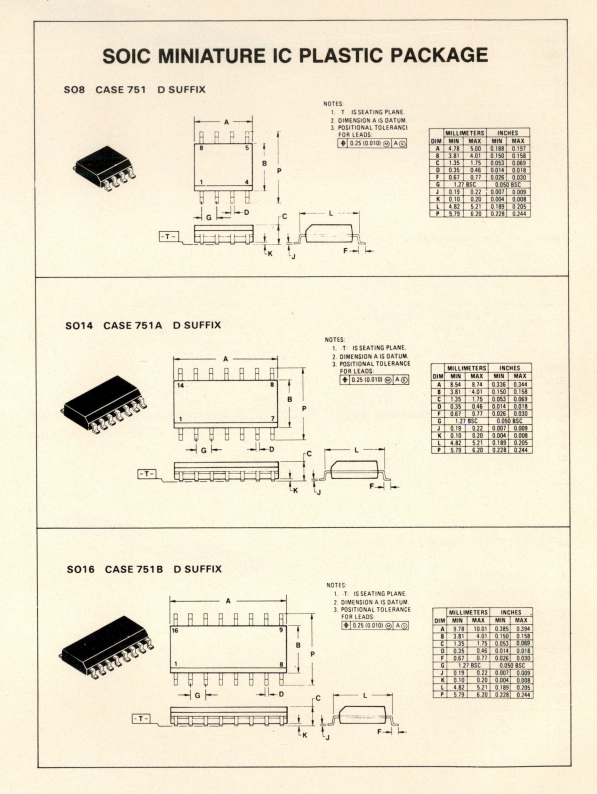

SOIC MINIATURE IC PLASTIC PACKAGE

SO8 CASE 751 D SUFFIX

NOTES:
1. T IS SEATING PLANE.
2. DIMENSION A IS DATUM.
3. POSITIONAL TOLERANCE FOR LEADS:
   [⊕] 0.25 (0.010) Ⓜ A Ⓢ

DIM	MILLIMETERS		INCHES	
	MIN	MAX	MIN	MAX
A	4.78	5.00	0.188	0.197
B	3.81	4.01	0.150	0.158
C	1.35	1.75	0.053	0.069
D	0.35	0.46	0.014	0.018
F	0.67	0.77	0.026	0.030
G	1.27 BSC		0.050 BSC	
J	0.19	0.22	0.007	0.009
K	0.10	0.20	0.004	0.008
L	4.82	5.21	0.189	0.205
P	5.79	6.20	0.228	0.244

SO14 CASE 751A D SUFFIX

NOTES:
1. T IS SEATING PLANE.
2. DIMENSION A IS DATUM.
3. POSITIONAL TOLERANCE FOR LEADS:
   [⊕] 0.25 (0.010) Ⓜ A Ⓢ

DIM	MILLIMETERS		INCHES	
	MIN	MAX	MIN	MAX
A	8.54	8.74	0.336	0.344
B	3.81	4.01	0.150	0.158
C	1.35	1.75	0.053	0.069
D	0.35	0.46	0.014	0.018
F	0.67	0.77	0.026	0.030
G	1.27 BSC		0.050 BSC	
J	0.19	0.22	0.007	0.009
K	0.10	0.20	0.004	0.008
L	4.82	5.21	0.189	0.205
P	5.79	6.20	0.228	0.244

SO16 CASE 751B D SUFFIX

NOTES:
1. T IS SEATING PLANE.
2. DIMENSION A IS DATUM.
3. POSITIONAL TOLERANCE FOR LEADS:
   [⊕] 0.25 (0.010) Ⓜ A Ⓢ

DIM	MILLIMETERS		INCHES	
	MIN	MAX	MIN	MAX
A	9.78	10.01	0.385	0.394
B	3.81	4.01	0.150	0.158
C	1.35	1.75	0.053	0.069
D	0.35	0.46	0.014	0.018
F	0.67	0.77	0.026	0.030
G	1.27 BSC		0.050 BSC	
J	0.19	0.22	0.007	0.009
K	0.10	0.20	0.004	0.008
L	4.82	5.21	0.189	0.205
P	5.79	6.20	0.228	0.244

APPENDIX XV Machine Screws

Screw Size	Maximum Thread Diameter (in.)	Thread System (threads/in.) UNC	Thread System (threads/in.) UNF	Clearance Drill Bit	Drill Bit Diameter (in.)	Clearance (in.)
No. 2	0.086	56	64	No. 38	0.1015	0.0155
No. 4	0.112	40	48	No. 30	0.1285	0.0165
No. 6	0.138	32	40	No. 23	0.1540	0.0160
No. 8	0.164	32	36	No. 15	0.1800	0.0160
No. 10	0.190	24	32	No. 5	0.2055	0.0165
No. 12	0.216	24	28	5.9 mm	0.2320	0.0160
¼″	0.250	20	28	¹⁷⁄₆₄	0.2656	0.0156

APPENDIX XVI Machine Screw Nut Dimensions (in.) (UNC Threads)

Screw Size	Width Across Flats, W	Thickness, T
No. 2	³⁄₁₆	¹⁄₁₆
No. 4	³⁄₁₆	¹⁄₁₆
	¼	³⁄₃₂
No. 6	¼	³⁄₃₂
	⁵⁄₁₆	⁷⁄₆₄
No. 8	¼	³⁄₃₂
	⁵⁄₁₆	⁷⁄₆₄
	¹¹⁄₃₂	⅛
No. 10	⁵⁄₁₆	⅛
	¹¹⁄₃₂	⅛
	⅛	⅛
No. 12	⁷⁄₁₆	⁵⁄₃₂
¼″	⁷⁄₁₆	³⁄₁₆

APPENDIX XVII Flat Washer Dimensions (in.)

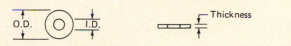

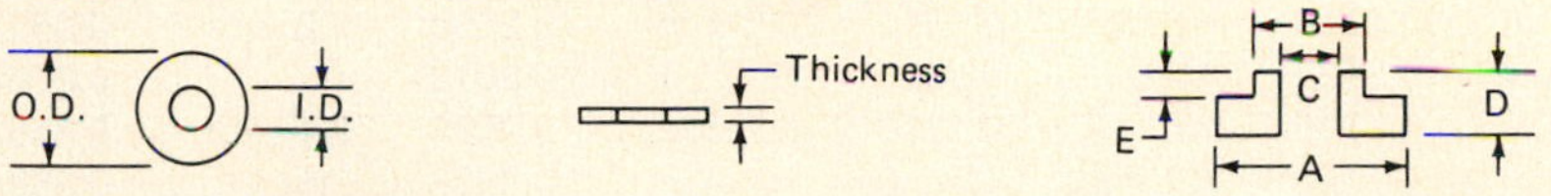

Screw Size	Flat Metal (Nickel-Plated Steel) OD	ID	Thickness	Flat Fiber OD	ID	Thickness	Flat Nylon OD	ID	Thickness	Shoulder (Fiber) Washer A	B	C	D	E
No. 2	¼	³⁄₃₂	¹⁄₃₂	¼	³⁄₃₂	¹⁄₃₂	¼	³⁄₃₂	¹⁄₃₂	—	—	—	—	—
No. 4	⁵⁄₁₆	⅛	¹⁄₃₂	¼	⅛	¹⁄₁₆	¼	⅛	¹⁄₁₆	¼	³⁄₁₆	⁹⁄₆₄	¹⁄₁₆	¹⁄₃₂
No. 6	⅜	⁵⁄₃₂	¹⁄₃₂	¼	⁹⁄₆₄	¹⁄₁₆	¼	⁹⁄₆₄	¹⁄₁₆	⁵⁄₁₆	³⁄₁₆	⁹⁄₆₄	¹⁄₁₆	¹⁄₃₂
No. 8	⅜	³⁄₁₆	¹⁄₃₂	⅜	¹¹⁄₆₄	¹⁄₁₆	⅜	¹¹⁄₆₄	¹⁄₁₆	⅜	¼	¹¹⁄₆₄	¹⁄₁₆	¹⁄₃₂
No. 10	⁷⁄₁₆	⁷⁄₃₂	¹⁄₃₂	⅜	¹³⁄₆₄	¹⁄₁₆	⅜	¹³⁄₆₄	¹⁄₁₆	⁷⁄₁₆	¼	¹³⁄₆₄	¹⁄₁₆	¹⁄₃₂
¼	⅝	⁹⁄₃₂	¹⁄₁₆	½	¹⁷⁄₆₄	¹⁄₁₆	½	¹⁷⁄₆₄	³⁄₃₂	½	⅜	¼	¹⁄₁₆	¹⁄₃₂

APPENDIX XVIII Lock Washer Dimensions (in.)

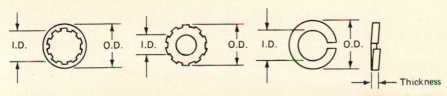

Screw Size	Internal OD	ID	Thickness	External OD	ID	Thickness	Split OD	ID	Thickness
No. 2	0.200	0.095	0.015	0.275	0.095	0.019	0.172	0.088	0.020
No. 4	0.270	0.123	0.019	0.260	0.123	0.019	0.209	0.115	0.025
No. 6	0.295	0.150	0.021	0.320	0.150	0.022	0.250	0.141	0.031
No. 8	0.340	0.176	0.023	0.381	0.176	0.023	0.293	0.168	0.040
No. 10	0.381	0.204	0.025	0.410	0.204	0.025	0.334	0.194	0.047
No. 12	0.410	0.231	0.025	0.475	0.231	0.025	0.377	0.221	0.056
¼	0.478	0.267	0.028	0.510	0.267	0.028	0.489	0.255	0.062

APPENDIX XIX   Drill Bit Sizes: Fractional, Letter, Metric, and Number

Drill Size	Decimal Equivalent (in.)	Drill Size	Decimal Equivalent (in.)	Drill Size	Decimal Equivalent (in.)	Drill Size	Decimal Equivalent (in.)	Drill Size	Decimal Equivalent (in.)	Drill Size	Decimal Equivalent (in.)
80	0.0135	1.25 mm	0.0492	2.55 mm	0.1004	4.1 mm	0.1614	B	0.2380	21/64	0.3281
79	0.0145	1.3 mm	0.0512	38	0.1015	4.2 mm	0.1654	6.1 mm	0.2402	8.4 mm	0.3307
1/64	0.0156	55	0.0520	2.6 mm	0.1024	19	0.1660	C	0.2420	Q	0.3320
0.4 mm	0.0157	1.35 mm	0.0531	37	0.1040	4.3 mm	0.1693	6.2 mm	0.2441	8.5 mm	0.3346
78	0.0160	54	0.0550	2.65 mm	0.1043	18	0.1695	D	0.2460	8.6 mm	0.3386
0.45 mm	0.0177	1.4 mm	0.0551	2.7 mm	0.1063	11/64	0.1719	6.3 mm	0.2480	R	0.3390
77	0.0180	1.45 mm	0.0571	36	0.1065	17	0.1730	E	0.2500	8.7 mm	0.3425
0.5 mm	0.0197	1.5 mm	0.0591	2.75 mm	0.1083	4.4 mm	0.1732	1/4	0.2500	11/32	0.3438
76	0.0200	53	0.0595	7/64	0.1094	16	0.1770	6.4 mm	0.2520	8.8 mm	0.3465
75	0.0210	1.55	0.0610	35	0.1100	4.5 mm	0.1772	6.5 mm	0.2559	S	0.3480
74	0.0225	1/16	0.0625	2.8 mm	0.1102	15	0.1800	F	0.2570	8.9 mm	0.3504
0.58 mm	0.0228	1.6 mm	0.0630	34	0.1110	4.6 mm	0.1811	6.6 mm	0.2598	9.0 mm	0.3543
0.6 mm	0.0236	52	0.0635	2.85 mm	0.1122	14	0.1820	G	0.2610	T	0.3580
73	0.0240	1.65 mm	0.0650	33	0.1130	13	0.1850	6.7 mm	0.2638	9.1 mm	0.3583
72	0.0250	1.7 mm	0.0669	2.9 mm	0.1142	4.7 mm	0.1850	17/64	0.2656	23/64	0.3594
0.65 mm	0.0256	51	0.0670	32	0.1160	3/16	0.1875	H	0.2660	9.2 mm	0.3622
71	0.0260	50	0.0700	2.95 mm	0.1161	12	0.1890	6.8 mm	0.2677	9.3 mm	0.3661
0.7 mm	0.0276	1.8 mm	0.0709	3.0 mm	0.1181	4.8 mm	0.1890	6.9 mm	0.2717	U	0.3680
70	0.0280	1.85 mm	0.0728	31	0.1200	11	0.1910	I	0.2720	9.4 mm	0.3701
69	0.0292	49	0.0730	3.1 mm	0.1220	4.9 mm	0.1929	7.0 mm	0.2756	9.5 mm	0.3740
0.75 mm	0.0295	1.9 mm	0.0748	1/8	0.1250	10	0.1935	J	0.2770	3/8	0.3750
68	0.0310	48	0.0760	3.2 mm	0.1260	9	0.1960	7.1 mm	0.2795	V	0.3770
1/32	0.0312	1.95 mm	0.0768	3.25 mm	0.1280	5.0 mm	0.1968	K	0.2810	9.6 mm	0.3780
0.8 mm	0.0315	5/64	0.0781	30	0.1285	8	0.1990	9/32	0.2812	9.7 mm	0.3819
67	0.0320	47	0.0785	3.3 mm	0.1299	5.1 mm	0.2008	7.2 mm	0.2835	W	0.3860
66	0.0330	2.0 mm	0.0787	3.4 mm	0.1339	7	0.2010	7.25 mm	0.2854	9.8 mm	0.3868
0.85 mm	0.0335	2.05 mm	0.0807	29	0.1360	13/64	0.2031	7.3 mm	0.2874	9.9 mm	0.3898
65	0.0350	46	0.0810	3.5 mm	0.1378	6	0.2040	L	0.2900	25/64	0.3906
0.9 mm	0.0354	45	0.0820	28	0.1405	5.2 mm	0.2047	7.4 mm	0.2913	10.0 mm	0.3937
64	0.0360	2.1 mm	0.0827	9/64	0.1406	5	0.2055	M	0.2950	X	0.3970
63	0.0370	2.15 mm	0.0846	3.6 mm	0.1417	5.3 mm	0.2087	7.5 mm	0.2953	10.1 mm	0.3976
0.95 mm	0.0374	44	0.0860	27	0.1440	4	0.2090	19/64	0.2969	10.2 mm	0.4016
62	0.0380	2.2 mm	0.0866	3.7 mm	0.1457	5.4 mm	0.2126	7.6 mm	0.2992	Y	0.4040
61	0.0390	2.25 mm	0.0886	26	0.1470	3	0.2130	N	0.3020	10.3 mm	0.4055
1.0 mm	0.0394	43	0.0890	3.75 mm	0.1476	5.5 mm	0.2165	7.7 mm	0.3031	13/32	0.4062
60	0.0400	2.3 mm	0.0906	25	0.1495	7/32	0.2187	7.8 mm	0.3071	10.4 mm	0.4094
59	0.0410	2.35 mm	0.0925	3.8 mm	0.1496	5.6 mm	0.2205	7.9 mm	0.3110	Z	0.4130
1.05 mm	0.0413	42	0.0935	24	0.1520	2	0.2210	5/16	0.3125	10.5 mm	0.4134
58	0.0420	3/32	0.0937	3.9 mm	0.1535	5.7 mm	0.2244	8.0 mm	0.3150	27/64	0.4219
57	0.0430	2.40 mm	0.0945	23	0.1540	1	0.2280	O	0.3160	7/16	0.4375
1.10 mm	0.0433	41	0.0960	5/32	0.1562	5.8 mm	0.2283	8.1 mm	0.3189	29/64	0.4531
1.15 mm	0.0453	2.45 mm	0.0965	22	0.1570	5.9 mm	0.2323	8.2 mm	0.3228	15/32	0.4687
56	0.0465	40	0.0980	4.0 mm	0.1575	A	0.2340	P	0.3230	31/64	0.4844
3/64	0.0469	2.5 mm	0.0984	21	0.1590	15/64	0.2344	8.3 mm	0.3268	1/2	0.5000
1.2 mm	0.0472	39	0.0995	20	0.1610	6.0 mm	0.2362				

APPENDIX XX   Determining Clearances for Automatic Insertion of Axial Components

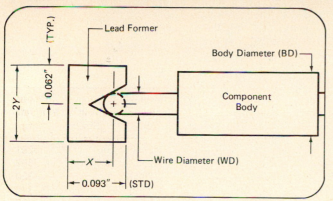

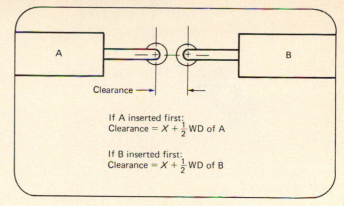

If A inserted first:
Clearance $= X + \frac{1}{2}$ WD of A

If B inserted first:
Clearance $= X + \frac{1}{2}$ WD of B

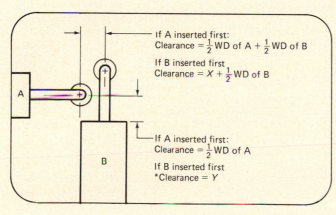

If A inserted first:
Clearance $= Y + \frac{1}{2}$ WD of A

If B inserted first:
Clearance $= X + \frac{1}{2}$ WD of B

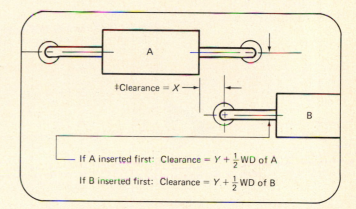

‡Clearance $= X$

If A inserted first: Clearance $= Y + \frac{1}{2}$ WD of A

If B inserted first: Clearance $= Y + \frac{1}{2}$ WD of B

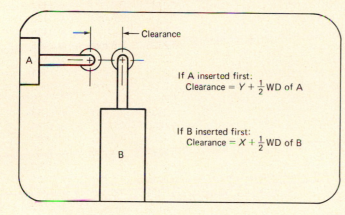

If A inserted first:
Clearance $= \frac{1}{2}$ WD of A $+ \frac{1}{2}$ WD of B

If B inserted first
Clearance $= X + \frac{1}{2}$ WD of B

If A inserted first:
Clearance $= \frac{1}{2}$ WD of A

If B inserted first
*Clearance $= Y$

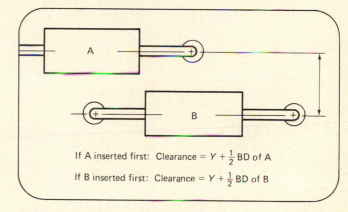

If A inserted first: Clearance $= Y + \frac{1}{2}$ BD of A

If B inserted first: Clearance $= Y + \frac{1}{2}$ BD of B

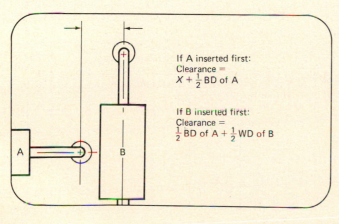

If A inserted first:
Clearance $=$
$X + \frac{1}{2}$ BD of A

If B inserted first:
Clearance $=$
$\frac{1}{2}$ BD of A $+ \frac{1}{2}$ WD of B

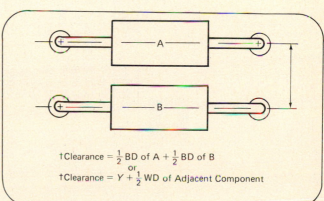

†Clearance $= \frac{1}{2}$ BD of A $+ \frac{1}{2}$ BD of B
or
†Clearance $= Y + \frac{1}{2}$ WD of Adjacent Component

*Check Y dimension for possible interference with body length of component B.
†Use whichever clearance value is larger.
‡Check X dimension for possible interference with body length of component A.

APPENDIX XXI   Reference Symbols

Square

Square (in.)	ID (in.)
0.090	0.025
0.100	0.025
0.110	0.025
0.125	0.025
0.150	0.040
0.180	0.030
0.187	0.040
0.200	0.080
0.230	0.031
	0.062
0.240	0.031
0.240	0.062
0.250	0.062
	0.093
0.260	0.062
0.280	0.062
0.300	0.062
0.312	0.062
	0.125
0.375	0.062
0.375	0.093
0.400	0.062
0.440	0.062
0.500	0.062
	0.093
0.625	0.093

Hexagonal

W (in.)	OD (in.)	ID (in.)
0.100	0.115	0.031
0.125	0.144	0.031
0.150	0.173	0.040
0.156	0.180	0.040
0.187	0.217	0.040
0.200	0.231	0.040
0.250	0.289	0.062
0.300	0.346	0.062
0.312	0.361	0.062
0.375	0.433	0.062

Triangular

W (in.)	H (in.)	ID (in.)
0.187	0.162	0.062
	0.217	0.062
0.250	0.217	0.093
0.375	0.325	0.093
0.500	0.433	0.062
	0.433	0.093
0.625	0.541	0.062
	0.541	0.093

Datum

| Diameter (in.) | | | ID (in.) |
A	B	C	
0.200	0.100	0.100	0.040
0.400	0.188	0.200	0.080

396

APPENDIX XXII Pads with Round Lead Access Holes

OD (in.)	ID (in.)	OD (in.)	ID (in.)	OD (in.)	ID (in.)	OD (in.)	ID (in.)	OD (in.)	ID (in.)
0.050	0.015	0.156	0.040		0.025		0.031	0.500	0.100
0.062	0.025		0.062		0.031		0.040		0.125
0.070	0.025	0.160	0.031		0.040		0.050		0.187
0.075	0.025		0.040		0.050		0.062		0.250
0.080	0.031	0.070	0.031	0.250	0.062	0.375	0.080	0.531	0.050
0.085	0.031	0.175	0.040		0.080		0.093		0.062
0.090	0.031	0.180	0.031		0.093		0.100	0.562	0.093
0.093	0.031		0.040		0.125		0.125		0.187
	0.025		0.025	0.260	0.062		0.187	0.600	0.062
0.100	0.031		0.031	0.270	0.062		0.031		0.080
	0.040		0.040	0.275	0.040		0.040		0.062
	0.050	0.187	0.050	0.280	0.050		0.050	0.625	0.080
0.103	0.039		0.062	0.281	0.031		0.062		0.093
0.104	0.032		0.093		0.062		0.080		0.125
0.106	0.032	0.190	0.135		0.031		0.093	0.650	0.031
0.110	0.031		0.025		0.040	0.400	0.100	0.687	0.187
0.111	0.032		0.031		0.050		0.125	0.700	0.031
0.113	0.040		0.040	0.300	0.062		0.200		0.031
0.115	0.031	0.200	0.050		0.080	0.404	0.062		0.062
0.120	0.031		0.062		0.093	0.416	0.080	0.750	0.080
	0.025		0.080		0.100		0.031		0.125
	0.031		0.093		0.031		0.050		0.320
0.125	0.040		0.125		0.040		0.062		0.031
	0.050	0.208	0.062		0.050	0.437	0.093	0.800	0.062
	0.062		0.031	0.312	0.062		0.100		0.125
0.130	0.032	0.218	0.040		0.080		0.125		0.062
0.140	0.031		0.050		0.093	0.468	0.050	0.875	0.125
	0.040		0.062	0.320	0.062	0.475	0.062		0.062
	0.031	0.220	0.093	0.340	0.062		0.031	1.000	0.093
	0.040	0.227	0.062	0.343	0.062		0.040		0.125
0.150	0.050	0.230	0.031	0.350	0.062		0.062	1.125	0.062
	0.062	0.240	0.040		0.180		0.080	1.250	0.125
	0.080		0.062	0.360	0.062	0.500	0.093	1.500	0.125
0.156	0.031	0.245	0.062						

APPENDIX XXIII Solid Pads

OD (in.)	OD (in.)	OD (in.)	OD (in.)
0.030	0.125	0.200	0.400
0.050	0.140	0.218	0.437
0.055	0.150	0.240	0.500
0.062	0.156	0.250	0.562
0.075	0.160	0.281	0.625
0.079	0.167	0.300	0.687
0.080	0.170	0.312	0.750
0.088	0.175	0.320	0.875
0.093	0.180	0.350	0.937
0.100	0.187	0.375	1.000
0.120			

APPENDIX XXIV Round Pads with Red Eyes

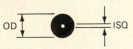

OD (in.)	ID (in.)
0.080	0.031
0.100	0.031
0.125	0.031
	0.040
0.150	0.031
	0.040
0.156	0.031
0.187	0.031
	0.040
0.200	0.031
	0.040
	0.050
0.250	0.040
	0.062
0.300	0.062

APPENDIX XXV Round Pads with Square Center Holes

OD (in.)	ISQ (in.)
0.100	0.031
0.125	0.031
0.150	0.040
0.156	0.040
0.187	0.040
0.200	0.062
0.250	0.031
0.300	0.062
0.312	0.062
0.375	0.062
0.400	0.062
0.500	0.062

APPENDIX XXVI Teardrop Pads

Radius Fillet

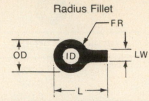

OD (in.)	LW (in.)	ID (in.)	L (in.)	FR (in.)
0.100	0.026	0.030	0.144	0.025
0.125	0.031	0.031	0.187	0.031
0.138	0.031	0.031	0.196	0.078
0.140	0.050	0.031	0.324	0.075
0.150	0.050	0.031	0.231	0.187
0.156	0.030	0.031	0.250	0.050
0.187	0.031	0.040	0.280	0.125
	0.062	0.047	0.280	0.125
	0.062	0.062	0.280	0.125
0.200	0.040	0.074	0.300	0.094
	0.050	0.040	0.300	0.094
	0.062	0.080	0.300	0.125
	0.080	0.032	0.300	0.062
0.218	0.093	0.062	0.312	0.094
0.219	0.080	0.064	0.260	0.062
0.240	0.062	0.062	0.312	0.141
0.250	0.050	0.040	0.375	0.094
	0.080	0.062	0.400	0.190
	0.100	0.040	0.375	0.062
	0.100	0.062	0.375	0.062
	0.125	0.040	0.344	0.125
0.281	0.125	0.062	0.375	0.062
0.300	0.100	0.040	0.500	0.125
	0.125	0.062	0.500	0.219
	0.125	0.080	0.500	0.219
0.312	0.100	0.040	0.500	0.125
	0.125	0.040	0.500	0.062
	0.125	0.062	0.500	0.062
0.375	0.100	0.062	0.562	0.250
	0.125	0.093	0.531	0.234
	0.125	0.047	0.562	0.250
	0.125	0.125	0.562	0.250
	0.250	0.093	0.531	0.422
0.400	0.100	0.062	0.562	0.250
	0.125	0.125	0.575	0.281
0.437	0.125	0.062	0.531	0.250
	0.125	0.125	0.531	0.250
	0.250	0.062	0.531	0.515
0.500	0.125	0.093	0.750	0.125
	0.125	0.062	0.812	0.406
	0.250	0.156	0.750	0.344
	0.250	0.062	0.812	0.422
0.562	0.125	0.080	0.718	0.141
0.625	0.125	0.093	0.875	0.125
	0.312	0.100	0.906	0.375
0.687	0.400	0.187	0.937	0.438
0.750	0.250	0.047	1.120	0.500

Straight-Side Fillets

OD (in.)	LW (in.)	ID (in.)	L (in.)
0.125	0.031	0.025	0.250
0.125	0.031	0.031	0.250
0.156	0.050	0.031	0.250
0.187	0.050	0.040	0.275
	0.050	0.062	0.275
0.200	0.050	0.040	0.350
	0.080	0.080	0.300
0.218	0.040	0.050	0.310
	0.062	0.062	0.350
0.250	0.050	0.040	0.375
	0.050	0.062	0.375
	0.062	0.062	0.500
	0.062	0.080	0.500
	0.062	0.062	0.625
0.281	0.040	0.062	0.420
	0.080	0.040	0.480
0.300	0.050	0.040	0.500
	0.062	0.040	0.500
	0.062	0.080	0.600
0.312	0.062	0.062	0.468
	0.080	0.040	0.500
0.375	0.040	0.062	0.560
	0.062	0.062	0.750
	0.125	0.062	1.030
0.400	0.100	0.093	0.900
	0.125	0.062	0.718
	0.125	0.080	0.718
0.437	0.125	0.062	1.000
0.500	0.125	0.125	0.750
	0.125	0.062	0.812
0.531	0.125	0.062	1.100
0.562	0.125	0.125	1.120
0.625	0.125	0.062	0.938
0.750	0.125	0.125	1.370
1.000	0.250	0.125	1.500

90° Elbow

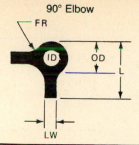

Double Ended

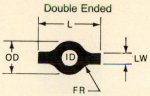

OD (in.)	LW (in.)	ID (in.)	L (in.)	FR (in.)
0.138	0.031	0.031	0.196	0.078
0.150	0.050	0.031	0.231	0.079
	0.062	0.047	0.280	0.250
0.187	0.093	0.062	0.275	0.125
	0.100	0.062	0.273	0.125
0.250	0.093	0.062	0.375	0.218
	0.100	0.062	0.375	0.125
	0.100	0.062	0.426	0.150
0.300	0.100	0.031	0.436	0.150
	0.125	0.062	0.462	0.437
0.312	0.093	0.062	0.406	0.125
0.375	0.125	0.093	0.500	0.468
0.400	0.125	0.125	0.575	0.562
0.437	0.125	0.062	0.487	0.500
	0.250	0.062	0.531	1.030
	0.125	0.093	0.750	0.250
0.500	0.200	0.125	0.750	0.281
	0.250	0.062	0.812	0.843

OD (in.)	LW (in.)	ID (in.)	L (in.)	FR (in.)
0.138	0.031	0.031	0.196	0.078
0.150	0.050	0.031	0.312	0.187
	0.062	0.047	0.375	0.250
0.187	0.093	0.062	0.360	0.125
	0.100	0.062	0.363	0.125
0.250	0.093	0.062	0.500	0.218
0.300	0.100	0.031	0.625	0.150
	0.125	0.062	0.700	0.437
0.375	0.125	0.093	0.688	0.468
	0.125	0.047	0.750	0.500
0.400	0.125	0.125	0.750	0.562
0.437	0.125	0.062	0.625	0.500
	0.250	0.062	0.625	0.250
	0.100	0.031	1.300	0.250
0.500	0.125	0.093	1.000	0.250
	0.250	0.156	1.000	0.687
	0.250	0.062	1.120	0.843

APPENDIX XXVII   Fillets

One Piece

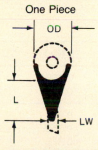

Two piece

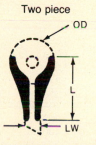

Use with OD (in.)	Use with LW (in.)	Use with L (in.)
0.200	0.020	0.200
0.250	0.025	0.250
0.300	0.030	0.300
0.375	0.040	0.375
0.400	0.050	0.400
0.500	0.062	0.500

Use with OD	Use with LW
0.250" to 0.600"	0.062" to 0.600"
0.100" to 0.375"	0.031" to 0.375"

One-Hole Offset

OD (in.)	L (in.)	C (in.)	ID (in.)	OD (in.)	L (in.)	C (in.)	ID (in.)	OD (in.)	L (in.)	C (in.)	ID (in.)
0.100	0.195	0.075	0.031	0.200	0.300	0.100	0.031	0.312	0.468	0.156	0.062
	0.200	0.050	0.031		0.300	0.100	0.062		0.487	0.156	0.093
	0.219	0.095	0.031		0.320	0.100	0.081	0.350	0.425	0.175	0.062
	0.245	0.075	0.031		0.340	0.100	0.031		0.450	0.175	0.031
0.125	0.175	0.062	0.031		0.350	0.100	0.040		0.500	0.175	0.031
	0.195	0.062	0.031		0.350	0.100	0.050	0.375	0.500	0.188	0.062
	0.200	0.062	0.031		0.440	0.100	0.031		0.750	0.188	0.062
	0.245	0.062	0.031	0.230	0.305	0.115	0.031	0.400	0.450	0.200	0.062
0.140	0.272	0.070	0.062	0.250	0.300	0.125	0.031		0.500	0.200	0.093
0.150	0.175	0.075	0.031		0.343	0.125	0.093		0.600	0.200	0.062
	0.200	0.075	0.031		0.350	0.125	0.062		0.650	0.200	0.093
	0.200	0.075	0.040		0.370	0.125	0.031		0.780	0.300	0.093
	0.225	0.075	0.031		0.375	0.125	0.031		0.980	0.300	0.093
	0.250	0.075	0.031		0.400	0.125	0.031	0.500	0.656	0.250	0.100
0.175	0.225	0.088	0.031		0.406	0.125	0.062		0.700	0.250	0.093
	0.250	0.088	0.062		0.425	0.125	0.093		0.775	0.250	0.062
0.187	0.312	0.093	0.040		0.450	0.125	0.031	0.600	0.800	0.300	0.062
	0.312	0.093	0.062		0.550	0.125	0.031		0.800	0.300	0.093
					0.656	0.125	0.093				
				0.300	0.400	0.150	0.062				
					0.400	0.150	0.093				
					0.420	0.150	0.031				
					0.450	0.150	0.031				

One-Hole Centered

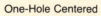
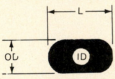

OD (in.)	L (in.)	ID (in.)
0.100	0.250	0.031
0.120	0.200	0.031
0.145	0.245	0.062
0.170	0.310	0.062
0.197	0.328	0.093
0.200	0.320	0.125
	0.400	0.093
0.225	0.300	0.093
0.232	0.400	0.078
0.240	0.400	0.078
0.250	0.400	0.062
0.375	0.500	0.062
	0.750	0.062
0.400	0.600	0.187
0.500	0.775	0.062

Two Holes

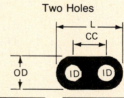

OD (in.)	L (in.)	CC (in.)	ID (in.)	OD (in.)	L (in.)	CC (in.)	ID (in.)
0.100	0.200	0.100	0.031	0.281	0.521	0.240	0.062
	0.300	0.200	0.031		0.751	0.470	0.062
0.125	0.225	0.100	0.031	0.300	0.625	0.325	0.062
0.150	0.350	0.200	0.050	0.312	0.500	0.187	0.062
0.187	0.375	0.187	0.062	0.345	0.750	0.405	0.062
0.200	0.400	0.200	0.062		0.830	0.485	0.062
0.250	0.500	0.250	0.062	0.375	0.750	0.375	0.062
	0.520	0.270	0.062	0.500	1.000	0.500	0.062
	0.550	0.300	0.062				0.093
	0.650	0.400	0.062				
	0.650	0.400	0.093				
	0.750	0.500	0.093				
	0.950	0.700	0.062				
	1.000	0.750	0.093				

APPENDIX XXIX   TO Patterns

8- and 10-Pin Round ICs

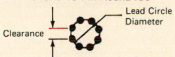

Lead Circle Diameter

Clearance

Patterns	Pattern Type	No. of Leads	Terminal Area (Pad) Dim. for Mounting Configurations			
			OD (in.)	ID (in.)	Clearance (in.)	Lead Cir. Dia. (in.)
	AA		0.046	0.010	0.031	0.200
	BB	8	0.054	0.015	0.024	0.200
	CC CUT PAD		0.062	0.015	0.025	0.200
			0.060	0.015	0.032	0.230
			0.078	0.015	0.037	0.300
	DD	8	0.093	0.015	0.040	0.350
			0.100	0.015	0.050	0.400
	EE*		0.093	0.015	0.057	0.400
	FF		0.046	0.010	0.025	0.230
	GG	10	0.046	0.015	0.015	0.200
			0.054	0.015	0.016	0.230
	HH CUT PAD		0.062	0.015	0.025	0.230
	II	10	0.070	0.015	0.040	0.350
			0.070	0.015	0.055	0.400
	JJ**	10	0.093	0.015	0.030	0.400

APPENDIX XXIX   **(continued)**

TO-5 Style

Pattern	No. of Leads	Terminal Area (Pad) Dim. for Mounting Configurations		
		OD (in.)	ID (in.)	Clearance (in.)
(3 leads triangular)	3	0.062	0.015	0.078
		0.075	0.015	0.066
		0.093	0.015	0.047
		0.100	0.015	0.041
		0.109	0.015	0.032
		0.125	0.015	0.016
(4 leads diamond)	4	0.093	0.015	0.047
		0.109	0.015	0.032
		0.125	0.015	0.016
Cut Pad (3 leads)	3	0.125	0.015	0.031
		0.125	0.023	0.031
Cut Pad (4 leads)	4	0.125	0.015	0.031
(3 leads with tab)	3	0.062	0.015	0.078
		0.075	0.015	0.066
		0.093	0.015	0.047
		0.100	0.015	0.041
		0.109	0.015	0.032
Cut Pad (3 leads with tab)	3	0.125	0.015	0.031
		0.125	0.023	0.031
(4 leads)	4	0.152—widest	0.015	0.022

	No.	OD	ID	Clearance
	6	0.058	0.015	0.020
	6	0.090—widest	0.015	0.025
	6	0.100	0.015	0.065

12-Pin Round ICs

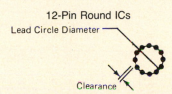

Lead Circle Diameter

Clearance

Patterns	Pattern Type	No. of Leads	Terminal Area (Pad) Dim. for Mounting Configurations			
			OD (in.)	ID (in.)	Clearance (in.)	Lead Cir. Dia. (in.)
	K		0.046	0.010	0.013	0.230
	L		0.037	0.010	0.015	0.200
			0.050	0.015	0.010	0.230
	M CUT PAD		0.062	0.018	0.018	0.230
		12	0.070	0.015	0.020	0.350
	N		0.080	0.015	0.023	0.400
	P		0.080	0.015	0.023	0.400

APPENDIX XXIX (continued)

TO-18 Style

Lead Circle Diameter — Clearance

Patterns	Terminal Area (Pad) Dimensions for Mounting Configurations				
	No. of Leads	OD (in.)	ID (in.)	Clearance (in.)	Lead Cir. Diameter (in.)
	3	0.087— widest	0.015	0.025	0.100
	4	0.093— widest	0.018	0.025	
	3	0.056	0.015	0.015	0.100
	4	0.056	0.015	0.015	
Cut Pad	3	0.062	0.015	0.025	0.100
	4	0.062	0.015	0.025	
	3	0.080	0.015	0.026	0.150
	4	0.080	0.015	0.026	
Cut Pad	3	0.100	0.015	0.025	0.150
	4	0.100	0.015	0.025	
	6	0.058	0.015	0.020	0.200*
	3	0.125	0.015	0.156	**
	3	0.125	0.015	0.037	**

*Pattern fits Thermalloy Spreader #7717-19
**Pattern fits Thermalloy Spreader #7717-21

APPENDIX XXX   Flat Pack Styles

Surface Mount

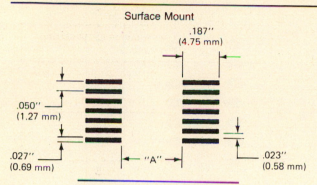

No. of Leads	"A" Dim. (in.)
14	0.187
14	0.312
16	0.312
24	0.312

Staggered Mount

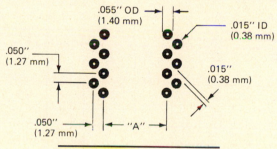

No. of Leads	"A" Dim. (in.)
14	0.325
16	0.400
20	0.800

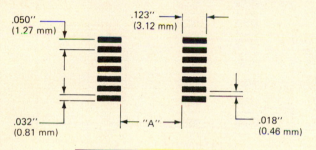

No. of Leads	"A" Dim. (in.)
14	0.250
14	0.312
16	0.312

No. of Leads	"A" Dim. (in.)
14	0.325

APPENDIX XXXI   Dual-in-Line Package Preforms

Oval

No. of Leads	"A" Dim. (in.)	"B" Dim. (in.)
8	0.300	0.175
14		
16		
18		
24	0.600	0.475
28		
36		
40		

Elliptical

No. of Leads	"A" Dim. (in.)	"B" Dim. (in.)
8	0.300	0.200
14		
16		
18		
24	0.600	0.500
40		

Round

No. of Leads	"A" Dim. (in.)	"B" Dim. (in.)
8	0.300	0.240
14		
16		
24	0.600	0.540
36		

Oval/Conductors

No. of Leads	"A" Dim. (in.)	"B" Dim. (in.)
8	0.300	0.175
14		
16		
24	0.600	0.475
36		

Cut

No. of Leads	"A" Dim. (in.)	"B" Dim. (in.)
8	0.300	0.175
10		
14		
16		
18		
20		
22	0.400	0.275
24	0.600	0.475
28		
36		
40		

APPENDIX XXXII   Insertion-Type Edge Connector Patterns

Center-to-Center Spacing (inches)	Pattern*
.050	.008" ID .050" .250" .031" .019"
0.100	.015" ID .050" .050" .100" .500"
0.100	.015" ID .062" .038" .100" .500"
0.125	.015" ID .062" .062" .125" .500"
0.150	.015" ID .080" .070" 150" .500"
0.156	.015" ID .093" .062" .156" .500"
0.200	.031" ID .125" .075" .200" 1.000"

*Available with or without plating bar.

APPENDIX XXXIII   Tape Widths (in.)

0.015
0.020
0.026
0.031
0.037
0.040
0.046
0.050
0.056
0.060
0.062
0.070
0.075
0.080
0.093
0.100
0.125
0.140
0.150
0.156
0.160
0.175
0.187
0.200
0.218
0.250
0.280
0.300
0.312
0.375
0.400
0.500
0.750
0.800
1.000
2.000

APPENDIX XXXIV Elbows and Universal Circles

30°	45°	60°	90°
Line Width, LW (in.)	Line Width, LW (in.)	Line Width, LW (in.)	Line Width, LW (in.)
0.031	0.031	0.031	0.031
0.040	0.040	0.040	0.040
0.050	0.050	0.050	0.050
0.062	0.062	0.062	0.062
0.080	0.080	0.080	0.070
0.093	0.093	0.093	0.075
0.100	0.100	0.100	0.080
0.125	0.125	0.125	0.093
0.150	0.150	0.150	0.100
0.156	0.156	0.156	0.125
0.160	0.160	0.160	0.150
0.187	0.187	0.187	0.156
0.200	0.200	0.200	0.160
0.250	0.250	0.250	0.187
0.300	0.300	0.300	0.200
0.375	0.375	0.375	0.250
			0.300
			0.375
			0.400
			0.500

Line Width, LW (in.)	OD of Nest (in.)	ID of Nest (in.)
0.031	0.373	0.125
0.040	0.445	0.125
0.050	0.525	0.125
0.062	0.625	0.125
0.080	0.890	0.250
0.093	1.062	0.312
0.100	1.100	0.300
0.125	1.375	0.375
0.150	1.640	0.437
0.156	1.680	0.437
0.160	1.720	0.437
0.187	2.000	0.500
0.200	2.100	0.500
0.250	2.500	0.500
0.300	2.900	0.500

APPENDIX XXXV Tees

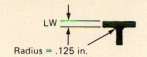

Radius = .125 in.

Line Width, LW (in.)
0.031
0.040
0.050
0.062
0.080
0.093
0.100
0.125
0.150
0.156
0.160
0.187
0.200
0.250
0.300
0.375
0.400
0.500

APPENDIX XXXVI Thermal Relief and Antipads

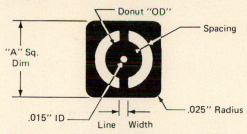

Donut "OD" (in.)	Spacing and Line Width (in.)	"A" Dim. (in.)
0.075	0.025	0.175
0.093	0.025	0.193
0.100	0.025	0.200
0.125	0.025	0.225

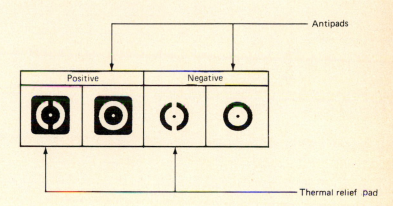

INDEX